Science and Religious Experience
Are they similar forms of knowledge?

To
ALISON & JULIAN,
SANDY & CHRISTOPHER,
BELINDA & SIMON
and their families,
with love.

Science and Religious Experience
Are they similar forms of knowledge?
GRAHAME MILES

sussex
ACADEMIC
PRESS
Brighton • Portland • Toronto

2 4 6 8 10 9 7 5 3 1

First published 2007, reprinted with corrections 2011, in Great Britain by
SUSSEX ACADEMIC PRESS
PO Box 139
Eastbourne BN24 9BP

and in the United States of America by
SUSSEX ACADEMIC PRESS
920 NE 58th Ave., Suite 300
Portland, OR 97213-3786

and in Canada by
SUSSEX ACADEMIC PRESS (CANADA)
90 Arnold Avenue, Thornhill, Ontario L4J 1B5

British Library Cataloguing in Publication Data
A CIP catalogue record for this book is available from the British Library.

Library of Congress Cataloging-in-Publication Data
Miles, Grahame.
Science and religious experience : are they similar forms of
    knowledge? / Grahame Miles.
    p. cm.
    Includes bibliographical references and index.
    ISBN 978-1-84519-116-0 (h/b : alk. paper) —
    ISBN 978-1-84519-117-7 (p/b : alk. paper)
    1. Religion and science. 2. Experience (Religion) I. Title.

BL240.3.M55 2006
215—dc22

                                              2005024376

Typeset & designed by Susex Academic Press, Brighton & Eastbourne.
Printed by TJ International, Padstow, Cornwall.
This book is printed on acid-free paper.

Contents

3   **How Do We Know What Knowledge Is?**   42
      An American Search for Personal Knowledge

4   **Are There Different Kinds of Knowledge?**   68
      Educationalists' Views of Knowledge after 1960

5 Changing Views of Scientific Knowledge   83

6 The Integrity of Science   99

Is Science Objective or Personal Knowledge?

Foreword by Janet Scott

Grahame's personal odyssey began when he was faced with a conflict between the Christian context of his home and school life and the assumptions in society that sense experience and science were the only sources of knowledge. This conflict influenced his career as a Religious Education teacher in Secondary Schools in Cleckheaton and then Letchworth. He began a research programme exploring the effects of two teaching programmes on the development of religious thinking, an interest which was to be developed further in his doctoral work.

In 1968, he was appointed Senior Lecturer in Religious Studies at Homerton College, Cambridge. Here he was greatly influenced by Jean Holm and became an enthusiastic supporter of the focus on World Faiths in her new college Religious Studies Course.

He went back into the classroom regularly, focusing his interest on how young peoples' understanding of religion develops. This led to two further research projects, one concerned with senior secondary and undergraduate students' understanding of religious experience. This work was later the basis of an undergraduate course. The other project was concerned with the development of religious thinking of primary school children.

I first met Grahame when we were both members of the Religious Education Council of England and Wales. Later I became a colleague at Homerton College in the religious studies department, where Grahame's generosity of spirit, keen enthusiasm for the subject, and boundless energy made a valued contribution to the work of staff as well as students.

Since he retired he has had the opportunity in this text to develop and set out his thinking about the different ways of knowing we all take for granted as well as the ways of knowing that have been rejected by many of us. A central concern has been to relate these ideas to an understanding of religious experience. This is of interest, not just to a teacher of religious education but to any senior school or undergraduate student, or any reader concerned with our different ways of knowing, and how they relate to each other.

Grahame tries to show how the structures of all kinds of knowledge involve what we grasp through our senses, how we interpret those impressions, first by description and then by an interpretive understanding of it. Such a theory or hypothesis is given support through evidence, argument, sufficient reasons, and then, if accepted, given credence by public acclaim. So, strictly speaking, one can no more prove scientific theory than religious belief, for each is upheld by a believing community.

It has been a pleasure to read this book, for in its blend of experience and explanation we can hear the authentic voice of a gifted teacher, like a householder who can produce from his store things new and old (Matt. 13: 52). Grahame can take difficult ideas and make them understandable; he links together knowledge from different fields and sources; and above all, he shares a sense of the excitement to be found in the processes of exploring and coming to know.

JANET SCOTT
Director of Religious Studies, Tutor and Praelector,
Homerton College, Cambridge

Acknowledgements

I have an enormous wish to thank so many people who have helped me in many different ways: My parents, Revd. William and Mrs Ruby Miles, for wonderful opportunities and loving support given to my brothers, Dr Malcolm Miles, Dr David Miles, and to me. And to:

Mr J. T. Lancaster, Mr C. N. Pleasance and Miss D. Cooper (Mrs Wickens, MBE) and all the Staff at Ashville College, Harrogate; Corpus Christi College, Cambridge: the Fellows and the Tutor, Mr M. W. McCrum, CBE; Lecturers in the Faculty of Geography and my Supervisor, Mr A. A. L. Caesar; Wesley House, Cambridge: the Principal, Revd. W. F. Flemington and Tutor, Revd. M. J. Skinner; Lecturers in the Faculty of Divinity; The University of London Institute of Education: Dr F. H. Hilliard; The University of Leeds School of Education: Professor Kenneth Lovell, Professor Margaret Sutherland; Directors of The Religious Experience Research Centre: Sir Alister Hardy, FRS, Mr Edward Robinson and Peggy Morgan.

It was my joy and privilege to be a teacher of Religious Education and Religious Studies all my professional life. I have an enormous thank-you to make to the pupils and students everywhere I taught, for their stimulating and fearless questioning, the fun and discussion, for I was never bored!

Everywhere I taught I enjoyed teaching and am very grateful to: Whitcliffe Mount Grammar School, Cleckheaton, and to Mr A. Morton and all Colleagues; Letchworth Grammar School and to Mr F. C. Lill, Miss Joan Williams, Dr Eric J. Macfarlane OBE and all Colleagues; Homerton College, Cambridge, and to four Principals: Dame Beryl Paston Brown DBE, Miss Alison Shrubsole CBE (Mrs Hilton Brown), Mr Alan Bamford CBE, and Dr Kate Pretty, Pro-Vice-Chancellor; to the Heads of the Religious Studies Department: Revd. Eric Lord, Jean Holm and Janet Scott; to all Religious Studies Colleagues: Revd. Vivian Buddle, Dr Jackie Hirst, Mr Peter Mitchell, Dr Louise Pirouet, Mrs Eleanor Scott, Revd. Dr David Stacey and all other Colleagues in Homerton, the Faculty of Education and the Faculty of Divinity of the University of Cambridge.

I enjoyed a range of educational opportunities, and am very grateful to:

All who helped me run an Easter Sixth Formers Sailing and Religious Discussion Broads Camp at Boundary Farm 1965–73, hosted by Mr Fred and Mrs Doris Cook, then Mr Don and Mrs Babs Cook, for Sixth Form (now Grades 12 & 13) students from Letchworth Grammar School; Heath Grammar School, Halifax with Mr Peter

Hand; Whitcliffe Mount Grammar School, Cleckheaton with Mr Keith Foreman OBE and Mr Ashley Petts.

The Principals, staff and students of The Sixth Form College, Luton, Hills Road Sixth Form College and Long Road Sixth Form College, Cambridge in 1977, for permission to teach a ten-week course on "Human Experience and Its Interpretation"; and to the Headteacher, staff and students at Stapleford Community Primary School for permission to teach a short course.

The President, Staff and Students at Stephens College, Columbia, and the University of Columbia, Missouri, for the invitation to give open and student lectures and seminars in the summer of 1985, and to my hosts, Professor Richard Gelwick and Revd. Don Johnson.

The Organizers of an annual Summer School for American Students at Homerton College, and then at Queens' College, Cambridge from 1988–93 in which I taught a course on "The Nature and Destiny of Human Beings".

Those who influenced this book: Dr Hilliard emphasized an educational approach to Religious Education, Revd. Eric Lord focused on relevance, and in the early 1970s Jean Holm broadened the educational approach, introducing the study of major World Religions besides Christianity, enriched by taking students to communities of Hindus, Muslims, Orthodox Jews and Sikhs, with an opportunity to share in their worship, festivals and home life; encouragement from Professor James Michael Lee, unstinting support from Janet Scott, close support from a former Homerton colleague, now Revd. Professor Michael Reiss who read the first draft of the text most helpfully, as did my recent R.S. colleagues including Mary Earl, and education colleague Dr Charles Bailey; and from others who have read parts of the script sent to them, Professor Michael Argyle, Dr Michael Bonnett, Revd. Douglas Brown, Professor Paul Hirst, Dr Will Large, Revd. Dr Arthur Peacocke, Revd. Dr John Polkinghorne, KBE, FRS, Dr John Shortt, Revd. Professor Adrian Thatcher, and Revd. Dr Fraser Watts; and to those who have given me references – Canon Professor Edward Bailey, Mr Roland Chaplain, Revd. Professor Leslie Francis, Professor Richard Gelwick, Mr Martyn Goss, Canon Brian Hebblethwaite, Professor Paul Hirst, Professor Terry McLaughlin, Dr Anne Miles, Christopher Miles, Mr Geoffrey Mizen, Revd. Dr Arthur Peacocke, Janet Scott, Professor Denys Turner, Dr John Waldram, Revd. Dr Fraser Watts, Dr David Whitebread and Dr Mark Wynn.

I am deeply grateful to the Religious Education Press, Birmingham, AL, for permission to quote from R. H. Hood Jr. (ed.), (1995), *The Handbook of Religious Experience* and to Mr Frank Clements, the Librarian at the University College of St Mark and St John in Plymouth, for permission to read in their library.

I am very grateful indeed for the support of Dr Barbara Miles, my joyous wife and closest friend, as well as her proof reading with Anne Shrewsbury and Patricia Fox. I deeply thank Malcolm Ward, the copy-editor at Sussex Academic Press, for his expert work, and other staff members who have contributed towards this publication. I am, in advance, sorry for all errors, which are mine alone.

GRAHAME MILES
Plympton, October 2006

Note to Students

Big words like *phenomenology* or *hermeneutics* used to frighten me and frighten me still, but now I have come to see that they are useful. They are shorthand once you get inside an area of study. Doctors use extraordinary language! But so do people who play sport. Have you ever met in the street a "square leg" or "a line backer"? Specialized words are necessary, but in the text that follows they will always be explained. Another matter also upset me as a student. I could not stand all this "name dropping". It made me feel so ignorant! Well again, names are useful because different ideas cluster around different names, and they too become a kind of short hand. As a student I hated footnotes. They made me feel so inadequate. Ignore them. So why do I use them now? If you are interested in or annoyed by a point made in the text, you may want to follow it up, so then the footnote may be useful. That is its only purpose.

How should you read this book? Different people read books like this one in different ways. Some start at the beginning and read to the end, or until they get bored! But if I picked up this book, I would go straight to the Chapters which dealt with subjects which interested or concerned me, like Freud's or Dawkins's reductionist arguments, to explore their understanding of different kinds of knowledge. Only as I wanted a fuller map of kinds of knowledge would I turn to other chapters. In the end I would read the lot as I would want to know the foundations of all kinds of knowledge and how they are related to each other. In that way I could be sure of the legitimacy of religious experience as the basis of a form of knowledge and of the argument that science and religious experience are reached in the same way, though the degrees of freedom of interpretation are different.

One more thing. It has taken me forty years to write this book; I am a slow learner. I like to be told what I am going to be told, to be told it, and then to be told what I have been told, because I am not sure at which part in the telling I will be asleep! This book is not just for highfliers but for those of us who grapple with ideas in stages. There will be repetition, because if I have only got hold of half an idea, I need to repeat it to get hold of the other half. The repetition is helpful for me, and hopefully helpful for many others. But it is easily skipped if you so wish.

Science and Religious Experience
Are they similar forms of knowledge?

Introduction

This book has six main threads. First, to collect together some of the main western contributions to the study of religious experience focused on the direct awareness or background awareness of God or "The Supreme Reality" of non-theistic World Faiths. Second, *are science and religious experience similar forms of knowledge?* Third, any such argument can only be well grounded in the context of an understanding of the nature of knowledge in general and scientific knowledge in particular, and the defusing of materialistic scientific criticism of religion. Fourth, we explore the thesis that *to become* a human being, a person, is a spiritual achievement. Fifth, the dangers of fundamentalism, an extrinsic or exclusivist form of religion and the opportunities of an intrinsic or inclusive form of religion invite reflection as we live in a global village. Sixth, do the final models of religious experience described offer a foundation to promote a greater understanding of the religious mode of discourse and dialogue between Humanists of every creed, religious, agnostic and atheist?

In Chapter 1 we begin our exploration on Lake Windermere, thinking about nature and then nature mysticism. This experience on Lake Windermere evokes Wordsworth's reflections on the subject. Then we explore the accounts of similar experiences that contemporary young people have had. I admit that in my early teens all these religious and mystical experiences in "my" world had to be questioned, because they were not "knowledge"; I had learnt that facts had to be "proved".

In Chapter 2 we explore "facts", only to discover that a perception – what we grasp through our five senses – needs to be described, and that involves an interpretation, which forms empirical knowledge. Observation, hypothesis or theory, and evidence are key elements in the formation of scientific knowledge, once regarded as proper knowledge. We then discuss European views of objective knowledge and the claims of the people who defined it, called logical positivists. Knowledge, according to the logical positivists, consists of material evidence and verified claims. Everything else is not knowledge, just expressions of emotion, opinions, or beliefs, which are unproven conjectures. We then discover how that theory collapsed, after which it became accepted that the meaning of language depends on the way it is *used*: the *use* test.

In Chapter 3 we read an account of the American approach to knowledge focused on personal knowledge, which claims that the person matters most and is irreducible. The humanistic psychological movement and a leading psychotherapist confirm that judgement. Being human involves knowledge of the dimensions of being human with the opportunity to work out an outlook on life, and freedom of action to pursue good or be tempted or entrapped by evil.

Chapter 4 helps to dismantle the view that only empirical knowledge and scientific knowledge could properly be called knowledge. Educational philosophers underpin the view that there are many forms of knowledge, such as empirical, scientific, aesthetic, moral, personal and religious.

Chapter 5 explores the conception of science as objective knowledge. This reveals how changes in the physical and biological sciences, as well as the development of chaos theory, show how systems, previously thought to be bottom-up closed systems, are in fact influenced by top-down causation and are open systems – "anything" can influence and change them.

In Chapter 6 we explore the integrity of science. Is science objective or personal knowledge? We read an account of Polanyi's analysis that scientific knowledge is based on personal discovery and, while scientific knowledge is tested for truth, results are not finally accepted as knowledge until the appropriate scientific community accepts both the theories and the results as knowledge. If the educational philosophers help to distinguish different forms and fields of knowledge, Peacocke helps to establish the emerging levels of knowledge. He shows how a higher form of knowledge is able to include a lower form of knowledge without any contradiction of its laws, encompassing them in a higher order of organizational complexity, highlighting that these transitions distinguish forms of knowledge which are different in kind. The revised nature of scientific knowledge shows us that it is provisional, revisable and full of unexpected surprises. This very fact, that these surprises cannot be predicted, underpins critical realism as the only defensible philosophical position in science. After further exploration, we shall discover that critical realism holds true in all forms of knowledge.

Chapter 7 focuses on the increasing recognition that emotion is more important than intellect ending the view that a human being is just an intellect on stilts. Macmurray opens up the discussion of different kinds of knowledge – aesthetic, empirical, scientific and personal – inverting the order of priority of kinds of knowledge and opinion held by the logical positivists. Personal knowledge, led by emotion, is at the apex of the hierarchy of knowledge for humanists, while the majority hold religious knowledge in that position. This leads to an exploration of how a biological human being, *Homo sapiens*, becomes a person – a spiritual achievement.

Is all knowledge relative? This question is explored in Chapter 8 by reviewing different contemporary views on the nature of knowledge and how these recent views of knowledge affect being a person, or, more accurately, *becoming* a person – a spiritual achievement. We then move on to explore hermeneutics, a frightening word which only means "methods of interpretation".

Chapter 9 offers a top-down account, beginning with a range of definitions of *religion*, then of *transcendence*, descending to different notions of *spirituality*, illustrating the overlapping uses of this notion between religionists and humanists, which may first have been formulated in children and held on to by some into adulthood in forms of relational consciousness between *I–Others, I–Self, I–World* and *I–God*. This chapter can be read as a bottom-up account, starting with the experience of children in which case I suggest that you begin with "Transcendence, Spirituality, Religion, and

Humanism", then read "Transcendence" followed by "Uses of Religion" and then "'To Be' or 'To Become'".

Turning to the subject of religious experience, we examine the main empirical, i.e. fact collecting, studies of religious experience in the twentieth century in Chapter 10.

Logically it follows that Chapter 11 explores humanist, empirical studies of mystical and peak experiences, and their accompanying theories. Theory provides the organizational words, or concepts and structures used to describe and classify experiences.

In Chapter 12 we identify the five model builders who give the most favoured accounts, from which eight models are selected that have the greatest acceptance.

In Chapter 13 we begin by exploring two writers who think that religious experience language is just picture language for purely human experiences, and argue for a non-realist account of religious experience. While this view is reductionist, it raises a question: how does one interpret religious experience? This leads to an exploration of levels of interpretation. How can one deal with one's suspicions when interpreting a religious experience? A method is explored to answer this question.

Having explored examples and models of religious experience, it is time to explore how Freud, Jung, and their successor in psychoanalytic writings, Object Relations Theory, treat religion and religious experience. The validity of their assessments are the focus of Chapter 14.

Chapter 15 explores the main psychological approaches and one feminist perspective about how religious experiences are formed and the effects these may have on the person who has the experience. The important "creative thinking" theory of Wallas is used, by Batson, Neitz and Sunden, to extend our understanding of religious experience.

Philosophical issues concerning religious experience begin the exploration of the integrity of religious experience in Chapter 16. Particular attention is given by Swinburne to the need to take seriously what people claim to have experienced, while Alston draws attention to the similar structure of basic beliefs in the everyday world and in religious experience. While being aware of the difficulties, Alston offers a strong case for the view that direct awareness of God provides an adequate defence of the existence of God based on testimony. Chapters 10, 12 and 16 together give a range of views from those who insist a religious experience is unique to its Faith to those who focus on shared characteristics between Faiths.

Chapter 17 notes the contributions of sociologists and phenomenologists – the people who study things as they are. We also take another look at a feminist perspective, another at views of the formation of religious experience from clinical psychologists, and another at "transitional space". New ideas about religious experience come from those reflecting on neuroscience and physical networks.

Chapter 18 analyses the Wallas models of religious experience within a contemporary context. Batson used a creative analogy of Wallas' model, successfully, showing that it could be applied to, and illuminate, our understanding of religious experiences. Miles used Wallas' model to analyse the elements in and structure of religious experience. Both models, which are complementary, represent a contribution to the contemporary psychological phenemenology of religion, using approved measurement techniques. It is argued that all forms of knowledge, except mathematics, and logic, have the same structure. Is religious experience a form of knowledge, and if so are religious experiences life enhancing?

Chapter 19 begins with facing a challenge to the two sets of philosophical assumptions of mind–body dualism, and of physicalism/materialism. It has recently been suggested that between mind–body dualism and materialism there can be a middle position called "non-reductive physicalism" which emphasizes that a person is a psychosomatic whole, with one physical system into which no external energy or spiritual substance is injected from without. How does intuition relate to different forms of knowledge? We reflect on the issue of religious experience within different Faith Communities. Then we turn to explore a sign-transformation model of religious experience, focusing on 'thought', noting the bottom-up and top-down influences, before exploring the religious experience of a child. How does the emotional intuition behind the experience become knowledge? Is emotional intuition behind all forms of knowledge?

We remember that Plantinga, Swinburne and Alston defend the pragmatic base of religious experience. Wynn's recent work focused on the emotional intuition – *a feeling (a value)*, central to religious experience, *bonded to a thought* – which could be linked to the sign-transformation model. The cumulative case advanced in this book gains its final support from Professor Robert Audi, an eminent epistemologist. So if all forms of knowledge (excluding Mathematics and Logic) are pragmatically established it seems that science and religious experience have similar forms of knowledge. Can the models of religious experience described help to provide a foundation for religious dialogue within our global village?

PART I

Uplifting Experiences
Religious and Humanist

Sources of Uplifting Experiences

LAKE WINDERMERE:
A MODERN EXPERIENCE OF UNION WITH NATURE

It was a glorious afternoon. The sun was shining on the lake and there was a clear blue sky. It was warm for early October – one of those lovely Indian summer days that sometimes come at this time of the year. A wind was driving the wavelets towards the shore of Lake Windermere in Bowness Bay where the steamers and pleasure boats clustered around the piers and jetties.

I parked my car, looked at my watch and found it was 2 p.m., so I had three hours before the first meeting I had to attend. I was alone, although the foreshore was crowded by trippers anxious to enjoy this wonderful day by the lakeside. Clouds can appear quickly in the English Lake District, so I took my plastic raincoat, rolled it up and walked towards the shore where rowing boats were available for hire. Fifty years ago I had not been allowed in one as a boy alone, but now I would remedy this by going for a row. The boatman pushed me out in a boat, after collecting his hour's hire fee.

I set off by rowing into the wind to get round the north side of Belle Isle and into the sheltered waters behind. The boat slid swiftly over the water and in no time I was around the island and approaching Thompson's Holme, a small island the shape of a banana. I stopped rowing. At that moment there was no sound or sight of any motor-boats, though there were two sailing boats in the distance, silently running downwind towards me. The leaves on the trees on the islands and around the shore line were tinted with their first autumn colours. It was a scene of classic lakeland beauty, and at that moment I was a part of it, right in the centre of it, in union and harmony with it all. The whole panorama was shot through with memories of this scene from fifty years ago, when I walked round the lake as a schoolboy.

"Forms of Knowledge" and Experiences of "Union with Nature"

Three kinds or forms of knowledge are experienced and expressed in the passage above. The first is often called the *empirical way of knowing* or *empirical knowledge*. This is sense experience, involving the perceiving of objects (I saw trees, hills and a lake), giving each object a concept, e.g. tree, involving interpretation and description (it is a tree and not a bush), and knowing your understanding is correct. Secondly, I experienced the view as beautiful. That involved almost instantaneous interpretation at the *aesthetic level of knowing*. Did I experience a sense of union with nature? Yes. That is the third level of meaning, the *sense of personal union with nature, a form of personal knowledge*. This is often called *nature mysticism*. The scene just described then involved at least these three ways of knowing mentioned.

Another way of knowing may be involved, whether mentioned or not; in the next experience, it is mentioned. Looking at the beautiful night sky above the hilly sheep pastures in Israel, the writer of Psalm 8 expressed his thoughts and feelings at a different level of meaning:

> O Lord, our Lord, your greatness is seen in all the world! . . .
> When I look at the sky which you have made,
> at the moon and the stars which you have set in their places –
> what are human beings that you think of them . . . ? (*Good News Bible* 1994)

These words also express my own thoughts and feelings. This fourth level of meaning is the *deeper, foundational or "metaphysical" meaning which, in this case, is religious*.

Do these four different levels or kinds of knowing conflict with one another? No. My third level of understanding was my feeling of union with the scene before me, which is often called nature mysticism and is a form of personal knowledge. The fourth level of meaning, which the psalmist expressed and with which I also agree, is stating the deep basis of my outlook on life, the metaphysical meaning, which is religious. By "metaphysical"[1] I mean that the psalmist states the *basic assumptions* which are the foundation of both our understandings of life, and express our "outlook on life".

Foundational Beliefs

I have stated that my foundational beliefs in life are religious. A non-religious person, a humanist, for example, might well say that "being in union with the scene" was both a statement of his thought and feeling, his personal meaning, and also his metaphysical meaning, which indicates the *basic assumptions* of his understanding of life as a humanist.

Simply, one can say that four basic assumptions distinguish the non-religious humanist from the religious person at the metaphysical level. The humanist's foundational assumptions are:

- ✧ nature is all there is;
- ✧ this life is all there is, the individual is alone in the world (i.e. there is no God);
- ✧ one is responsible alone for one's own life and for influencing life around one;
- ✧ death is the end. (Blackham 1968: 13)

For the humanist, naturalism expresses the view that the world you see is all that there is, and nature mysticism is a purely human relationship with nature. The humanist, if an atheist, will say 'there is no God' or supernatural agency, as above, or, if an agnostic, may say they see no convincing argument for believing there is a God.

The central difference between the religious and humanist is that each *interprets* life at the metaphysical level differently, holding their basic assumptions as foundational beliefs which they claim to be true.

So, in a religious "outlook on life", the existence of 'God' as creator is the minimal foundational, or metaphysical, assumption, while in a humanist outlook, 'human life' is the foundational, or metaphysical, assumption underlying their "outlook on life". There are of course other foundational assumptions, held by other people. Some people are "materialists", believing that all that exists fundamentally is "matter". Some wildly claim that "football is their religion", though most people who say that are really indicating that football is their greatest interest in life, an empirical meaning rather than a foundational or metaphysical meaning.

BACK TO LAKE WINDERMERE:
WORDSWORTH AND NATURE MYSTICISM

I looked up at the ring of hills that rose up around the northern end of Lake Windermere and around Grasmere and Ullswater to the north, then I rowed towards the island ahead, landed and walked the length of Thompson's Holme, with the wind rustling through the trees. This lovely little island is looked after by the pupils of a local school.

As I walked I recollected how William Wordsworth (1770–1850), 200 years ago, had drunk in the beauty and strangeness of nature at Grasmere, Ullswater and other parts of the Lake District and frequently fallen into trances, or transports of delight, wonder and sometimes terror. In his autobiographical poem *The Prelude*, he describes the impact of nature on him after the age of ten. He describes how, while growing up, he had been influenced by many experiences of beauty and fear through his contact with nature – the lakeland scene around him. They fostered the *aesthetic* and *moral* dimensions of his personality:

> Fair seed-time had my soul, and I grew up
> Fostered alike by beauty and by fear . . . [2]

A number of these experiences were particularly significant, e.g. the stolen boat at Ullswater (Wordsworth, *The Prelude*, 1850: bk I, lines 357–400). One source of "fear" which worried him was his conscience. Wordsworth describes this incident in his poem *The Prelude* quoted above. One summer evening, without permission, he took a little rowing boat from its usual mooring in a rocky cave. His powerful rowing made the boat ride through the water "like a swan". As he rowed out, a craggy peak increasingly towered over him. He thought that it stood above him in judgement, separating him from the stars. Conscience stricken, he quickly turned the boat and rowed swiftly back to its mooring.

The "huge peak, black and huge" is a picture of his conscience, which follows him, accusing him of stealing the boat. He notes that this association of large land forms and "Judgement from Above" led to experiences in his mind of "huge and mighty forms . . . [which were] a trouble to his dreams".[3] He goes on to say that through such visions he found that the wisdom and spirit of the universe sanctified pain and fear, as well as beauty, ennobling the significance of both nature and human nature.

When he was eighteen years old, a critical moment was reached in this long series of joyous and mysterious moments when, early one morning during the long vacation, he crossed a moor at sunrise and saw the mountains transfigured "grain-tinctured, drenched in empyrean light" (*The Prelude*, p. 143, bk 4, line 328) ("empyrean" refers to the sphere of pure fire – the abode of God). In the presence of this divine light, like Moses in the presence of the burning bush, he felt a call and responded with a dedication:

> I made no vows, but vows
> Were made for me; bond unknown to me
> Was given, that I should be, else sinning greatly,
> A dedicated Spirit. (*The Prelude*, 1850: p. 143, bk 4, lines 334–37)

This religious calling, a religious experience, and his dedication in response to it, committed him to become a poet both of nature and of human beings.

Wordsworth, in these extracts, reports experiences in which the dominant levels of meaning and modes of knowing were *aesthetic, moral* and *religious* respectively.

Wordsworth, Personal Union and Rapport

Wordworth's concern with human beings and relationships is well illustrated in "Michael, a Pastoral Poem", which he wrote when he was thirty years of age. He first expresses his own response to the power of Nature, then turns to the importance of human life and relationships:

> while I was yet a Boy,
> Careless of books, yet having felt the power
> Of Nature, by the gentle agency
> Of natural objects, led me on to feel
> For passions that were not my own, and think
> (At random and imperfectly indeed)
> On man, the heart of man, and human life. (De Selincourt (1965), Vol. 2, p. 81, lines 27–33)

In particular he focuses on human commitment as he talks about Michael the Shepherd:

> in his shepherd's calling he was prompt
> And watchful more than ordinary men. (De Selincourt (1965), Vol. 2, p. 82, lines 46–47)

Until the age of eighty, Michael's work involved him going out to rescue his sheep in blizzards, a time when the hills were places of great danger:

> hills, which with vigorous step
> He had so often climbed; which had impressed
> So many incidents upon his mind
> Of hardship, skill or courage, joy or fear;
> Which, like a book, preserved the memory
> Of the dumb animals whom he had saved,
> Had fed or sheltered. (De Selincourt 1965, Vol. 2, p. 82, lines 66–72)

And yet for Michael

> Those fields . . . [which] had laid
> Strong hold on his affections, were to him
> A pleasurable feeling of blind love,
> The pleasure which there is in life itself. (Vol. 2, p. 83, lines 74–77)

The poem continues Michael's story, telling of his faithful wife and Luke, the child of their old age, who had worked with his father until bad news came of an unknown and unexpected debt which had to be paid. Many years earlier, as surety to start his brother's son in business, Michael had guaranteed to meet his brother's son's debts if necessary. It was necessary and the debt could be met only by selling half of all his property. The mortgage on the farmland had only been paid off three weeks earlier.

So Luke left home for the city to earn money to try and meet this demand. He began well but then fell into bad ways; in shame he was driven "to seek a hiding place beyond the seas". Even after receiving this news, the old shepherd faithfully continued his work until his death. The farm was sold three years later after the wife died. After Michael received the news of his son's disappearance overseas and his reasons for going, Wordsworth says in the poem:

> There is a comfort in the strength of love;
> 'Twill make a thing endurable, which else
> Would overset the brain, or break the heart. (p. 93, lines 448–50)

Similar feelings can be found in the poem which follows "Michael", "The Widow on Windermere Side" (De Selincourt 1965: Vol. 2).

Sometimes these experiences are described by Wordsworth as a "visitation from the living God",[4] but often such an explicit reference is omitted, which is why he is sometimes regarded as an exponent of nature mysticism or nature ecstasy (Butler 1967: 228). He gave expression to these "spots of time" (to which he would constantly repair in memory) often years later, drawing from them creative and moral power, often as religious experiences in the presence of God, who dwells in the light of fire and setting suns.[5]

Comment on Wordsworth's Work

First, in the extracts in the sections above, Wordsworth focuses on what it means to be human and engage in personal relationships. At the centre of relationships, he suggests, only love is able to make disastrous work situations, disastrous human situations or disastrous relationships bearable. They would otherwise break one's heart and send one mad.

The supreme example of this was Jesus of Nazareth, whose love for the people around him helped him to forgive from the cross all those whose actions or inaction had placed him there. Knowing what it means to be human and engage in personal relationships is a distinct form of knowing that is different from empirical, aesthetic or moral knowledge. It is *personal knowing*.

Here the mode of knowing, or kind of discourse used, is *personal knowledge*. As we have seen previously, personal knowledge often *includes these other forms of knowledge*, and as such is called a "higher" form of knowing, because its inclusive and personal nature gives it greater value. So far we have noted two forms of personal knowledge: knowing which involves a sense of mystical union with nature, and knowing which involves relationships with one's work and with other people.

Secondly, the experiences of nature mysticism Wordsworth describes often seem to be infused with Wisdom and Spirit, hinting at the idea of pantheism, i.e. God within nature. These references have sometimes been interpreted as referring to the Spirit of nature, in other words secular naturalism. However, the mention of "empyrean light", "Wisdom" and "Spirit" makes this unlikely, as they are three of the images referring to God in the brief extracts mentioned above. Wordsworth accepted the calling to be a poet rather than that of a clergyman, though he strongly supported the Church (Byatt 1997: 13, 72).

Thirdly, Wordsworth was intuitively aware of, and distinguishes, the different ways of knowing in the experiences he describes. He knows that nature mysticism involves empirical, aesthetic, personal, moral and religious ways of knowing in different combinations. As has been noted, he knows that personal knowledge, as explored in his poem "Michael", often involves all these kinds of knowledge. And also that when this knowledge is underpinned by love, whether of work, of people or both, it is a source of faith and hope in the worthwhileness of human life. This insight is at the heart of the great world religions, e.g. Judaism and Christianity, which was the religious cultural source which Wordsworth knew. This underpinned his personal experiences and insights (e.g. Leviticus 19: 18; Deuteronomy 6: 4; Mark 12: 29–31; 1 John 4: 7–9a.

Fourthly, nature mysticism and personal mysticism were frequently *sources of spiritual energy and moral encouragement*.

This leads to the question, were Wordsworth and the other poets of his time unique in having these sensitivities, or do many of us today sometimes feel this way?

MICHAEL PAFFARD'S *INGLORIOUS WORDSWORTHS*

In 1973 Paffard published a book called *Inglorious Wordsworths*. He taught sixth form pupils and later undergraduates. He says: "I found myself wondering how many of them (a small and not very interested class of sixth form boys) knew at first hand what Wordsworth was talking about in 'Tintern Abbey' and the 'Intimations' ode" (Paffard 1973: 21). The method he used to explore this question was to quote a description of a typical transcendental experience in the questionnaire he designed and asked the respondents to write about any experiences of their own which they felt were in any way similar to the one quoted (Paffard 1973: 85). The transcendental experience he quoted contained no God-talk; it was an example of nature mysticism.

Paffard defined a *'transcendental' experience*,[6] indicating that it was an uplifting experience 'outside of' or 'greater than' normal experience of an intimate spiritual relationship with nature or with a person. Paffard suggests that such an experience must have some of the following ten characteristics:[7]

1   be inexpressible – not put into words easily
2   be brief
3   be rare
4   lead towards contact or union
5   seem outside one's body
6   be valuable or important
7   give knowledge or insight
8   have a sense of being outside time or space
9   be experienced as having a divine or supernatural origin
10   involve an intense aesthetic experience (e.g. of beauty, harmony etc.).

This is the passage from W. H. Hudson's *From Far Away and Long Ago* (1936) that Paffard gave the pupils to read:

> It was not, I think, till my eighth year that I began to be distinctly conscious of something more than this mere childish delight in nature. It may have been there all the time from infancy – I don't know; but when I began to know it consciously it was as if some hand had surreptitiously dropped something into the honeyed cup which gave it at times a new flavour. It gave me little thrills, at times purely pleasurable, at other times startling, and there were occasions when it became so poignant as to frighten me. The sight of a magnificent sunset was sometimes more than I could endure and made me wish to hide myself away. The feeling, however, was evoked more powerfully by trees than by any other sight; it varied in power according to the time and place and the appearance of the tree or trees, and always affected me most on moonlit nights. Frequently after I had first begun to experience it consciously, I would go out of my way to meet it, and I used to steal out of the house alone when the moon was at its full to stand, silent and motionless, near some group of large trees, gazing at the dusky green foliage silvered by the beams; and at such times the sense of mystery would grow until a sensation of delight would change to fear, and the fear increase until it was no longer to be borne, and I would hastily escape to recover the sense of reality and safety indoors, where there was light and company. (Hudson 1936, ch. 8; Paffard 1973, ch. 1; Miles 1983: 435)

The pupils were asked: "Does this remind you of anything you have ever felt? If you have ever had an experience which you feel is in any way similar to the ones the writer of this passage is describing, please try to write about it on the blank page overleaf. (It does not matter in the least where you were or what you were doing at the time; for instance, writers have described having experiences like this in a train, or ill in bed or while reading a book, etc.)"

Before proceeding any further it is necessary to make some simple distinctions to avoid confusion in our examination of the experiences which will be quoted later in our exploration.

Transcendental experience has been defined earlier by Paffard as an uplifting experience of an intimate spiritual relationship with nature or with a person "outside of" or "greater than" normal experience. Unless otherwise stated, it will be assumed that a transcendental experience can also refer to an experience of God's Presence. Transcendental experience may be either unitary or bipolar. However, Paffard and Wordsworth do not distinguish between bipolar and union forms of mysticism. Other experiences, like religious experiences or aesthetic experiences can be bipolar. For example, I may enjoy a beautiful scene, but the scene and I can remain distinct. That is aesthetic experience. Likewise, I may have an experience of the Presence of God, or like a great friend very much indeed, but in both cases I can remain distinct from what is the object of my experience. However, it is possible for a bipolar experience, e.g. relating to God, to develop into a union experience, e.g. union with God, but we need not explore this at present.

Wordsworth is mainly concerned with *mystical experience*. Mystical experience has three main forms. It can be defined as union between nature and an individual, in the case of nature mysticism; it may refer to union between two people, so that they see themselves as one, as suggested in the Christian marriage service, where two become one 'personal mystical union'; and it may indicate religious mystical union, as when a religious devotee claims they are one with God.

Paffard talks about transcendental experience, but it will be necessary to divide up the experiences he quotes into ten subcategories clustered into four groups (nms = nature mysticism secular; pms = personal mysticism secular, and so on):

1   nature or person mysticism – secular; (*nms, pms*)
2   nature or person mysticism – religious; (*nmr, pmr*)
     and mystical union with God; (*mG*)
3   aesthetic, moral or personal experience – secular; (*aes, pes*)
4   aesthetic, moral or personal experience – religious; (*aer, per*)
     and religious experience meaning experience of God's Presence. (*eG*)

Paffard's Transcendental Experiences: Religious and Secular

Here are four examples of the experiences collected from sixth form and undergraduate students by Paffard. The subject SM18, was a *S*ixth-form *M*ale aged *18*.

1   SM18 "Ever since early childhood I have been moved by occasional overwhelming sensations of the smallness of myself, and the greatness and magnificence of the universe. Thus, when about ten, I arrived at St. Pancras Station on my first journey to London, and stepped out of the train carriage and stood amazed at this enormous structure of stone and glass. I was in London, *London*, the capital of England! Since then I have felt the same, upon a larger scale, when reading or talking about the infinite size of the universe.

In direct contrast to the 'smallness' I feel on occasions, such as described above, on other occasions I feel a sensation of greatness, exhilaration or joy. For example when listening to a moving piece of music, or loving and being loved. These incidents . . . remain something outside, and superior to, my everyday life, indefinite and indefinable." (*nm, pm*) (Paffard 1973: 173; Miles 1983: 422)

Emotions evoked here by nature and majestic buildings include awe and amazement – the contrast between his smallness and the greatness of the universe or great buildings. This experience is not just at the level of sense experience, the empirical level, but also at the personal level, involving his emotions in a personal response greater than just aesthetic appreciation, but involving personal knowing as a result of a symbiotic union with nature (*nm*).

Emotions evoked by music and relationships of loving and being loved evoke feelings of greatness, exhilaration and joy, which indicate personal knowing through communion, or union, with the other (*pm*).

These two experiences of the emotions of amazement and awe, evoked first by nature and second by loving and being loved by persons, can both provoke an interpretive response at a foundational or metaphysical level which may be religious or secular. This particular experience contains no report of such reflections. Paffard's question is centrally concerned with whether a pupil has experienced similar *feelings* to those expressed in the passage he quoted from Hudson as a stimulus for their thought. The next sixth-form male answers in the way requested:

2   SM16: "Often when out alone at night, when the sky is cloudless, I cannot help gazing at the universe beyond our tiny sphere. It provokes in me a great feeling of wonder and makes me realize how relatively worthless we are when considered side by side with all the work of the great Creator. I cannot help trying to think of a solution as to how space can stop, the mere thought of it going on endlessly almost giving me a brain storm. I am filled by utmost awe by the spectacle." (*per*) (Paffard 1973: 174; Miles 1983: 422)

This sixth-form male experiences emotions of wonder, insignificance and awe at the immensity of the universe. For him it was a religious experience of an interpretive kind. It was not an experience of the Presence of God, but rather an experience of the world he believed was *created* by God; his prior religious beliefs were essential to his interpretation of what he experienced. 'UF' means undergraduate female etc.

3   UF19: "I have had feelings similar to those suggested quite a few times, particularly in my own garden at home. It is surrounded on two sides by tall poplar trees which at times have a menacing effect. The garden is very damp and everything grows luxuriously. Often standing in it towards the evening, I have had the sensation of meeting an overpowering spirit, far stronger than anything that could be experienced from encountering another person, and the sight of the delicate blackness of these trees against the heavy blue of the evening sky has thrilled me to a point when I could stand the eerieness no longer and must go indoors. Often in open country, too, I have had a feeling of loneliness and again the feeling of something far stronger than myself was present. This 'something' I prefer to call God." (*eG*) (Paffard 1973: 175)

15

This nineteen-year-old undergraduate female experiences being threatened and thrilled with the beauty of nature, evoking feelings of eerieness and loneliness, and also the sensation of meeting an overpowering spirit, a strong sense of 'presence', which she called God. This religious experience includes an appreciation of the empirical, aesthetic and personal relational dimensions of this experience of the Presence of God.

> 4   UF21: "Industrial scenes, particularly in calm surroundings with a sugges-
> tion of mystery, can be very awe-inspiring, particularly huge cranes, chimneys
> or cooling towers but I suspect that these feelings are merely the normal reac-
> tions of most people to such things . . . these feelings are predominantly
> awe-inspiring, making me realize how very powerless and insignificant I am
> and yet they are included, perhaps predominant among the things which
> make me feel how worthwhile it is to be alive." (*nm*) (Paffard 1973: 190)

This undergraduate experiences emotions of mystery, awe, powerlessness, insignificance in an industrial landscape, which, paradoxically, gives her the feeling of the worthwhileness of life. This passage hints at mystical union as the source of 'worth-whileness of life', which she seems to deduce intuitively from the fact that she is part of the scene.

These kinds of responses were checked and confirmed in a replication study with sixth form students. Here are four examples from that project (Miles 1983: 474–8):[8]

> 5   EF39: "When I travelled to America with my family last year, one night whilst
> we were camping in a large picturesque mountainous region, my father,
> brother, uncle and myself went out for a walk in the early evening. It was I
> think the most beautiful sight that I have ever seen. We tended to separate
> slightly, and I wandered off on my own towards an enormous river. I sat
> myself on a large rock next to the river, and drank in the surroundings. It was
> an exhilarating experience. As I looked up to the oncoming dusk, I could see
> stars appearing from behind the mountain tops. It was so very quiet and
> peaceful there that I could almost have believed that I was the only person in
> the world at that moment. The landscape was so enormous and awesome
> that I felt quite overwhelmed, and began to have deep thoughts about the
> meaning of life, who I was, what I was doing there, what I was doing with my
> life, and many other such feelings. I felt that I could have sat there for hours
> on end. Although I have had many such thoughts as that since then, never
> have they been quite the same, or stuck in my memory so easily as the evening
> I spent in America." (*nm*)

This young sixth-form woman experienced feelings of exhilaration, peacefulness, awe, of being overwhelmed. This is an experience of nature mysticism, because she felt herself to be part of the scene, which provoked deep reflection about the meaning and significance of life and her place in the world in which she found herself.

> 6   CF155: "I often admire the beauty of God's world . . . When I was small we
> lived in a block of terraces and there was . . . a very large park nearby. This

park was freedom for me and I was never happier than there – at all seasons
. . . " (*aer*)

This female enjoyed the beauty of nature and the freedom it gave her, but it all
speaks to her of God's world. Here her prior belief in God as Creator was foundational in her interpretation of her experience, but she could well say that she was simply
describing her experience. For this reason it will later be necessary to explore the relationship between "description" and "interpretation".

7   EM21: "The passage evokes an experience I felt when looking over the
     Severn Estuary from Portishead to Wales. The sunset was fading but was
     still magnificently coloured, and in the uncanny darkness, the lights on the
     other side of the estuary were reflected in the water. This gave a view of
     colour which was quite magnificent. For a while I stood and looked, quite
     stunned." (*ae*)

This male expressed emotions of eeriness, overpoweringness and awe in the presence of the beauty of nature: pure aesthetic appreciation. There is not quite enough
evidence to suggest that it is an example of nature mysticism. He is simply answering
the question he was asked, namely, for an account of his feelings, and gives an account
of an intense aesthetic experience.

8   EM 059: "I like in the day to be all alone in a large field and lie on my back
     and see nothing but the large blue cloudless sky, so deep, mystical, religious,
     so powerful . . ." (*nmr*).

This male, who experiences emotions of immensity, mystery and power, comes
close to expressing the sentiments of the writer of Psalm 8: 3–4, quoted earlier on page
8. "When I look at the sky which you have made . . . what are human beings that you
think of them." Both the student and the Psalmist describe a religious experience in
the sense that each assumes the existence of God, a basic belief, and each interprets
his experience of nature religiously.
   The experiences quoted above provide examples of nature mysticism in young
people. Examples 1, 4 and 5 cannot be interpreted as evidence of the underlying foundational or metaphysical position of these students, either religious or humanist,
because they were not asked for their foundational beliefs. So they could be religious people
giving an account of an experience of nature mysticism, or humanists doing the same.
Others, unasked, did give a religious metaphysical base to their experience, because,
for them, it was an intrinsic aspect of the experience. We can nevertheless assume that
some of those who did not give a metaphysical base to their experience would be religious and that some would be secular and probably humanists. The humanists, then,
were giving a statement of their metaphysical foundations, as for them only the natural
world exists.
   The three examples of pure nature mysticism cited above, examples 1, 4 and 5, all
affirmed personal value and the meaning of life, while the four religious examples, 2,
3, 6 and 8, affirm God's Presence in different ways. Example 3 refers to an experi-

ence of God, while the other three interpret their experiences in the light of their belief in God. All four unconsciously invoke the estimate of human importance affirmed in Genesis 1: 26 and 2: 7, that humans are the climax of creation and its most important creatures, because they are "made in the image of God". Another important point has already been made by Wordsworth above, namely, that these experiences of nature, whether interpreted as a humanistic or a religious event, were frequently recalled and were *a source of creative and moral power*. This could apply to example 7 above.

Another example taken from the follow-up study will suffice as further support for Wordsworth's claim:

> 9   EF084: "It was just this year that I felt a similar experience. On holiday to get away from the hustle and bustle I would go for solitary walks along the beach in the evening. Here with the wind blowing through my hair I felt at peace. My mind seemed to be cleared of its cobwebs, and I could straighten out any worries. To make the evening complete was to see the sun setting against the sky and the sea. It was a really magnificent sight and my heart would lurch and pound faster just as it sank out of sight. I then felt set to meet the demands of the modern world." (Miles 1983: 475)

Paffard's Classification

Paffard collected experiences from 200 pupils (and 200 undergraduates) whom he tested. He classified them into fifty-five possible categories of experience, using the criterion of one or two of the ten emotions: *joy, pleasure, enhancement, calm, trance, longing, melancholy, awe, pain, fear* (e.g. "joy" or "awe" or "pleasure-calm", or "pleasure-fear", etc.). In completing his piece of research Paffard expressed four concerns:[9]

1   He wanted 'the spiritual' expressed in non-religious concepts and stories when relating
    (a) our "highest" experiences of ecstasy and awe;
    (b) the values we hold of supreme importance;
    (c) and for all experiences "in another dimension".
2   Education should not only concern facts, ideas and arguments but also feelings of ecstasy and awe which may result from reflection and contemplation.
3   Transcendental experiences explored in the curriculum can be a source of strength and joy.
4   Transcendental experiences could be the start of religious, moral and aesthetic life.

Comments on Paffard's Work

First, Paffard, carefully avoiding any religious language or stimulus, starts from agnostic assumptions in the passage he gave to his students to begin their own reflections about nature mysticism. This has the enormous advantage of helping the reader recognize that any references to religion in the answers students gave indicated the religious, and therefore metaphysical, basis of their outlook on life. *Secondly*, an answer in the form of an expression of nature mysticism *alone*, without qualifying comment, e.g. "nature is all that exists", may simply be a statement of aesthetic harmony with nature, which does not of itself indicate whether the student in question is a humanist or a religious person. It may be a statement of the metaphysical assumptions of a humanist, but unless that is clearly indicated, it cannot be assumed. *Thirdly*, in Paffard's four final comments, there is inconsistency between his first and fourth comments, for if he really means that a transcendental experience may be the basis of religious insight, it then follows that this will be expressed in religious language. It is rarely possible to express a religious interpretation of the spiritual in non-religious language. *Fourthly*, Paffard's study is a model of how to enquire into Wordsworthian experiences of people today, and accurately reflected Wordsworth's own position in interpreting the personal relationships he describes with nature or other people in terms, *first*, of nature mysticism or personal mysticism *or*, *second*, in sometimes indicating their underlying foundational beliefs, describing them then as religious nature mysticism or religious personal mysticism.

SOURCES OF UPLIFTING EXPERIENCES: AN OVERVIEW

First, so far the exploration of transcendental experiences (Paffard) or nature mysticism and personal mysticism (Wordsworth) has identified many aspects of nature and personal relationships as sources, or, more accurately, as triggers of these kinds of experience. There are many other triggers for such experiences. Some mentioned above include music, pictures, the appearance of design in the world, the reality of death, experiences in church.

Secondly, attention has been drawn to the presence of specific and powerful emotions which are integral elements of these experiences.

Thirdly, we have explored how examples of nature mysticism 'feeling in union with nature' or personal mysticism 'feeling in union with another person' can be experienced in at least five different levels of meaning:

- ✧ empirical – a landscape of trees; its beauty – aesthetic;
- ✧ personal union with nature or with a person, two forms of personal knowledge;
- ✧ at a metaphysical level or foundational level as religious or as humanist nature mysticism.

Wordsworth notes that such experiences were a source of creative and moral power to him.[10]

Looking Back

All this makes sense to me now, but when I was in my teens, while I secretly enjoyed the feelings described, I would not discuss them because they were thought not to be knowledge. After all, in school we only dealt with knowledge in such subjects as Maths,

Physics, Bible Knowledge, English Literature, Geography and History. Attention focused on facts in each case, so I would not have considered experiences like those described earlier in this chapter as having any important meaning or value.

Indeed, the crunch came when I was about thirteen. I suddenly discovered that a statement I believed to be true, i.e. "Jesus is the Son of God", wasn't "true" any more. I felt cheated and annoyed. But I have not yet made myself clear. What I mean is that I thought then that all statements which were true were *literally* or *empirically* true.

I thought that "Jesus is the son of God" made the same kind of biological or empirical claim as "Jim is the son of Joan", as the sentences are identical in structure. The second sentence is factually or empirically true: it can be supported by evidence available to the senses. We can outline the steps by which it can be tested: look at Jim's birth certificate, ask the midwife, do a DNA test. It can be shown to be factually true, i.e. is is *knowledge*. This discovery of mine took place about 1947. I didn't know what kind of statement the first sentence was and didn't know how some people could claim it to be true. How could it be true? Was it knowledge? If not, what kind of sentence was it? If it was knowledge, what kind of knowledge was it? It just seemed to me to be nonsense. What could I do?[11]

The only answer was to set out on a quest to find out what 'knowledge' was. It has taken many decades of exploration. So, before I can return to an exploration of 'religious experience', I need to know what 'knowledge' means, and what it is.

NOTES

1 'Metaphysics' is used in this work to indicate the statement of first principles or assumptions behind all knowledge claims. All structures of thought, e.g. logic, mathematics, history or religion, make assumptions as the foundation of any statement or argument. Every knowledge claim is based on underlying metaphysical assumptions or presuppositions. These absolute presuppositions may vary from one culture to another, or more obviously, from one discipline or form of knowledge to another. Even if a metaphysic is correctly deemed to refer to a holistic outlook on life, the underlying foundation statements or assumptions lay down the metaphysical base of such a view. This book will not be concerned with other uses of the term "metaphysics".

2 Wordsworth *The Prelude* (1850: bk I, lines 301–2 in J. Wordsworth, M. H. Abrams and S. Gill (eds), (1979) *The Prelude*. New York, London: Norton, p. 45). While references are given to the 1850 edition, the first published text, the 1805–6 edition, which was read to Coleridge, has the immediacy of the original inspiration, which I frequently prefer for this reason. It is printed on the left-hand page of this edition (cf. Moorman 1963: 23.589–9).

3 One summer evening . . . I found
    A little boat tied to a willow tree
    Within a rocky cave, its usual home.
    Straight I unloosed her chain, and stepping in
    Pushed from the shore. It was an act of stealth
    And troubled pleasure, nor without the voice
    Of mountain-echoes did my boat move on;
    Leaving behind her still, on either side,
    Small circles glittering idly in the moon,
    Until they melted into one track
    Of sparkling light . . . lustily
    I dipped my oars into the silent lake,

And as I rose upon the stroke, my boat
Went heaving through the water like a swan;
When, from behind that craggy steep till then
The horizon's bound, a huge peak, black and huge,
As if with voluntary power instinct
Upreared its head. I struck and struck again,
And growing still in stature the grim shape
Towered up between me and the stars, and still,
For so it seemed, with purpose of its own
And measured motion like a living thing,
Strode after me. With trembling oars I turned,
And through the silent water stole my way
Back to the covert of the willow tree:
There in her mooring place I left my bark . . . (*The Prelude*, 1850, p. 49, bk I, lines
357–67, 373–88)

4   W. Wordsworth (1814), 'The Excursion', bk I, line 212 in E. De Selincourt (1965),
Wordsworth's Collected Works, Vol. 5, p. 15.

5   For example, 'Lines Composed a Few Miles above Tintern Abbey' (1798), in De
Selincourt (1965), *Wordsworth's Collected Works*, Vol. 2, pp. 259–62:

Reflecting on the pastoral view at Tintern Abbey, he said:

These beauteous forms,
Through a long absence, have not been to me
As in a landscape to a blind man's eye:
But oft, in lonely rooms, and 'mid the din
Of towns and cities, I have owed to them
In hours of weariness, sensations sweet,
Felt in the blood, and felt along the heart;
And passing even into my purer mind,
With tranquil restoration . . . (lines 22–30)

Later, Wordsworth hints at religious presence:

And I have felt
A presence that disturbs me with the joy
Of elevated thoughts; a sense sublime
Of something far more deeply interfused,
Whose dwelling is the light of setting sun (lines 93–97)
. . . a spirit, that impels
All thinking things
And rolls through all things. (lines 100–2)

Finally, the poet says that in nature, thus transformed, he found a source of moral
encouragement:

The guide the guardian of my heart, and soul
Of all my moral being. (lines 110–11)

6   A transcendental experience is "characterized not so much by the external stimuli or occasion, which may or may not be in any way extraordinary, as by the subjective response. In

other words, the subject must feel that it transcends the ordinary modes of his consciousness, thought, feeling or perception. *If* the experience is somehow 'in another dimension', outside the frontiers of ordinary experience, a necessary concomitant will be that it cannot be adequately described in ordinary language: it will be felt to be ineffable" (Paffard 1973: 33).

7 Paffard (1973: 35): ". . . ineffable; transitory; rare; unitive or desire for union or contact; loss of bodily sensation, of being; outside the body valuable or important (but not necessarily joyful); giving knowledge or insight; having a divine or supernatural origin; exceptionally intense form of aesthetic experience."

8 The experiences quoted are labelled CF (control female) or EM (experimental male) etc. The experimental group was, and the control group was not, taught the programme of lessons in the 1983 project. The number is the student's reference number.

9 First, he feels: "We need a category of the 'spiritual', but divested of myth and supernaturalistic connotations, for our highest experiences of ecstasy and awe, our deepest intuitions of value, for those experiences which seem somehow 'another dimension'" (Paffard 1973: 226).

   Secondly, he feels: "Education reflects knowledge that is not propositional . . . it cannot comprehend reason of the heart as well as of the head." Intellectual education "cannot tolerate browsing, rambling . . . contemplation, and yet it is at those times when we are least self consciously purposive that we are most likely to know transcendent ecstasy and awe" (227).

   Thirdly, if we accept transcendental experiences as an element in the curriculum "we can help ourselves and others to share their source of strength and joy" (231).

   Fourthly, "Transcendental experience . . . could be the essential germ from which religious, moral and aesthetic life begins to grow" (231).

10 This account of awe and wonder experiences takes no account of our common experiences of suffering, illness, torture, war, disaster, cruelty, estrangement and similar negative aspects of our relationship with the world as it is, e.g. earthquake, flood. This is not our focus here. See Chapter 3, Personal Knowledge of Human Evil.

11 Cf. the young Pip's response to "language" in Dickens's *Great Expectations* (Soskice 1997: 197–203).

REFERENCES

Blackham, H. (1968), *Humanism,* Harmondsworth: Penguin.
Butler, D. C. (1967), *Western Mysticism,* London: Constable.
Byatt, A. S. (1997), *Unruly Times: Wordsworth and Coleridge in Their Time,* London: Vintage.
De Selincourt, E. (ed.), (1965), *Wordsworth's Collected Works,* Vols. 2 & 5, Oxford: Oxford University Press.
Hudson, W. H. (1936), *From Far Away and Long Ago,* London: Dent.
Miles, G. B. (1983), "A Critical and Experimental Study of Adolescents' Attitudes to and Understanding of Transcendental Experience" (PhD dissertation, University of Leeds), Vol. 2, p. 435.
Moorman, M. (1963), "Wordsworth, William", in *Collier's Encyclopaedia,* New York: Crowell-Collier, vol. 23, pp. 589–90.
Paffard, M. (1973), *Inglorious Wordsworths,* London: Hodder & Stoughton.
Soskice, J. M. (1997), "Religious Language", in P. L. Quinn and C. Taliaferro (eds), (1999), *A Companion to the Philosophy of Religion,* Oxford: Blackwell, pp. 197–203.
Wordsworth, W. (1850), *The Prelude* (1979), in J. Wordsworth, M. H. Abrams and S. Gill (eds), New York and London: Norton, p. 45.

PART II

What is Knowledge?

2 | How Do We Know What Knowledge Is?

A European Search for Objective Knowledge

For more than 200 years science has been an assured source of accurate knowledge. It is seen to be objective, i.e. it is not based on subjective opinion, but on evidence, and theory related to it and tested by it. Some European philosophers set out to establish similar clear tests for truth. From about 1920 to 1953[1] they analyzed and refined the understanding of language and how to test what was knowledge. The movement, called 'logical positivism', focused their attention on the meaning of sentences, before outlining their view of the nature of knowledge.

KINDS OF KNOWLEDGE

The concern here is not the history of different theories of knowledge, as this can be found elsewhere (Chisholm 1966; Hamlyn 1970; Scheffler 1978; Trusted 1997). We only need to know that the three main theories of knowledge at the beginning of the twentieth century were Rationalism, Empiricism and Pragmatism. No discussion of pragmatism is necessary here as it was from empiricism and the success of scientific methodology that logical positivism emerged. Brief mention will be made of rationalism, as empiricism also uses reason.

The positivists began with facts and with theories which tried to explain them and their relationship to one another. They believed that knowledge was established only through the process of verification and testing, to see if the facts and theories were true or false. Logical positivism, and philosophical analysis which grew out of it, can be summarized in three questions:

1   Are there such things as facts, or is our knowledge of the world *interpreted* experience?
2   Is science the only kind of knowledge, or are there other kinds?
3   Is religion a form of knowledge, and is religious experience first-hand knowledge?

These questions need exploring and answering in different ways. They provide the agenda for our exploration of religious experience, as we need to know how religious experience fits into an inclusive picture of knowledge.

We begin this exploration by making use of one of the achievements of logical positivism and philosophical analysis, both of which will be defined later when we explore what the terms mean. The achievement is the demand that language is used as carefully and as precisely as possible.

Different Kinds of Sentences

When I was about thirteen I suddenly discovered that a statement I believed to be true, "Jesus is the Son of God" was not *literally* or *empirically* true. I thought that it meant the same as "Jim is the son of Joan", which is literally true, as the sentences are identical in structure. I did not know what kind of statement the first sentence was and did not know how some people could claim it to be true. How could it be true? Was it knowledge? If it was not, what kind of sentence was it? If it was knowledge, what kind of knowledge was it? Over a long period of time I found out there are other kinds of sentences associated with different ways of knowing. Here is a list of some:

✧ That is a beautiful building.
✧ She is an honest policewoman.
✧ I have several times felt the so-called 'consciousness of a presence' (James 1968: 34).
✧ He is a good friend.
✧ There are five of them in the group.
✧ I know that my Redeemer lives.

What kinds of sentences are they? Are they true? Are they pieces of knowledge?

We will explore these questions slowly and carefully, one step at a time, looking for help from specialists, exploring each central idea or concept carefully.

Using Language

Later on, we will find that the philosophers and other writers we must consult talk about "statements" and "propositions", so we need to know what they mean by those words.

In everyday speech the terms "sentence", "statement" and "proposition" are used rather loosely. To avoid confusion, philosophers often give these terms precise meanings.

A *sentence* will refer to any series of words, in connected speech or writing, forming a grammatically complete expression of a single thought, contained between two full stops. Some examples are:

✧ Chitty Chitty Bang Bang (a flying car in a children's story) flew over the sea.
✧ The lion chased the unicorn.
✧ The moon is made of green cheese.
✧ The moon is beautiful.
✧ There is a chair there.

These examples show that sentences can be fiction, fantasy, feeling, fact, or other kinds of utterance (Ayer 1982: 10).

A *statement* indicates a state of affairs which can be expressed in different sentences. For example:

- ✧ Joan is his wife = he is married to Joan.
- ✧ The moon is beautiful = the moon is aesthetically pleasing.
- ✧ God is omnipotent = God is all powerful.

Statements, like sentences, can be fiction, fantasy, feeling, fact, or other kinds of utterance (Ayer 1982: 11).

The word *proposition* is reserved for a claim to be stating the truth. That claim is only accepted as true when it has been tested through verification procedures involving a series of identifiable steps. (According to Ayer [1982: 11], 'propositions' form a subclass of 'statements'.) For example:

- ✧ There are three chairs here.   Test: count them.
- ✧ This church has no font.   Test: ask if the church has a font. If yes, ask where it is, then check.

EMPIRICAL KNOWLEDGE

Our interest in this topic is to try to discover the answers to three questions. The first is, are we dealing with facts, or observations which need *interpretation*? (This question takes us into the theory of knowledge, called epistemology.)

Let us begin with the world we all know today, the world of sense experience, the world of trees, houses, animals, shops, tables and chairs – the empirical world. Empirical knowledge is the knowledge which is tested by sense experience.

The second question which concerns us is, what is the source of knowledge? The third question we must try to answer is what is the relation of *percepts* and things, described by *concepts*?

Let us start with an example. If I look out of the window and see the front of a square house at an angle to me, what I actually *see* is the front and side walls meeting at an angle capped by a portion of roof. I cannot see the rear walls from where I am. What I deduce from the image that prints itself on my retina (the "photographic plate" at the back of my eye) is the result of a complex process. Light falls on the house and is reflected to my eyes, which stimulates the rod cells and the cone cells in them to form an inverted picture, which is transmitted to my brain – a *perception*.

Perception

Perception is the name given to what I have seen, which is stored by the brain and *interpreted* by it. The brain initially seeks to co-ordinate two walls and a bit of roof. No sense, no success. Then the brain appeals to the *memory*. The memory has a record of such a pattern of two walls and a bit of roof being evidence of a house because of a previous similar experience. It is not a two-dimensional painting of a house on the backcloth on a stage for a play. The memory checks both, and co-ordinates all available data in order to suggest it is a house. Now "house" is a *concept* we use to classify lots of dwelling places. Look at photographs of different houses. Notice that the judgement that what I see is a house is not based on complete evidence, but on the evidence available, which is two walls and a bit of roof. Those clues are checked against the memory of such data, and the brain *guesses* or *intuits* that the object is a house. But if the dwelling is only one storey high, you will call it a bungalow, the *concept* used to describe one-storey dwellings.

Let me repeat this in a different way. The brain has had to collect together a lot of data: two walls and a bit of roof which has been seen, two walls and some more roof which cannot be seen but are remembered, and then has had to fit them together *as a whole* to make a house. The key point to notice about the act of seeing a house, which we do in a split second, so quickly does the brain work, is that even the data presented to the brain has had to be *interpreted*. Then, using our brain, we choose a *concept*, here either a "house" or a "bungalow" to describe the dwelling.

Let us explore some more examples. I am walking across a field with you on a foggy day. I stop walking and say, "There is a man over there." You strain your eyes to see into the fog and say, "I think it is a bush." As we walk closer, it becomes clear that you are right. The process of *perception* and *interpretation* can clearly be seen at work here, becoming more adequate as more information becomes available.

These two processes are also involved when one attempts to represent a three-dimensional figure on a two-dimensional surface; here a piece of paper. Two famous examples which illustrate this double process are Neckar's cube (figure 2.1), and Rubin's profile and vase illustration (figure 2.2; opposite). If you look at the former, choose one of the corners in the centre and 'pull' it, then 'push' it and the corner will be the near corner and then the distant corner of the cube. If you look at the latter you can see either two faces or a vase.

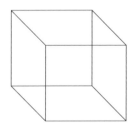

Figure 2.1   Neckar's Cube

The final example (figure 2.3) involves matching an object, *the perception*, to a concept, *the interpretation*. There is a four-legged square coffee table 60 cm square and 45 cm high at one end of five pieces of furniture, and at the other end is a stool 20 cm square and 45 cm high (figure 2.3; opposite). In between are three pieces of furniture all 45 cm high but 30 cm, 40 cm and 50 cm square.

The question you must answer is, What is the name of each of the intervening pieces of furniture? Somewhere in the middle you will reach a crisis: is this a coffee table or a stool? Never mind that question now. The point to grasp is that the mind is moving from perception of the object in front of us to the memory's account of, or concept of, what constitutes a stool or a coffee table. How do we *interpret* the object in front of us, as a coffee table or as a stool? Which concept do we allocate to the object we are looking at?

Figure 2.2   Rubins's Profile

So far we have established that when a person looks at an object, whatever the object of reference is, that person engages in one way of obtaining knowledge: through the twin activities, carried out almost simultaneously, of *perception* and *interpretation*. That answers our first question: Are we dealing with facts or *observations* and *interpretations*? We can therefore conclude that there are no such things as *uninterpreted* facts.

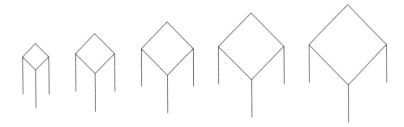

Figure 2.3   Which are stools or coffee tables?

In answer to our second question, What is the source of knowledge? we see that *perception* is one *source* of knowledge, but it is not *knowledge* until the perception is *interpreted*. The answer to our third question, What is the relationship between perceptions and things? is that we allocate a *concept* to an object, and that involves *interpretation*.

Intuition

Intuition is a 'connecting or creative faculty' which makes the link between the object, or perception of the object, and the interpretation of the perception or concept which comes into one's mind. We know, through intuition, that the universe exists. We know, through intuition, that material things exist. We know, through intuition, that persons exist. One might have intuitive awareness of God, but there is no proof (Strawson 1959; Baelz 1968). Intuition is what connects our minds to 'what is outside of them'. Intuition is not based on a perception, a fact, a sentence, an object, or a memory. It operates in various ways. For example, it tells you that 1+1=2, which is an eternal mathematical truth; it helps you recognize a colour – "That's red"! (Russell 1995: 382).

Memory

Memory is a second source of knowledge we have mentioned. It was needed in some of the examples explored above. Empirical knowledge is established through using *perception, interpretation*, and, frequently, *memory*. We do not need to spend much time on memory, except to be quite clear that it is not a static photograph album.

Two examples will do to illustrate this point. Do you remember an occasion in company when perhaps someone made a rude remark about you but you were very busy explaining something of importance to a group of people at the time? You probably ignored the comment as unimportant, for you were concentrating on your exposition. The comment was forgotten until a few days later, when you passed the person who made it, exchanged greetings with him and walked on, you remembered his comment. You were furious; it really was rude.

A second example: Mary was really upset when she did not get the job in Bristol and ended up in a similar job in Exeter. However, there she met a wonderful person, a member of the worshipping community. She fell in love, got married and then felt that God had guided her to go to Exeter.

Has there not been a change in the *interpretation* of the initial events in these two situations in terms of their *significance,* and in the second case in the mode of knowing?

Inner Consciousness

Many contemporary philosophers now think there are two other sources of knowledge, besides *perception, memory* and *intuition,* which we need to record at this stage of our explorations.

The first can be called *inner consciousness*. How do I *know* my own state of mind? For example, "I feel like a cup of coffee now" or "I would like a sail in that boat" or "I would like to get to know Joan". The discussion here is not about coffee, a boat or Joan, but about my feelings in relation to each of them. The argument here is that I can only *know* my feelings through my inner *consciousness*. My self-knowledge is obtained by the ability of my mind to reflect on my feelings, and to evaluate and conceptualize them. Such knowledge can be seen as parallel, in some sense, to perception, which has no "knowledge" content without evaluation and conceptualization.

In other words, we see an object and apply the concept "table" to it. So also in consciousness we have such feelings as pain, delight, attraction and repulsion, which need evaluating and conceptualizing as in the three examples above. There is a second area of concern: a person's consciousness of their own mental states or dispositions. Here are some examples:

- ✧ She favours this political party.
- ✧ They dislike this music.
- ✧ I love writing this book.

Perception, evaluation and interpretation all seem to be involved here.

Reason

Reason as the fifth source of knowledge, *reason* is included because, while Descartes listed "understanding, imagination, sense and memory", Thomas Reid clearly lists as of equal importance "the faculties of consciousness, memory, external sense, and reason". Reason is self-evidently important in any form of thinking, action, exploration or human activity. The account given, however, provides a basic framework from which to explore the structure of knowledge. Philosophers are not all agreed on this account of five sources of knowledge (Chisholm 1966: 57).

EMPIRICISM

Empiricism is the philosophical view or school of thought which holds that *sense experience* has primacy in human knowledge and justified belief. *A posteriori* is a phrase used to characterize reasoning from effects to causes, from experience (not axioms) *inductively* to solutions to problems. *Induction* is the method involved in the process of proposing a model or theory from the assembling of particular observations.

RATIONALISM

Rationalism is the philosophical view or school of thought which holds that "reason takes precedence over other ways of acquiring knowledge, others think that it is the unique path to knowledge" (Garber 1995: 673). Use is made of a number of forms of reasoning like *a priori* (lit. "from first things"), a phrase used to characterize reasoning from causes to effects, from abstract notions to their consequences, from assumed axioms, basic assumptions or accepted laws of a subject (and not from experience) *deductively*. *Deduction* can be defined as "any drawing of conclusions from stated propositions using the laws of logic" (Onions 1975). *A priori* applies to two kinds of propositions, those which are self-evident propositions, and those which are derived from propositions which are self-evident. Here are some examples of self-evident propositions. The first two cited are examples of the so-called "law of non-contradiction" for necessary truths.

- ✧ I am in Birmingham in Alabama, therefore I cannot at the same time be in Plympton in Devon.

✧ The person you spoke to was Joan, therefore it cannot be another person, June or Judy.

Here is an example of *a priori* reasoning in which *deduction* from two true propositions yields new information which is true:

> There was only one Wayfarer, a 16ft sailing dinghy, for hire on the lake and Jim tells you he has had a lesson in it today, alone with his instructor friend. A day later you meet Mark who says he has been sailing in a hired Wayfarer which has only been on the water once in the last two days in the estuary.

You can *deduce* that Mark was teaching Jim how to sail yesterday from the two statements above, assumption (a) and assumption (b), from which (c) is deduced. This inference is true because it has been worked out from *reason* applied to two true propositions. In other words, it follows logically that (a) if Jim had a lesson in the only hired Wayfarer in the lake yesterday, alone with his instructor friend, and (b) if Mark had been sailing in the hired Wayfarer in the lake on the one occasion it has been out in the last two days, then (c) he must have given Jim his lesson, as they must both have been in the one and only hired Wayfarer on the lake on that day.

The Relationship of Mind and Matter

How do we know how mind and matter are related? I have in my mind an "inner picture" of the house I saw in the first example discussed. Dualists are people who say that the "inner picture" in the mind serves as a representation of the matter across the road – the house. In other words, dualists are so called because they say that everything in the world consists of *mind* or *matter*. They assume that the house was a physical object – an empirical object – which you can recognize with your five senses, but that the picture in your mind was psychical or mental: the one is *matter,* the other is *mind.* So they affirm "that the mind is independent of matter, that the brain is not just a machine" (Watson 1995: 210–11). There are other forms of dualism, but they do not concern us here.

NINETEENTH-CENTURY SCIENTIFIC KNOWLEDGE

We need briefly to explore the nature of Newtonian scientific knowledge, which is a major area of empirical knowledge, before we explore how this affected the way the logical positivist group of philosophers defined all knowledge.

The understanding of science for most of us is based on the ideas of Newton (Barbour 1998: 11). His main ideas, for our purpose, are these: first, the world is understood as and pictured as a *machine* (18). Secondly, it consists of *bits of matter,* unchanging fundamental particles or bits of nature which can be combined in many different ways. Thirdly, the world, being a machine, is organized by *mechanical laws* or *causes* which are responsible for states and changes in the arrangement of matter. However, "Newton used *induction, i.e.* he could propound models which fitted his observations. He could not formally *deduce* the models from the facts available. Employing Occam's razor, that is avoiding unnecessary hypotheses, Newton's theory was extremely economical with assumptions, which explained a huge number of facts".[2] Fourthly, Newton adopted the *dualist* view of the world, following Descartes' view that the world is made up of two kinds of substances – *mind* and *matter* (Watson 1995: 210). That is, the mind is independent of matter: the brain is matter but not a machine, because the mind has limited freedom in making decisions, which is sometimes called "sanctioned choice" (Walsh 1960). Newton regarded God and human minds as the only exceptions in a materialist world. Newton himself believed that God created the world and was the First Cause. This simply means that God created it with all its bits of *matter* and the *minds* of people and other creatures. Science was the study of the secondary causes which governed the universe; "God" was used to explain *why* the world was created. Science seeks to explain *how* the world was created.

However, some adopted the view that the notion of God was now redundant: the world could be completely explained without God. When Napoleon asked Laplace, who had explained the world to him in Newtonian terms, where God fitted into his scheme of things, he replied: "Sire, I had no need of that hypothesis" (Barbour 1998: 35). This latter materialistic view of the universe and of science rejected any notion of the existence of *consciousness* or *inwardness* as being subjective illusion and held that the world could be understood through the processes of *observation, theory,* and *deduction.*

One example of an early experiment illustrates *the method.* A biologist wanted to know what elements made up a tree. Apparatus: plant pot, sapling, soil. Each was weighed. The plant was watered and the weight of all the water recorded. Sometime later, everything was taken apart and weighed again. Weights: soil – the same, plant pot – the same; sapling – heavier; water – disappeared. Those were the *observations.* The *deduction* became the *theory:* the tree was made of water! This was a valid *interpretation* based on *induction* from available data.

This experiment occurred before photosynthesis was understood and looks silly from today's position of more comprehensive knowledge. But that is not the point; the point is that the scientific method of *observation, induction, theory, deduction* was properly followed, given that the limits were what was thought to be relevant to the experiment then. Those limits still obtain today: scientists can only include in their work factors they think are relevant, but they might miss something out. That is one reason why there is very often extended scientific debate after a new theory is advanced: it needs extensive testing, or *verification.*

Let us summarize the scientific method which obtained up until about the beginning of the twenty-first century: it involved *observation, induction, theory and deduction, and verification.* The motivation in science of many was often scientific curiosity leading to understanding, leaving prediction and control to technology.

If you want to improve your roses, you seek to understand, predict and control what happens to them. At present they have no scent, the blooms do not last very long and they are easily destroyed by rain. So you search out a rose to suit your requirements. The number of scented white roses is extremely limited. You want a scented Hybrid tea white rose in which blooms last a long time, which withstands rainfall well and which can be used for exhibition purposes. Recently bred in America, the *Evening*

Star is well worth considering, for it has been developed to possess all these features (Hessayon 1981).

Before we end our discussion of Newton and Laplace we must clarify the difference between them. They both agreed that the *methods* of exploring the world were mechanistic, because the world was governed by unchanging laws of nature. However, Newton believed in God as the First Cause. He was a deist, i.e. he pictured God as a clock-maker, and the clock he made would be kept going through the laws which governed it. He believed in a remote, non-interventionist God.

Laplace was an atheist who elevated the *methods* of exploring and studying the universe into an "atheist religion" or a metaphysical stance called *scientism*. Such a person thinks, for example, that a person is *nothing but* a collection of atoms, of bits of matter. "Nothing but" is a clue that this is a *reductionist* view of a human being.

So, for example, if I was taken to a laboratory and all the bits were separated out and sent to my home in test tubes, with each constituent element correctly labelled, would that still be me? Of course not. But am I not more than the mechanical sum of the parts which chemically constitute me? This is a question we must address in more detail later.[3] Here it is enough to answer "Yes".

Empirical knowledge is established through the use of *inductive thinking*. This method is used in all sciences, practical thinking and in most other kinds of thinking. Darwin's *Origin of Species* (1859), the book in which he advances the *hypothesis* or *theory* of evolution, uses induction as the principle method of argument, though Darwin also uses deductive argument as well.

Here is one example of a deductive argument he uses. Darwin found that the domestic duck has a wing which weighs less that the wing of a wild duck, while the legs of the wild duck weigh less that the legs of the domestic duck. He goes on: "I presume that this change may be safely attributed to the domestic duck flying much less, and walking more, than its wild parent" (Darwin 1968: 74). An example of inductive thinking appears after Darwin has reviewed a large number of plants and animals with their slight differences and then makes his *hypothesis* or *theory*. Darwin says: "I attribute the passage of a variety, from a state in which it differs very slightly from its parent to one in which it differs more, to the action of *natural selection*" (Darwin 1968: 107).

The theory of evolution by natural selection contained three elements: random variation, the struggle for survival and the survival of the fittest (Barbour 1966: 84–5). The crucial elements in the scientific method here are the interplay between *observation* and *theory*, which is then *tested or verified* by further *observation*, or by *experimentation* (Darwin 1968: 71–2). The creative imagination is central to the development of a theory or hypothesis.

The theory of evolution by natural selection is still widely accepted as the best inductive account available to date. It is based on lots of slightly different variations between members of, for example, a species of butterfly which was white in the unpolluted countryside but became black in the industrialized conurbations up to the 1950s. Darker variations in the polluted towns survived.

The theory of evolution by natural selection, with its three elements, is the inductive hypothesis which best explained the different coloured population of "white" butterflies in town and country. It is believed to be true. However, one needs to be clear about the *status* of that truth, and to hold it "provisionally" or *heuristically*. This simply means that the theory is confidently held by a very large number of scientists in agreement, but provisionally in the sense that one may need to revise one's belief in evolution if presented with another hypothesis that offers a more adequate theory to account for existing and perhaps new data.[4] But do note that a theory is the result of *interpreting* the evidence examined.[5]

While there may be agreement amongst scientists about methods of investigation and kinds of explanation or theories to be explored and tested, there is no necessary agreement on the nature of science. This debate we must briefly explore.

The popular view of science outlined above, which is said to involve *precise observation* yielding *pure facts* (Barbour 1966: 138–9) leading to the creative formulation of *theory or hypothesis* to account for them, was embraced in the first quarter of the last century by a group of philosophers called *logical positivists* as the model for the nature of knowledge. This view dominated philosophical discussion in Britain until about 1953, but still lingers in the popular imagination, as Revd. Dr. Fraser Watts, the University of Cambridge Starbridge Lecturer in Science and Religion, has indicated: "There is a widespread perception that science and religion are in conflict; that science has disproved religion; that religious faith is intellectually disreputable. The perception comes from the latter part of the nineteenth century. It doesn't date back any further than that, but it is the received wisdom. . ." (Stannard 1996: 189). This is so even though changing views of knowledge have been explored by the scientific and philosophical communities for more than forty years.

LOGICAL POSITIVISM'S VIEW OF KNOWLEDGE c.1920 – c. 1953[6]

The spectacular advances in the various sciences which were explored in the nineteenth and twentieth centuries attracted a group of philosophers, called logical positivists, to science as a model of what knowledge is: they stated that empirical knowledge is the only kind of knowledge. The positivists generalized from their understanding of the scientific method, suggesting that all real *knowledge* had to meet the following specification. All concepts used must have been tested. All sentences are either *meaningful* or *meaningless* (note carefully the definitions which the positivists give these words, which are much "narrower" than the way you probably use them, as we will see later). If it is meaningful, it is also either *true* or *false*. In other words:

1   There are only two main kinds of meaningful sentence:
     (a) an analytic sentence, or definition, e.g. all bachelors are unmarried men;
     (b) a synthetic sentence, or observational sentence, e.g. there are three chairs in this room.
2   The meaning of a proposition depends on the operations involved in verifying it (Ashby 1985: 497), or more simply, the meaning depends on the steps needed to test it.

Let us explore what this means through examining some examples. If I said "all bachelors are balloons", that clearly is "meaningless". It does not define a bachelor, so it cannot be an analytic statement or definition. I will take some examples of definitions from the dictionary: "A lion is a large tawny carnivorous animal with tufted tail and in the male a flowing shaggy mane"; "A unicorn is a fabulous animal with a horse's body and single straight horn". Both definitions are meaningful and true.

Let us use them in synthetic sentences, remembering that a synthetic sentence is usually a statement of what can be observed. "There is a lion in the zoo"; "There is a unicorn in the zoo." The first we can verify because we can go and see the lion, so the sentence is both *meaningful and true* (if a lion is present) or *meaningful and false* (if absent). We cannot see a unicorn because it is an imaginary animal found in fables, so the sentence "there is a unicorn in the zoo" is *meaningless* because we cannot outline the steps by which we can verify it, and so the question of it being true or false *strictly* does not arise, if you are rigorously following the positivists' method of establishing knowledge. But of course, in common-sense terms, it is false!

One more pair of examples: "The moon is beautiful"; "The moon is made of green cheese." The first is *meaningless* because we cannot outline the steps by which it can be proved to be true or false. The second is *meaningful* because someone has been to the moon, but *false.*

Another pair of examples: "The boy is six foot tall"; "The boy is honest." The positivists say the first sentence is *meaningful* and the second *meaningless* because we know the steps to take to verify the first, but there are no empirical tests to verify the second.

John wants a bar of chocolate, so he takes fifty pence out of the pocket in his coat in the cloakroom. Peter wants a bar of chocolate, so he takes fifty pence out of the pocket of John's coat in the cloakroom. The positivists argue that all that can be established is where the fifty pence came from (provided witnesses agree). To say Peter was dishonest is a *value judgement,* and that cannot be empirically established, for value judgements are regarded as meaningless because one cannot outline the steps by which they can be verified. A value judgement fails to meet the positivists' definition of what counts as knowledge.

Another pair of examples: "The boy is six foot tall"; "I like the boy." The same pair of verdicts apply to "the boy is honest" and "I like the boy". Both are value judgements, the first is moral and the second is personal.

Last examples: "God created the earth"; "John created that globe." As synthetic sentences the first is *meaningless* and the second *meaningful,* and may be true or false depending on supporting evidence. We cannot outline the steps by which we can verify "God created the earth," so it is technically *meaningless,* its truth or falsity therefore does not arise.

Look again at the examples given and you can see that the logical positivists say empirical statements are both *meaningful* and *true* or *false* depending on the evidence.

But aesthetic (beauty), moral (honest), personal (I like you), or religious (God loves you) sentences are *meaningless* or technically nonsense, and cannot be *knowledge* claims. It becomes evident that we do use sentences like those dismissed by the logical positivists, so they must have some "meaning".

Some positivists made the concession that they are not knowledge, but emotive or expressive statements, i.e. they express my feelings but tell me nothing about what can count as knowledge.

KNOWLEDGE AND PHILOSOPHICAL ANALYSIS:
THE USE OF LANGUAGE AFTER 1953

As mentioned above, this theory dominated philosophical thinking in Britain for the period from the 1920s until 1953 when Wittgenstein's book *Philosophical Investigations* was published two years after his death.

The rule or verification principle of the logical positivists stated that the meaning of a sentence depends on the steps by which it can be verified. Wittgenstein replaced this by a second rule or verification principle: "For a large class of cases – though not for all – in which we employ 'meaning' it can be defined thus: the meaning of a word is its use in the language" (Putnam 1998: 150). For our general purposes here we can simplify this and say that *the meaning of a word in a sentence depends on the* use *to which it is put.*

Wittgenstein indicated that words often have family resemblances (Quinton 1964: 540–1), which, when used in family groups, he described as "language games" (541). Here are four examples:

- ✧ The *boy* ran around the *pool.*
- ✧ *Repenting*, he was *saved* from his *sins* and *restored* to a *right relationship* with *God.*
- ✧ He *apologized* for smashing the window and *promised* not to do it again.
- ✧ The *beauty* of the painting lies in the use of *proportion*, *perspective* and *colour.*

The "language games" here are empirical, religious, moral, and aesthetic respectively.

Notice that you can see all the objects indicated by the words in italics in the first sentence, but that the words in the other sentences indicate nothing visual. But you think you can see "colour"? Think again. Do you not see "red", or "yellow"? "Colour" is an abstract aesthetic word referring to all colours. Later these ideas were developed by, amongst others, some philosophers of education like Professor Paul Hirst (see Chapter 4).

You can discover something of what was said for yourself. Classify the following list of words in table 2.1 into their family groups with appropriate names for the different families of words.

Ignore whether they are nouns or adjectives.

Table 2.1

beautiful	holy	mud-flat	well-proportioned		ravine	ought
valley	sacred	duty	ugly	river	constable	ox-bow
horse-stable	salvation	right	confession		sinful	timetable
pretty	turntable	wrong	perspective		table	promise

Groups of students faced with this task, and using their common sense, always came up with the same number of word-groups, or "families of words", often with exploratory names. There was extended discussion over some words. Was "confession" moral or religious? Likewise "sinful" and some others. As a result of their discussions, they discovered words could be used literally or metaphorically, and that helped to distinguish the precise *use* of a word in a particular family group. Through this exercise those students reached one of the central conclusions that Wittgenstein reached in his 1953 book. To see the groups see the note.[7]

Why Did Logical Positivism Collapse?

Before turning to those developments, we need to turn back to ask why logical positivism collapsed. We have seen that testing the meaning of *language by identifying its* use *focused attention on language games, i.e. family resemblances between different clusters of words, such as those discussed above, group them into moral, religious, aesthetic, personal and empirical families.*

Sometimes each family of concepts, when used in clusters to outline a particular kind of meaning in a narrative, is regarded as a form of life, a mode of discourse, a form of knowledge.

Metaphysics can be considered as a mode of discourse and as a language game. The key aspect of metaphysics is that it explores the limits of understanding beyond the world of experience into the world of the supra-sensible, and thus is concerned to set out and to lay down the first principles as a foundation for all knowledge. It is often concerned with "limit" questions like the limits of reality or the limits of the finite world.

Here are three metaphysical sentences which state the necessary presuppositions which underlie a materialist, humanist and religious outlook on life respectively: – "The world only consists of bits of matter in different combinations"; "What matters most is what is human"; "God created and continues to create the Universe." But the first principle is accepted by the humanist and religious person as acceptable as an empirical statement, and the second as a principle within the mode of personal discourse for the religious person. They are all true in the different senses just outlined.

Q   What did logical positivists count as knowledge, and how did their view of knowledge collapse?

A   Logical positivism held that there were only two main kinds of sentences containing necessary presuppositions which can claim to be knowledge: analytic and synthetic. The first is a definition, – the one kind of knowledge positivists accepted, e.g. "A bachelor is an unmarried man." The second is a knowledge claim tested by the verification principle: "The meaning of a sentence depends on the steps by which it is verified or tested for its truth." For example, "there are three girls in this room." Test: count them.

Q   Is the "verification principle" analytic or synthetic?

A   Neither. So, according to their criteria, this question is meaningless.

Q   So does logical positivism fail according to its own criteria?

A   Yes.

Q   So what kind of sentence was "the verification principle"?

A   A metaphysical sentence which had attempted to limit the nature of knowledge to analytic and synthetic statements. Its system sought to abolish all metaphysical, moral, personal, aesthetic and religious *knowledge*.

Metaphysical sentences outline the necessary presuppositions which underpin any theory of knowledge. So it follows that metaphysical sentences are the 'foundational' statements of an outlook on life, like logical positivism, Christianity, humanism, Judaism or materialism.

Wittgenstein's new verification principle, we have noted already, states that the meaning of a word in a sentence depends on the *use* to which it is put. Attend to the use words in sentences, then you will understand the message or meaning.

SUMMARY

So far in our exploration of the nature of knowledge as modern philosophers describe it, we have noted four possible sources of knowledge: *perception, leading to empirical knowledge; memory; inner consciousness;* and *reason*.

We have seen that empiricism is a theory of knowledge which is grounded in and derived from experience, whose truth claims are subject to testing or verification in theory or practice. In other words, the steps of a verification or testing process can be identified, which leads to a result which shows it is either true or false.

Rationalists assert that reason establishes necessary connections, i.e. it discerns a necessary basic structure in the world through intuition. For example, reason assumes the *order* in the universe, illustrated by the scientist's enterprise, which is meaningless unless there is order to be discovered.

Rationalists also use rational analysis, illustrated by, for example, necessary truths of non-contradiction and inference, for which examples have already been given, and causality, for which an example follows:

"The cup lay broken on the floor because the man pushed the shopping basket away from where he had placed the newspaper, so that the shopping basket pushed the unseen cup onto the floor."

However, Kant, the father of modern philosophy, saw that there were good points in both empiricism and rationalism, and sought to combine them. Kant held that two elements make up our knowledge of the world. External data, which we perceive through our senses, he called the *materials* of knowledge. This was the contribution from empiricism. The second element, from rationalism, included not only the methods of using reason which we have discussed, but also the intuitions of human reason so that humans always perceive within the dimensions of *time* and *space*, and assume *causality*. To put this another way, humans cannot know what the world is like in itself; we can only "see it through the glasses" of "time", "space", "cause and effect" (Gaarder 1995: 248–9).

However our main conclusion is that the main search for a test of "what knowledge is" in Europe from the 1920s to 1953 was that of logical positivism, which used science as the model for truth. Its test was called the "verification principle", which stated that the meaning of a proposition depends on the steps by which it can be verified. It accepted two main kinds of statements, an analytic statement or definition, and a synthetic statement which was an empirical truth claim e.g. there are four chairs in this room. As the verification principle is neither an analytic statement nor an empirical statement, but a metaphysical statement, the whole theory collapsed.

Wittgenstein showed the fatal flaws of logical positivism and replaced their verification principle with his new rule for testing meaning: the meaning of words in a statement depends on the *use* to which they are put. In other words, meaning is established through how people *use* language.

When you classified the groups of words earlier, you discovered that there were five kinds of word groups – moral, religious, aesthetic, empirical and geographical. Wittgenstein called these similar word groups, once they were in sentences, different language games, different forms or kinds of knowledge, or different modes of discourse, by which he meant different kinds of talk, moral, religious and the other examples quoted above. But this brillant insight will be explored further in Chapter 4.

NOTES

1   This is a selective account of epistemology, or the theory of knowledge. The criterion of selection of movements discussed is the relevance of a particular movement to the study of transcendental experiences, including religious and mystical experiences. So, e.g., in discussing the logical positivists' view of knowledge from *c.* 1920 to 1953 on p. 35, no account is given of the work of Russell or the early Wittgenstein, which preceded logical positivism, as the latter movement pervaded national consciousness in Britain throughout the period discussed.

2   Personal communication from Dr J. R. Waldram.

3   A fuller account of scientific methodology can be found in Barbour (1998).

4   Those who argue that the theory is "true" because it is "useful' are adopting a pragmatic theory of truth, i.e. they assume a theory is "true" when it fits in with other aspects of their experience (cf. Speake 1981: 265).

5   Popper (1974: 80–93). He challenged the established view that the inductive method could *prove* the truth of any scientific theory true. Science had to live by faith. He advanced the principle of falsifiability, so that scientific claims were tested by deductive consequences. Hypotheses then could not be verified, only falsified. So if you find one exception to a "rule" or "theory" it may well be wrong. Existing theories still stand because they withstand attacks on them, not because they are proved to be true.

6   In 1953, Ludwig Wittgenstein's *Philosophical Investigations* was published, which formally ended logical positivism.

7   The groups of words indicating their primary or literal meaning were:

moral	religious	aesthetic	empirical	geographical
ought	holy	beautiful	table	mud-flat
duty	sacred	well-proportioned	turntable	ravine
right	salvation	pretty	timetable	valley
confession	sinful	ugly	horse-stable	river
wrong	promise	perspective	constable	ox-bow

REFERENCES

Ashby, R. W. (1985), "Logical Positivism", in O'Connor, D. J. (ed.), (1985) *A Critical History of Western Philosophy*, (1964) Glencoe, NY: Free Press.

Ayer, A. J. (1982), *Language, Truth and Logic* (1946), Harmondsworth: Pelican.

Baelz, P. R. (1968), *Christian Theology and Metaphysics*, London: Epworth.

Barbour, I. G. (1966), *Issues in Science and Religion*, Englewood Cliffs, NJ: Prentice-Hall.

Barbour, I. G. (1974), *Myths, Models and Paradigms*, London: SCM Press.

Barbour, I. G. (1998), *Religion and Science*, London: SCM Press.

Chisholm, R. M. (1966), *Theory of Knowledge*, Englewood Cliffs, NJ: Prentice-Hall.

Darwin, C. (1968), *Origin of Species* (1859), London: Penguin.

Gaarder, J. (1995), *Sophie's World*, London: Phoenix House.

Garber, D. (1995), "Rationalism", in R. Audi (ed.), (1995), *The Cambridge Dictionary of Philosophy*, Cambridge: Cambridge University Press.

Hamlyn, D. W. (1970), *The Theory of Knowledge*, London: Macmillan.

Hessayon, D. G. (1981), *The Rose Expert*, Waltham Cross: PBI Publications.

James, W. (1968), *Varieties of Religious Experiences* (1902), London: Fontana.

Onions, C. T. (ed.), (1975), *Shorter Oxford Dictionary*, Oxford: Oxford University Press.

Popper, K. (1974), *Unending Quest*, London: Fontana.

Putnam, H. (1998), *Renewing Philosophy* (1992), Cambridge, MA: Harvard University Press.

Quinton, A. M. (1985), "Contemporary British Philosophy", in D. J. O'Connor (1985), (ed.), *A Critical History of Western Philosophy* (1964), Glencoe, NY: Free Press.

Russell, B. (1995), "Intuition", in R. Audi (ed.), (1995).

Scheffler, I. (1978), *Conditions of Knowledge* (1965), Chicago: University of Chicago Press.

Speake, J. (ed.), (1981), *A Dictionary of Philosophy*, London: Pan Books.

Stannard, R. (1996), *Science and Wonders*, London: Faber & Faber.

Strawson, P. F. (1959), *Individuals*, London: Methuen.

Trusted, J. (1997), *An Introduction to the Philosophy of Knowledge*, 2nd edn, London: Macmillan.

Walsh, W. (1960), *The Use of Imagination*, London: Chatto & Windus.

Watson, R. A. (1995), "Dualism", in R. Audi (ed.), (1995).

Wittgenstein, L. (1953), *Philosophical Investigation*, Oxford: Blackwell.

3 How Do We Know What Knowledge Is?

An American Search for Personal Knowledge

In the first half of the twentieth century the exploration of the nature of knowledge developed from empiricism in both the UK and the United States, but in different ways. In the UK we have seen that the focus gradually centred on the nature of science as the model of true knowledge, but in the United States two new movement's sought to explore personal knowledge, Pragmatism and Personalism. These were built on by later movements in the disciplines of psychology, psychiatry and psychotherapy.

PERSONAL KNOWLEDGE FOR HUMANS

Pragmatism

Pragmatism[1] is an instrumental view of knowledge, which simply means "what works is knowledge"; and therefore is true. It is tested by the way it fits in with our other experience. Its focus is on *how* we know, though pragmatists claimed to give an account of *what* we know as well.

For pragmatists the search for knowledge started from observations and experiences, but then involved the experimental manipulation of the situation in response to a problem which needed to be solved. Basically, then, pragmatism is a two-part theory. The first part is a 'theory which concerns the nature of believing' (Chisholm 1966: 110). The second part asserts that if you experience it, it is part of your *being*. It is basically a belief in what you hope will happen, that is made true when you experience it. So, if you expect that you will enjoy playing tennis, then play tennis and have the satisfaction of enjoying the game; the result of the game is the test of the truth that tennis happens and is enjoyable! This means that if you try out an idea in practice and it works, it is true.

William James was a leading pragmatist. In his essay "Pragmatism", written in 1907, he said, "Truth *happens* to be an idea. It *becomes* true, is *made* true by events" (Audi 1995: 386). And again in the same essay: "Ideas become true just so far as they help us to get into satisfactory relations with other parts of our experience" (Speake 1979: 265). Paul Tillich provided a valuable summary of James's position when he said, "Pragmatism, as developed by William James . . . reveals the philosophical motive behind this elevation of experience to the highest ontological rank, viz. reality is identified with experience" (Roberts 1991: 156). In other words, all being is identified with experience: our experiences reveal to us the true nature of all being, including what it is *to be* ourselves. Ontology is here used to mean the study of being.[2] This assumes that *being*, e.g. a human being, is irreducible. If it is reduced to its *material* elements, it is no longer a *being: being* itself is irreducible.

This theory of knowledge is distinguished from, for example, the *materialists* or *empiricists*, for whom *matter* provides the irreducible elements of *knowledge*. Materialism tends to imply that a *being* can be *reduced* to the bits of *matter* of which it is composed. We will later discuss how these two theories of knowledge can be seen to be complementary when we study the hierarchy of ways of knowing.

James is also the scholar who led the revival of the study of religious experience in the twentieth century. His brilliant book written in 1902, *The Varieties of Religious Experience*, will be studied later. His defence of religious experience as knowledge is pragmatic:

> When we think certain states of mind superior to others, is it ever because of what we know concerning their organic antecedence? No! It is always for two entirely different reasons. It is either because we take an immediate delight in them; or else it is because we believe them to bring us good consequential fruits for life. (James 1968: 36–7)

What he is saying is this: the value or truth of religious experience is not due to its *origins*. For example, it is not true to say that Winston Churchill was great because he had an American mother, or that Jesus revealed the will of God because of the virgin birth. Indeed, the story of the virgin birth emerged, like so many similar biblical birth stories, on which it is modelled, to use the fallacious argument that truth can be established from origins. Many a biography employs the fallacy of trying to justify the greatness of a person by their origins. The truth of religious experience, argues James, has two tests: Did it fill you with deep joy? Did your religious experience bear fruit in your newly changed or inspired life? If you can say "yes" to both questions, then your experience was real: you *know* what you experienced.

The emphasis on *being* led to the development and flourishing of *personalism* in America between the late nineteenth century and the mid-twentieth century.

Personalism

Personalism[3] grew out of idealism, which was the dominant movement in America in 1900. Idealism argued that reality is basically spiritual, and that our higher human characteristics indicate its nature, for without mind or spirit the material body of a person is but a corpse. It clearly provided a philosophical foundation for religion.

Personalism developed from idealism to incorporate American democratic liberalism and individualism. It affirms that "all reality is ultimately personal. God is the transcendent person and the ground or creator of all other persons" (Delaney 1995: 575; Roberts 1991: 162–3). The natural world is regarded as "an order within or as a function of the mind of God. Finite persons are created by the uncreated Person. Human persons are therefore centres of intrinsic value . . . God is the source and conserver of all values" (Roberts 1991: 164). Basically, human persons are individual, irreducible, finite but indestructible. Persons are manifestations of Infinite Person who are united in a co-operating community. This community was an "Eternal Republic" bound together in God.

The vision of personalism, that individual persons are united in community, has had a widespread effect on philosophical and theological thinking in, for example, the broadening of the outlook of the Methodist Church in the United States, and is having a considerable effect on the black churches. The thought of Martin Luther King Jr, as well as his leadership, underlines the importance of this movement.

PERSONAL KNOWLEDGE ABOUT BEING HUMAN:
HUMANISTIC PSYCHOLOGY

The views of humanistic psychologists that are explored here are of interest to us because of the philosophical assumptions they make about the nature of a human being and the way they use personal discourse to express a form of knowledge.

Humanist psychology is a movement of thought which rejects *behaviourism* and *Freudian psychoanalysis* for different reasons.

Behaviourism flourished until the 1940s, focusing attention on the study of behaviour rather than on the mind. This was because behaviourists believed that the environment initiated an unbroken chain of cause and effect and was the exclusive determinant of behaviour. The central concepts used were mechanistic, like *reflex, stimulus, response, conditioning*. Thus a child, or a pigeon, could be trained by offering it a stimulus to perform an action and following the stimulus by a reward to reinforce it, thus implanting a conditioned response. Such treatment reduces a human being to that of a programmed robot.

Psychoanalysis, on the other hand, also treated the human as a machine which could be manipulated. A range of mechanisms were identified which Freud used to explain behaviour. *Identification* is the original form of emotional attachment to an object or a person, e.g. "That's my mother there!" *Introjection* involves taking into oneself required behaviours or values, e.g. "It's wrong to take money from Mum's purse without permission to buy sweets!" *Projection* is placing on another person one's own undesirable feelings or ideas as a defence mechanism to avoid self-guilt and to blame the other. For example, I may despise a person who is late for work to avoid facing my own failure to be at work on time. *Displacement* has a similar meaning. *Substitution* is another form of the displacement and projection mechanisms, where a characteristic or emotion is transferred from the subject to another person or object. It may be the replacement of an unsafe object or emotion with one which is personally acceptable. *Sublimation*, say followers of Freud, channels repressed sexual needs into socially acceptable non-sexual activities, such as playing a musical instrument well. *Rationalisation* describes acceptable reasons which conceal the true and unacceptable basis of the behaviour. An imaginary example: to an interview panel "he" argued that the female was the best-qualified candidate for the job in the office, but actually "he" was very much attracted to her!

Both behaviourism and psychoanalysis, though giving valuable insights of particular kinds, are reductionist in two ways. First, attention is focused away from the whole person to a particular action in isolation, as if a person were a determined mechanism. Secondly, the explanation is mechanistic, not least because of the desire of both schools of thought to be "scientific", as it was understood then, i.e. impersonal and objective. Both kinds of analysis centres on the component parts of persons as mechanisms.

The humanist psychological movement reacted against both these approaches. It focused on the nature of a whole person, and what it is to *be* a person, without seeking to reduce a human being to a mechanistic collection of parts, or looking for the atomistic basic elements of what constitute a person. The humanistic psychologists attended to the nature of a person in their wholeness, engaging in freely chosen and willed actions.

The key thread linking this group of American psychologists is their concern to promote a humanistic faith, often through an existentialist approach, in the wholeness and irreducibility of *being* human. This faith may be expressed through ethical humanism or non-dogmatic religion. An important achievement of this community was to outline two ways of being religious or humanist which we need to explore, as it will affect the way in which you react to the analysis of religious experience later.

Humanistic Psychology's Models of Being Human

A distinction is made between *authoritarian religion* and *humanistic religion* (Fromm 1971: 36–7), but the terms are used to include humanists in both models. The same distinction is made by a religious psychologist in the humanist psychology camp who talks of *intrinsic and extrinsic* religious orientations, so their insights will be combined (Allport and Ross 1967: 432–43; Wright 1967; Allport 1969). Before formally describing the distinction, a practical religious example of an extrinsic or authoritarian orientation will be given.

Q   Who is Jesus?
A   He is the Son of God.
Q   What do you mean by that?
A   Well, he is the Messiah who has come to save us.
Q   What is a Messiah?
A   He is God's Anointed One. 'Messiah' is the Jewish word which means the same as the Greek word, 'Christ'.
Q   Do you or I need a Messiah?
A   Yes.
Q   What for?
A   We all need to be saved from our sins to have a right relationship with God.

This verbal learning is, for a Christian, doctrinally "orthodox". But the speaker answering the questions cannot get out of the verbal game of chess with the kind of pieces being used – Messiah, Son of God, Christ, Saved, Sin and God. The answers fail to get to grips with everyday life and relationships, which is what religion is really about, and is the base from which the questions came. The questions are not given answers relevant to daily life in this exchange.

An authoritarian or extrinsic religious orientation is life restricting. This approach is often characteristic of members of conservative, literalistic, religious communities and is dogmatic and divisive because you must either join them or remain outside. Any faith which claims a mandate to be the sole means of salvation is adopting an authoritarian stance that was largely unknown to Jesus in the New Testament, but widely known to many of his followers who wrote about him in the New Testament. Members of other stances for living or world beliefs, including humanism, can be either an extrinsic or intrinsic religious orientation (Bowker 1987).

Freud seems only to have known religion in the extrinsic mode, and in his context many of his insights are appropriate to the authoritarian religious stance, which can become a collective neurosis. It can involve the projection of a powerful father figure as a psychological protection against the dangers of nature and the tribulations of life. It can defend an ego identity within the boundaries of a "saved" in-group. Such a stance can be guilt ridden or reward seeking. It can be used to attain personal ends, like peace of mind, power, self-justification, status, security, support for one's racial, ethnic and religious prejudices, and for rigid moral beliefs and moral prejudices. It provides a basis for dogmatism, political conservatism, any or all of which, depending on the person, are more important than his religion which he uses, often unconsciously, as a means to his ends.

The distinction between the authoritarian religious orientation and the humanistic religious orientation is well illustrated, inadvertently, by Piaget, in the distinctions he has made between children's stages of development.[4] Young children at the *heteronomous* stage, i.e. the stage of being under the law or rule of another, such as parents or teachers, live under external rules without full understanding. For example, look at this conversation between two young children before a meal:

> "You can't have a sweet before lunch."
> "Why not?"
> "Because Mummy says so."

Older children at the *autonomous* stage of development have internalized sensible rules or principles. So the answer of an older child at the later stage would be "Because it will spoil your appetite".

A person with an authoritarian outlook might be horrified to be told that they are at an immature stage of religious development, such as following religious rules without understanding. Altruism is sometimes confusingly mixed up with deep personal needs. Sometimes a person intuitively, and perhaps unconsciously, "recognizes" the deep focus on love. However, that person, at a verbal level, quotes and accepts the received doctrinal formulas because that is the only language they know, and because, like many of us, they have delegated the "expertise" in the areas of religion to the "specialist", e.g. the imam, minister, priest or rabbi.

Most of us delegate the expertise of repairing the car or electrical or gas breakdowns in the home to the appropriate professionals, whom we trust. Life is too short for us to investigate and be knowledgeable in all areas, so we all delegate different areas of life in this way. Such persons may use the mythological language of religion in good faith for that which is deeper, and their quality of life may well be considerably greater than that of the person who holds their faith rather casually, in the mode of an intrinsic religious orientation. These comments are made to make it clear that people do not necessarily fit exactly into one model or the other. The models are used to make a distinction between an authoritarian/extrinsic and a humanitarian/intrinsic outlook, which may be either religious or humanist.

The humanitarian intrinsic religious orientation indicates, as with the developing child, that a person holds their religious faith autonomously, i.e. they have made rules or principles their own. Such a person has internalized their beliefs and values and, after appropriate analysis and reflection, made them their own. This orientation can be life enhancing and expand horizons, and is inclusive of all humanity, nature and the universe. It is reflective, open to modification and constructive. Such a person perceives God "within" and "without", and experiences a sense of oneness with all. Such a person is guided by principles and norms, motivated by love, and faces life gratefully, joyfully and with courage.

The mature religious sentiment has been defined as "a disposition, built up through experience, to respond favourably, and in certain habitual ways, to conceptual objects and principles that the individual regards as of ultimate importance in his own life, and as having to do with what he regards as permanent or central in the nature of things" (Allport 1969: 64–5). Such a mature religious outlook has six characteristics (Allport 1969: 64–5; cf. Macmurray 1972: 248–9)

1   It is well differentiated in outlook, rich and complex. It includes all aspects of life, taking account of views like those of Macmurray, mentioned earlier, that "the aim of religion is human perfection in relation with others" and the view of Whitehead, who defined religion as "What a man does with his solitariness". A mature person has worked out, or is working out, his beliefs in relation to God and to other people, and the values and situations involved in that programme and the actions which need to be taken to implement it.

2   It is derived from many sources, is functionally autonomous, and is dynamic in the sense that it is continually attempting to promote situations for wholesome living without fanaticism or prejudice. A person autonomously decides what to think or do.

3   It has, or develops, a consistent morality.

4   It is comprehensive, taking account of all aspects of the world, life and relationships, and identifying the facts, emotions and values needed in the pursuit of perfection. Being comprehensive, it is tolerant. "Truth is one, men call it by many names."

5   It is concerned to integrate different aspects of experience, like science and a modern religious outlook, in a way which takes account of the limiting factors of life, which include social and environmental factors, including evil in various forms, and which determine the context of life and the space in which freedom can be exercised.

6   It is heuristic, that is, a faith is held with confidence and conviction, but provisionally. For example, I believe in Darwin's theory of evolution, which is derived from his biological data. But if another biologist were to advance another theory, and the fraternity of biological scholars accepted the new theory after a few years of debate, research and perhaps experimentation if

that were necessary, then I would accept the new theory. Another example: at one stage in life I had only heard of the idea of God "up there". When John Robinson (1963) wrote *Honest to God* and talked about Him existing "in the depth of your being", my idea of God changed to include that second picture or image.

This model of the intrinsic, humanistic or mature religious orientation, which has been contrasted with the model of the extrinsic, authoritarian 'rule following without understanding' orientation, has been brilliantly illuminated by Martyn Goss in his *Faith Attitudinal Spectrum*, which he designed and uses as a basis for group discussion.

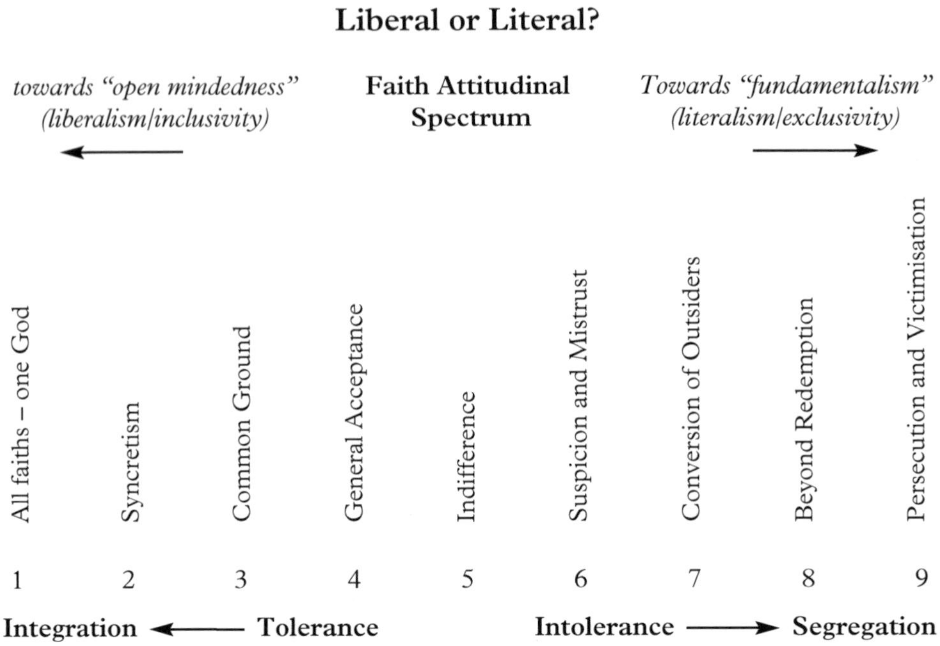

Liberal or Literal?

towards "open mindedness" (*liberalism/inclusivity*)	**Faith Attitudinal Spectrum**	*Towards "fundamentalism"* (*literalism/exclusivity*)
←		→

All faiths – one God	Syncretism	Common Ground	General Acceptance	Indifference	Suspicion and Mistrust	Conversion of Outsiders	Beyond Redemption	Persecution and Victimisation
1	2	3	4	5	6	7	8	9

Integration ← Tolerance   Intolerance → Segregation

[N.B. Such a spectrum can be found within Faith Traditions as well as across them]
Source: Reproduced by permission of Martyn Goss.

Figure 3.1   Liberal or Literal?

We have looked at the model of an intrinsic, humanistic or mature religious orientation and at two examples of answers in this vein. Now we can take the student questions asked earlier, starting with the last question. In this example, this student shows she has an intrinsic religious orientation:

Q   Do you or I need a Messiah?
A   I do need help to restore my broken relationships. I think that I need to tune into the teaching of Jesus about the primacy of love, and how to attempt to heal upsets I have caused. Then when I've done all I can to mend those rela-

tionships, I think I'll be more at peace with God. In that way I do think that Jesus brings me freedom from my guilt and helps me to restore personal harmony as far as it is possible in my situation.

This response has involved internalizing the teaching of Jesus, summarized so well in 1 John 4:20: "If we say we love God, but hate our brothers and sisters, we are liars. For people cannot love God, whom they have not seen, if they do not love the brothers and sisters, whom they have seen."

The Atheistic Humanist Group

This group focuses on the need of individuals to embrace their human freedom. They believe they have the power to choose whether or not to strive towards maturity by *freeing* themselves *from* the constraints of the social and physical environment in order to be *free to* develop their creative talents in areas suited to them. These people aspire towards the higher values of human beings, which are often called the spiritual values.

If there are five levels of human need, the basic three are:

✧ physiological needs: food, clothing and shelter;
✧ safety and security: usually provided in the home through parents and other social institutions;
✧ belonging, being loved and loving others.

The higher levels are:

✧ developing high personal esteem and the high esteem of others, which provides a foundation for the fifth level of need, the development of
✧ higher values and virtues like truth, beauty, goodness, perfection, justice, wholeness, aliveness, uniqueness, completion, simplicity. Those who aspire to this level are sometimes called 'self-actualizers'. That simply means that they strive to fulfil the inward urge to achieve their wholeness of *being* (Maslow 1970).

The enabling attitudes and emotions which must be cultivated in order to develop the higher values,which are understood by them in a *purely humanist* sense, include, for the atheistic humanist group, awe, sense of mystery, adoration, worship, gratitude, love, humility, approval, devotion, union, bliss and ecstasy. These values and feelings constitute a *non-personal unity*. To attain these values is to reach spiritual maturity, though there is no spiritual realm outside human beings. If a person does not follow their conscience in aspiring to these ends, and fails to make the appropriate moral choices, they fail to fulfil their biological essence. Then neurosis, unhappiness, destructiveness and evil are the expressions and behaviours which thwart a person's potential to become true being.

Many share the feeling that

there is a formative directional tendency in the universe which can be traced and observed in . . . complex organic life and in human beings. This is an evolutionary tendency towards greater order, greater complexity, greater inter-relatedness. In human kind the tendency exhibits itself in individual moves from a single cell origin to complex organic functioning, to knowing and sensing below the level of consciousness, to a conscious awareness of the organism and the external world, to a transcendent awareness of the harmony and unity in the cosmic system, including mankind. (Rogers 1980: 133)

For the religious person personal values can find full coherence and expression in a personal God. However, the atheistic humanist psychologists focus on non-personal values, as they are called, in order to deny the existence of a personal God. Nevertheless, they do accept the philosophical legitimacy of the religious humanists. This finds expression in their writings in two ways. First, they accept the aims of the religious humanists if they share their own focus, which is on living a life of love, and a concern for the pursuit of truth. The atheist focus includes a recognition that the pursuit of personal power or happiness are illusory goals, while the pursuit of love and reason towards others leads to deeper self-fulfilment. The second way religion attracts their interest is in analyzing two *ways* of being for a religious person, which are *psychologically identical* to the two ways in which an atheistic person may be a person. "The question is not *religion or not*, but *which kind of religion*, whether it is one furthering man's development, the unfolding of his specifically human powers, or one paralysing them" (Fromm 1971: 26).

The Religious Humanist Group

A central focus of this group is to search for meaning in life and, in some cases, help people recover meaning in life, for life has meaning even in Auschwitz, the Nazi extermination camp in which countless numbers of Jews died (Frankl 1963). Evil either deepens faith or destroys it. Like the atheist humanists, attention focuses on what it means to *be* fully human, in other words what is the nature of *being* as far as humans are concerned. The study of *being* is called ontology.

Both groups affirm that while a human person has a biochemical constitution and a social background and personal history, all of which can be analyzed into the elements which constitute the whole, *being* itself is the result of a phase change after its constitution and is irreducible. Being is just being, just as morals are morals, music is music, and molecules are molecules. Once you try to dissect or discover how each is constituted, you are no longer dealing with the original but with ingredients and the phase changes through which they go before becoming molecules, being, music or morals.

Life becomes meaningful in three ways:

1   through what we create or give to life – our life's work;
2   through our response to the higher values which challenge us in life, like the good, the true and the beautiful;
3   through our response to guilt, suffering and death, for guilt and failure are incentives to do better.

Religious humanists, like their atheistic colleagues, focus on the higher or spiritual values, making a clear distinction between the instinctual and bodily aspects of a person, like the basic levels of human need noted by the atheistic humanists above, and the higher or specifically human needs of the human mind, which are spiritual. Self-transcendence is directed to all realms of meaning. These religious humanists focused on the "immediate data of actual life experience", "bracketing out" preconceived patterns of interpretation held in order to get back to the basic uninterpreted experience.[5]

Human beings are committed to the search for meaning, and ultimate meaning is found in the ultimate being, God. This is because human life is related to something or someone other than self, in the form of a meaning to fulfil or a human being to encounter. This is driven by a deep personal willing or intention which provides the *reason for* not the *cause of* the pursuit of desired goals.

The human will is free to choose, though it is obviously limited by biological, social and psychological constraints. The person is not conditioned to behave in a particular way, but has "freedom within limits", which is real freedom. I am responsible for "willing what I ought to will". Reasons pull us towards our goal in contrast to causes, which push from behind. Love is the highest goal towards which we can aspire and is sought through intentional action. Conscience leads us towards this goal. A transhuman agent speaks to us through our conscience, and this is a human source of transcendence which many call God, although the atheists find no reasons for this naming.

The religious humanists recognize that two different levels of knowing are involved. Human knowing is personal knowledge; transcendent disclosure is religious knowing, which the believer claims as knowledge of God, who is "hidden" even though revealed. Unconscious religious awareness emerges through a pool of images from the heart of a person's being. These images are shaped by existing and previous cultural forms, so that when we are born we enter a world of symbols and images which we can "adopt and adapt". Repressed transcendence finds expression as "unrest of the heart". Ultimate meaning, expressed through transcendence, can only be grasped from the depths of our being through faith, for it is discovered not through reason, but through commitment, and it separates the life of ultimate absurdity from the life of ultimate meaning. The attraction of the disclosure of transcendence is offered without compulsion (Fuller 1994: 241–2; Frankl 1977).

Ultimate Being can only be freely approached and responded to in faith, but then that is true of the way you approach and respond to the one you hope to spend your life with. Here's one version from a sixteen-year-old school girl, written in 1959, called:

"Life Is Not So Simple When You Fall In Love":

How strange and complex are the rules of love.
The boy mustn't show he likes the girl even if he's head over heels in love with her,
    in case she take his love for granted and goes out with other boys.
The girl mustn't show she loves the boy lest he thinks she's chasing him.
The parents of both must not be allowed to know that the two are in love lest they
    disapprove.
The boy must be neither too fast nor too slow in case he loses the girl,

and she must not encourage him in case of being thought a flirt,
or show too little interest in case he ignores her.
With all these rules to keep, I'm surprised there are so many marriages!
Somehow, it seems love wins through in the end![6]

PERSONAL KNOWLEDGE ABOUT BEING HUMAN:
PSYCHIATRY AND PSYCHOTHERAPY

The concern of science to understand "psychologically healthy living" finds expression not only in the humanistic psychology movement, but also from psychiatrists involved in psychotherapy. Psychiatrists are concerned to help people face difficulties in life which they find emotionally crippling. Psychiatry is a medical speciality concerned with the prevention, diagnosis and treatment of mental illness; a psychiatrist is a doctor trained to recognize disturbance of mood, behaviour and intellect (Campbell 1987: 222).

Psychotherapy involves discussion sessions between the healer, who may be a psychoanalyst or a psychologist, and the patient. Attention focuses on experiences in earlier life which created a repressed emotional reaction. The healer tries to help the patient to recognize and express repressed earlier experiences which threatened them in the past, and may still threaten, in order to express any repressed emotions, such as anger, shame and guilt. Having identified the emotional prison repressed in the unconscious, and showing the patient that the emotions can be released without damage to self, the patient is then helped to find different ways of reacting in situations which previously caused such stress (Campbell 1987: 227).

Psychiatrists, like other sensitive human beings, recognize that life is difficult. Dr M. Scott Peck offers a very valuable account of confronting and solving painful problems.[7]

Scott Peck

He outlines a new psychology of love, traditional values and spiritual growth. The book he wrote is in four sections: Discipline, Love, Growth and Religion, and Grace.

Discipline   Because life is difficult we need to develop *a discipline*, which involves the constructive use of what Peck calls the four "techniques of suffering".

I   Work before play. This increases the pleasure of leisure or recreation as an earned reward for work done. Playing first may leave inadequate time to attend to work which needs doing and increases a person's problems.

2   Problems don't go away: they need facing. Time is needed to attend to them, otherwise they become a barrier to spiritual growth. *Responsibility* must be exercised: start, continue, finish and then play! Rather than say "I can't" or "I couldn't", say "I have to, there is no choice".

3   We need to be committed to truth. Truth is what is real. The problem is not intellectual truth – in learning data in subjects like decorating a room,

plumbing a bathroom, understanding mathematics or geography. The problem is usually emotional truth. For example, if, as a child, you could not trust and were afraid of, were hurt by, a parent, teacher or a bully in school, it is inappropriate to *transfer* this distrust onto your adult employers or colleagues. This is often done unwittingly, because a "childhood trust map" is *transferred* to an adult world because no other map is known. Thought, and sometimes help, is required in testing the trustworthiness of people in new situations and consciously avoiding the inappropriate *transference of outdated maps*.

4   *Balancing* is the discipline which gives us flexibility. For example, we should not become angry automatically if someone "puts us down", for that may not be the *intention* of the speaker. The intention might be to correct a point of information or to end a meeting on time. In such a situation, anger is misplaced, and one needs to avoid a rigid, unthinking response in order to be understanding and flexible. Give and take, including forgiveness, are required.

To summarize, we each need to develop a disciplined approach to facing the problems of life, whether it is relational (requiring emotional skills) or intellectual (requiring learning skills). Most problems require both sets of skills. This discipline involves developing and using four techniques:

1   *Delaying gratification*, which is work before play;
2   Taking *responsibility* for dealing with a situation, rather than being a helpless victim;
3   Being committed to the *truth*. This is less about dishonesty and more about *awareness* of the inappropriateness of *transferring* outdated "trust maps" from childhood to adult situations without re-evaluation and testing.
4   *Balancing* one's emotional responses through a disciplined analysis of situation, intentions, ameliorating factors affecting a speaker and the *flexibility* of give and take, including forgiveness.

Love   Psychologists looking at human life, its purpose or purposes and problems, quickly centre on *love* as the heart of the spiritual growth of a person. Love has many meanings. The ancient Greeks had separate words for three meanings, and in exploring the meanings of love, it is helpful to refer to them so that we can later use them as shorthand. They are *eros, philia* and *agape*. The third word, in particular, we must explore more fully than Peck does, as it is central to spiritual health.

Eros has a statue in Piccadilly Circus in London. It refers to the sensuous and passionate desire between a man and a woman which, when combined with friendship, is usually very positive, leading to a permanent relationship and marriage. Without friendship it can be very evil. Alone it is animal desire for sexual gratification.

Philia refers to family affection and natural friendship in normal human relationships. It is a healthy working mode of convivial relationships at home, work or play.

Agape refers to the highest form of love, which involves respect and reverence, and which, for religious people, includes God's love for humans, humans' love for God and love of neighbour, which is a disposition of the will, not an emotion, directed to all in need, whether likeable neighbour or hostile enemy. Love as *agape* is defined as "the *will* to extend one's self for the purpose of nurturing one's own or another's spiritual growth".

This definition of love, *agape*, has six characteristics:

1   It is concerned with an end or a goal, the well-being of another person or group.
2   The definition is circular in the sense that you "grow bigger" as you extend yourself. This in turn gives you the opportunity for further growth in an evolutionary upward spiral.
3   Love includes self-love and love of others, friend or foe.
4   To extend one's limits of loving requires an act of *will.*
5   Love is an expression of freedom: we choose to love.
6   Love, for the religious person, is emotionally driven by God's love, love for God, love for the human family, the needs of others, or any combination of these or other philanthropic emotions.

Love, *agape*, implies loving attention, effort, commitment, even in the face of the risks of confrontation, material or situational loss, loss of independence. Love implies self-discipline translated into action. Love, when including all three forms, *eros, philia and agape*, even in marriage implies the separateness and independence of persons who, paradoxically, are simultaneously committed to interdependence and the support of others. True love in any of the three forms is, when life affirming, shot through with awe, reverence and beauty, and is "given away" if one treads the pathway to true joy.

Agape is a key term in Christian religious talk about God. This love is associated with loving-kindness, grace, mercy, forgiveness and fidelity or loyalty. These qualities are features of God's love when pictured as the love of a faithful husband for an erring wife (cf. God's love for Israel in the Book of Hosea). The lure or effect of God's love "calls for" or evokes love in response which is expressed in worship and through love of neighbour (Leviticus 19:18; Deuteronomy 6:4–5; Mark 12:28–31; 1 John 4:19–21). Christians find the model of God's love, the model of *agape*, in the life and teaching of Jesus of Nazareth. So now love as *agape* in human life is the creative intention of a person to enable another to fulfil their potential by giving him or her the space, freedom or encouragement to become his or her authentic self, overcoming obstacles and being healed of injuries. This involves the letting be of the other to enable, through loving, the realization of the potential of being one's mature, whole and authentic self (Richardson and Bowden 1991: 341).

Growth and Religion   As human beings grow in discipline, love and life experience, their understanding of the world and their place in it grows. Everyone has a set of ideas which constitutes their *world-view*. It is their life stance, their philosophical outlook, their way of living. It is their religion, (*religio* is the Latin word meaning "bond" or "commitment"), says Peck, meaning it is their world-*view*, which may be religious (e.g. Christian or Hindu) or secular (e.g. humanist or communist).

Spiritual growth from childhood to adulthood is fraught with growing pains and problems. Spiritual growth involves growth in knowledge. Love, for the growing individual, involves an expansion of self, a move into unknown experience. This involves the developing of skills in expanding or rebuilding schemes of knowledge, and assessing the trustworthiness of people. The former we discussed earlier (Campbell 1987: 222; see also section on Piaget in Chapter 4).

Assessing trustworthiness is often affected by earlier experiences. It might have been the case that we were surrounded by authoritarian, possessive and abusive parents as children, and that we *transferred* these intuitively learned attitudes to honest and friendly colleagues in our adulthood. Or it might have been the case that we were surrounded by loving and enabling parents in childhood and, in adulthood, unconsciously *transferred* these intuitively learned ways of responding to those around us, who were dishonest rogues. *Transference* is the psychiatric insight that a set of ways of perceiving and responding to the world which is developed in childhood is inappropriately transferred to the adult environment.

We have to accept much second-hand information in life, because life is too short for us to discover everything ourselves, first-hand. So I trust the ticket collectors who tell me the time of the next train, or the career adviser on how to apply for a job, or the doctor who diagnoses my illness and tells me which pills to take. Life is too short for me to train myself in every career and profession, to discover all the information I need to live successfully.

Trust in second-hand knowledge is necessary for successful living, even though it is sometimes wrong. We all have to consider ultimate questions, like, Why am I alive? What really matters in life? Which *world-view* shall I adopt? Which values do I embrace? Is death the end? I begin life, as a child by accepting second-hand answers – those of my parents. We start life by inheriting and accepting a second-hand faith. But in that form it is an authoritarian or extrinsic faith. We have to turn this into a humanistic or intrinsic faith. That is, we have to adopt or build a *world-view* for ourselves.

If we begin with a childhood religious *world-view*, with God in the sky looking down lovingly on all the human family below, and then become aware that some with wonderful careers ahead die of cancer in their teens and others are killed in avalanches. when skiing, this may cause painful reflection. Scott Peck suggests that science is a religion, a *world-view* that is broader than a childhood religious view in which experience, if it can be repeated, can be tested for truth. He sees science as a religion of scepticism, and Peck defends its status as a religion because of its international character, though it has difficulty dealing with the reality of God.

Peck finds that religion has a bad record. It is restricted by dogmatism, i.e. it holds beliefs which it will not change. Some examples from conservative churches are: no women can become priests; no abortion; no salvation except through Christ alone. The consequences of dogmatism have included inquisitions, torture, persecutions, wars and hypocrisy. Peck says that religion has a multiplicity of images without agreement. So is God an illusion and belief in Him a nervous sickness to be dealt with by the psychiatrist? Peck's case book shows that the answer to this question is sometimes "yes" and sometimes "no". The kind of religious orientation will be a key factor affecting the psychiatrist's judgement.

Peck suggests that many scientists have two problems when dealing with religion. First, they throw the baby out with the bath water. The bath water includes cruelty, torture, persecution, religious wars, ignorance, superstition, dogmatism, rigidity, self-righteousness and hypocrisy. But that is no reason for throwing out God (Scott Peck 1990a: 238–9). Secondly, Peck accuses scientists of tunnel vision. It is true that both science and religion can be boxed in by tunnel vision. Religion is boxed in by its dogmas. Science, says Peck, is boxed in by two things: measurement and paradox. From the beginning of science as we know it, the scientific method has relied on experience, observation, verification. It is boxed in by measurement, which has been an idol: if we cannot measure something, we cannot know about it. However, he now feels that is no longer the case, for, since the advent of relativity, quantum theory and the exploration of complex systems, sometimes called chaos theory, measurement is more problematic. For example, we cannot know the location of a particle of light at the same time as its momentum. At any given moment we can only know one or the other, but not both together at the same time. Measurement is no longer an idol.

Secondly, argued Peck, scientists could not accept paradox, while religion did. Now scientists accept paradox: "light is both a wave and a particle at the same time". Religions accept paradox: "a human is both mortal and eternal at the same time". Peck concludes by saying that science and religion "have begun to speak the same language. Is it possible that the path of spiritual growth that proceeds from religious superstition to scientific scepticism may indeed ultimately lead us to a genuine religious reality?" (1990a: 243).

Peck's hope for a re-establishment of harmonious relationships between science and religion seems well founded. It does seem to be the case that science largely asks questions focused on how things change, while religion is concerned with why, e.g. Why is there a universe? Why are there human beings in the universe? While my distinction here is too simple, it does point the way to recognizing that science and religion are complementary forms of knowledge, as we have previously noted. Peck seems to be unaware of this.

Grace What strikes us as amazing is a sign of grace, suggests Peck. These moments or situations which amaze us are examples of what theologians call uncreated grace, examples of God's gift of himself to us. God's grace is unconditional acceptance of human beings, even though we are self-centred, pursuing styles of life far from those which promote our authentic being. Prevenient grace, or unexpected help before facing difficult situations, is recognized in the incidents or situations in which humans discern the hand of God: that which attracts us to God. Created grace is the result of humans responding to grace – the undeserved love of God – leading to the changed nature of humans as they aspire towards a divine nature. This is the accepting of unexpected help after we have got ourselves into difficult or unwholesome situations from which we need to extract ourselves.

Peck's thesis is that grace "manifested in part by 'valuable or agreeable things not sought for', is available to everyone, but while some take advantage of it, others do not" (Scott Peck 1990a: 275). From a psychiatrist's perspective he notes the miracle of health, the miracle of the unconscious mediating help through dreams, idle thoughts, even Freudian slips of the tongue. The miracle of serendipity is the gift of finding valuable and agreeable things not sought for. All these can be seen and grasped as signs of grace. They have some common characteristics:

- ✧ they serve to nurture, support, protect and enhance human life and spiritual growth;
- ✧ the mechanism of how these things happen is not fully understood;
- ✧ occurrence is frequent, routine and commonplace;
- ✧ their origin is outside the conscious will.

Peck sees these events as the expression of a single powerful force outside human consciousness which nurtures the growth of human beings. Religious people call this grace.

Theology talks of two traditions which seek to locate the origin of grace. Emanence is the doctrine "which holds that grace emanates down from an external God to men". Immanence is the doctrine "which holds that grace immanates out from the God within the centre of man's being" (Scott Peck 1990a: 279). This is a paradox, due to the fact that we tend to regard "God" and "grace" as entities which have a location.[8]

Evolution: a Miracle of Grace?   Peck then focuses on another miracle (1990a: 282), the growth process of life: evolution. If the second law of thermodynamics states that energy flows naturally from a state of greater organization to a state of lesser organization, the universe is winding down. A stream flows downhill. Evolution is a process of development of organisms from lower to higher states of complexity, differentiation and organization, e.g. from a virus, little more than a molecule, to a human with an enormous cerebral cortex and complex behaviour patterns. The first miracle here is that the force of entropy is overcome, and the second is that humans push themselves to improve themselves and their cultures, as in, for example, the growth in scientific understanding, and in technological and creative achievement. Peck then poses the key question: "But what is this force which pushes us as individuals and as a whole species to grow against the natural resistance of our own lethargy? We have already labelled it. It is love."

This is the miraculous force which defies the law of entropy.

If love is the human expression of evolution, where does the whole force of evolution come from? Love is conscious; grace is not. Where does this "powerful force originating outside of human consciousness which nurtures the spiritual growth of human beings" come from? We do not know. But to explain the miracles of grace, which initiated evolution, and love which enables humans to continue spiritual growth in self and society, "we hypothesise the existence of a God who wants us to grow – a God who loves us". God wants each of us to become himself or herself, for God is female and male in harmonious union. The impediment to this aim is that which spoils the relationship of humans with God.

Laziness, or Original Sin   If evolution points "upwards", does laziness point "downwards"? It is human entropy, laziness, one form of which is fear, which theologians have called original sin. Laziness, or original sin, is the form of our sick self. Even though the science of psychology has acted as if evil did not exist, Peck concludes from his work as a psychiatrist that it is real. The sight of, and therefore the effect of, the outrageousness of evil is a signal to us to purify ourselves: "It was evil, for instance, that raised Christ to the cross, thereby enabling us to see him from afar. Our personal involvement in the fight against evil in the world is one of the ways we grow" (Scott Peck 1990: 286–99).

The Evolution of Consciousness   Consciousness is a central characteristic of a human being. To be conscious means "to know with" the mind. The unconscious "mind" possesses extraordinary knowledge. The development of consciousness involves developing awareness in our conscious mind along with our unconscious mind, with which it comes into synchrony. Peck hypothesizes that our unconscious is "God within us" (1990a: 301). The Christian concept of the Holy Spirit is largely focused on God active in human hearts and does not necessarily support this hypothesis (Wiles 1994: 96; Bowker 1998: 416).

Peck's central concern as an experienced psychiatrist is to demonstrate that people's capacity to love, and their will to grow, are nurtured not only by the love of their parents during childhood but also through their lives by grace, or God's love (Scott Peck 1990a: 321). Failure to respond is sometimes laziness, but sometimes selfishness which may become pure evil.

Welcoming Grace?   Peck reflects on his view that our human myths, the stories which give us the vision of life by which we live, have been replaced by scientific information. This has caused us to suffer a sense of personal meaninglessness. "Yet it is that same science that has in certain ways assisted me to perceive . . . [that] the reality of grace indicates humanity to be at the centre of the universe" (Scott Peck 1990a: 333).

Summary:
Psychiatric Insights into Being Fully Human

First, it was only when reading Peck's book that readers discovered that this "best-selling author" was a Christian. Through outlining insights gleaned from treating his patients, he found that the Christian way of life perfectly prescribes an existential health plan for the fullness of life. The central details of his prescription are:

✧   Self-consciousness operates through awareness when dreaming, when awake, and when responding to a transcendent voice.

✧   Discipline involves work before pleasure, facing problems responsibly, commitment to truth in word and deed, making balanced judgements and flexible responses where appropriate.

✧   Altruistic love is at the heart of spiritual growth, learned from persons surrounding one, and practised as concern for all around one.

✧   Spiritual growth involves developing and committing oneself to a personal world-view – religious or humanist – in which altruistic love is central. This grows out of the cultivation of the higher spiritual values (1 Corinthians 13:1–13; Galatians 5:22). A materialist world-view with its basic beliefs in the primacy of matter and mechanism is simply sub-human reductionism.

✧ Spiritual knowledge contains and complements scientific knowledge by completing it.

✧ Clinical experience leads to the recognition of the reality of grace. This follows from seeing life as a gift: we hardly earn it or have a right to it before we have it. It follows from recognizing unexpected or undeserved help from without, a source most people recognize as God;

✧ Recognition of God's presence provides opportunity for a conversational relationship with God who is Mother and Father.

✧ The conversational relationship of friendship with God can be damaged by laziness or selfishness, leading to ill health and, at worst, a downward spiral into evil.

Peck recognized, in the last point in the summary above, that he had to write a book about human evil (Scott Peck 1990b: 10).

PERSONAL KNOWLEDGE OF HUMAN EVIL

A Psychiatric View: Scott Peck

In *People of the Lie*, Peck (1990b) explores the evil side of human nature from his perspective as a practising psychiatrist. He makes the point that the most basic scientific investigation of evil can be made only by a therapist: "There is no method of looking into the core of a person's being that can approach psychoanalysis for its depth and discernment" (Scott Peck 1990b: 299). The scientific exploration of human nature, however, cannot be detached from moral questions. There is therefore a need for a psychology of evil. Using his case-book experiences as a psychiatrist as evidence, he has met the evil side of human nature and offers a description of it. Finally, a psychology of evil needs to relate to the religious models of evil and religious psychology (Scott Peck 1990b: 50).

Peck defends his position that a scientific study of evil cannot be a value-free study of facts; it is also a matter for moral and religious judgements, for two reasons. First, humans are moral beings with world-views. Secondly, the recognition that it is no longer possible to defend the autonomy of intellectual disciplines, which he finds one of the most interesting developments in the intellectual history of the later half of the twentieth century. Thus morality and religion can no longer be separated from science (Scott Peck 1990b: 44). The scientific uses of nuclear energy as well as "the arms race" are two of many scientific and technological questions which pose moral and ultimate questions, from which they cannot be detached (302).

He distinguishes between most people, who occasionally do evil deeds from time to time, those who feel possessed by evil, a state frequently ended by exorcism, and those who are wholly evil people. Those possessed seek healing, for healing is impossible without free will assent. Those wholly evil seek no healing; they only want their own way. His insights as a psychiatrist led to, and later informed, his decision to become a committed Christian in 1980, at the age of forty-three, after first exploring Buddhism and Islamic mysticism (Scott Peck 1990b: 11).

Psychiatrists understand humans in terms of health and disease – a medical model – which parallels the Christian model of good versus evil: God versus the Devil (Scott Peck 1990b: 236). The Devil or Satan is a spirit, yet Satan or the Devil is just a name for a pattern of demonic or evil spiritual behaviour. The Devil can only be observed in a human body, and has no power except in a human body. Satan's powerhouse is *lies*. Satan's weapons are greed, pride, vanity, seduction, flattery, intellectual argument and especially *fear* (238). Evil is that which kills the spirit of another person, i.e. it deeply damages their awareness, feelings, mobility, growth, will or autonomy. Evil also leads to intense hatred and murder (47).

Peck notes three models for evil. First, the Hindu and Buddhist view that good and evil are two sides of a coin, and evil is illusion. This, he feels, is not very far from Augustine's unsatisfactory view that evil was the absence of good. A second model regards evil as distinct from good but still part of God's creation. To endow humans with free will, they must have the choice to do good or to do evil. God, whose will is good, does not want evil but has to permit it as the evidence of real free choice for human beings. The third view holds that evil is not God's creation, but a cancer beyond his control. Peck regards this as the only model which adequately deals with murder and the murderer.[9]

Parents are evil when they allow children to assume responsibility for some evil incident, like the suicide or accidental death of a friend or relative (Scott Peck 1990b: 67). Evil people are consistently destructive of others, manipulative, and frequently lie. They are covert, intolerant of criticism and of any attempt to puncture their fictitious image of good character and respectable life style. Lyingly denying that they have hateful feelings or vengeful motives, they are intellectually devious (146).

Evil people are always morally aware, and so seek scapegoats when things go wrong. Evil people are self-absorbed, narcissistic, i.e. they worship themselves, assert their will, but meticulously cultivate their appearance of "being good", thus living a lie. The lie is designed to deceive themselves, since they cannot accept criticism and therefore cannot criticize themselves. They are dedicated to preserving their image of moral perfection. They lack all motive to be good, but intensely desire to look good. A good disguise is as a deacon in a church. Evil patients Scott Peck met were dominated by the will to power, based on a foundation of lies. He regards evil as a mental illness. Evil is the ultimate disease. "Despite their pretence of sanity, the evil are the most insane of all" (Scott Peck 1990b: 84–5 footnotes, 304).

Because of self-centredness, Satan, evil, has no understanding of love. Love is to be fought, and lies are its powerhouse. Free will always takes precedence over healing. Satan, the truly evil person, rejects the humiliating hand of friendship and maintains an icy solitariness (Scott Peck 1990b: 240, 288).

The healing of evil, scientific or otherwise, can only be effected by the love of individuals. A willing sacrifice is required. The individual healer must allow their own soul to *absorb* the evil. When this happens there is sometimes a strange reversal of roles. When an innocent victim is killed or disempowered, guilt can reverse the roles, making the original persecutor the victim and the original victim the forgiving "victor", except that the aim of the "victor" was the absorption of evil, not personal "victory". We do not know how this reversal of roles happens.

Peck has known good people who deliberately allowed themselves to be pierced by the evil of others, broken by injury or humiliation, yet were unbroken in spirit. They are "killed in some sense and yet still survive and not succumb. Whenever this happens there is a slight shift in the balance of power in the world" (Scott Peck 1990b: 309). That is, the good in the world increases slightly and the evil correspondingly diminishes (cf. Luke 11:20).

A Religious View: Judaeo-Christian Insights

Peck chose the third of the three religious models he described in order to give an ultimate frame of reference to his discussion of evil. In exploring Peck's work through his eyes as a psychiatrist, we have looked at a picture of the nature of evil and how it may be contained.

Philosophy is concerned to apply reason to problems like evil. The Christian philosopher Plantinga sketches a helpful picture of God and evil (Peterson *et al.* 1991: ch. 6). He suggests that it is not inconsistent to believe that God exists and is omnipotent or all powerful and perfectly good, *and* that evil exists. Evil exists because of the actions of free, rational, fallible creatures – human beings. So humans have free will, and free will can cause human evil, which harms the spiritual growth of an individual, in the sense that it restricts, arrests, violates or ends their spiritual development as persons. It follows that God has freely limited his own power in order to allow a human freedom that is not restricted by a divine Presence. God did not have to do this; he chose to do it (Luke 15:1–2, 11–32). This account of God and evil is sometimes called the "freewill defence of a good God in the presence of evil in the world".

The defence, as we have said, is that God allows human freewill so that humans can choose to do good or evil without divine constraint. There are, of course, other constraints on freewill, like time, space, personal ability, the personal values of the one intending to do an evil deed, and situational opportunities for doing evil deeds. The distinction made by Peck between doing an evil deed, which most people do sometime in their lives, being unwillingly possessed by evil, and evil people has already been mentioned. The locations of purposeless suffering provide a place where true human goodness can occur.

God's willingness to allow humans freewill must always be understood in relation to his providence. That is, God shows loving foresight in dealing with human beings, and gives them opportunities for dealing with Him in person. Humans can thwart God's good designs, and it may be that, like a good chess player, He can turn some of our evil actions to good effect (Genesis 37–47, esp. 37:3–28, 45:4).

Reflection on experience leads many to recognize that for the most part, God seeks to persuade, as a husband seeks to win back a wife or a father receives back an erring son, for we all need to know that God is love and wants to see us mature as loving human beings (Hosea, esp. 11 and 14; Luke 15:1–2, 11–32). His chosen method is the lure or attraction of His selfless love towards us as communities and as individuals (Pittenger 1970). Healing can only occur where evil is absorbed by an accepting victim. This is supremely demonstrated in the life of Jesus of Nazareth.

Some may see life as the arena for character building, or the vale of soul making, which provide the opportunity for humans freely to chose to respond to the call of their inner being (religious people call this God's call) to a life of love or to pursue selfish ends. This does not, however, deal adequately with the innocent suffering of the very young (Hick 1968: 264).

The suffering and death of an innocent child is the occasion for two opposite responses. One is provided by the powerful fictional situation in Dostoevsky's *The Brothers Karamazov*, where Ivan rejected belief in God (Dostoevsky 1982: 284–7). He cannot accept that a loving God would allow the suffering and death of an innocent child. He *knows* that this is a logical contradiction. This is the verdict of one who thinks he *knows* the whole situation. This is the voice of reason. He chose atheism. Camus gives his own view through the mouth of Dr Rieux in his novel *The Plague*. He shares Ivan's view, for he too refused to accept a God who allows the suffering and death of an innocent child (Camus 1984: 174–8).

In a real-life situation, Elie Wiesel witnessed a child being hanged beside two men on one of three gallows when they returned from their day's slave labour at Auschwitz. Behind him he heard a man asking: "'Where is God now?' And I heard a voice within me answer him: 'Where is he? Here he is – He is hanging here on this gallows' . . ."[10] If God is a co-sufferer, then theodicy, the defence of the love of God in the face of innocent suffering, becomes theophany, the appearance of God at work. The paradox of a God of love coexisting with evil is understood.

Wiesel's profound reflection on this and other atrocious events he witnessed needs reporting. "Deep down, I thought man is not only an executioner, not only a victim, not only a spectator: he is all three at once" (Surin 1986: 120). For Wiesel, the question, Where is God? can only be answered, if at all, by a person "imprisoned together with his God in the abyss – it can be done only by him who finds himself in that hell where God and humankind, full of terror look into each other's eyes" (122).

For a victim to accept and absorb this evil without behavioural signs of vitriol or hatred can be interpreted as a sign of prevenient grace i.e. that in some mysterious way God is caringly present in advance of, and during, the suffering to be endured. The focus of the victim has already been transformed from a concern with self-will to an acceptance of another will (cf. Luke 23: 46). Here is a "*hoping love*" which can do nothing but submit. Where this has the power to encourage others to share in suffering love in the fight against evil, it turns into an "*active love*", and therein lies the power of the suffering and death of Christ, leading to the disciples' shared religious experience of the risen Christ on Easter Day. On that day "a hoping love" of Jesus became "an active love" of the disciples.

The book of Job has an important contribution to make to this discussion. The writer of it is concerned to show two things. First, the innocent Job's suffering was not God's punishment for sins unseen, as his friends suggested with faultless logic (Job 2:11; see also Brown *et al.* 1995: 466–7). This view, called retributive justice, widespread then, at Jesus' time, and since, assumes that God exacts "payment for" or "revenge for" sins committed. Camus assumed that this was the current Christian view, as he placed this argument in the mouth of Father Paneloux in his novel *The Plague* (1984: 80–1). This idea of retributive justice is based on a very primitive and unacceptable picture of God as a legalistic monarch, not a loving father. Jesus faced the same problem with his disciples in John 9: 1–3. The disciples *assumed* that the man was born blind because of past sin. Their only question was whether the cause of blindness was the man's sin or that of his parents before his birth. Jesus corrects them by saying this was an opportunity for healing.

Second, more importantly, the writer of Job poses a question. Do people love God and their neighbour altruistically, i.e. for their own sake, or do they love God and their neighbour instrumentally, i.e. in order to gain a reward? The author of *Job* states that God assumes that a good person loves and avoids evil altruistically (Job 1:8–12). In any case, there is no reward except that one has kept faith with one's calling from a loving father to love one's neighbour. Job corrects his friends about retributive justice but makes a mistake with God by insisting on his integrity, failing to recognize that that implies no obligation on God (Job 31). Jesus makes this clear when he says that when we have done all that we should do "we have simply done our duty" (Luke 17:10).

Job is portrayed as a saint to establish that his suffering *cannot be* a punishment for sins committed. God has to allow Satan to test Job to show that, even though his understanding is at fault, his goodness is altruistic. God trusts those who serve him; that is one central point of the book of Job (Brown *et al.* 1995: 466–7). The writer of Job puts this judgement into the mouth of God: "Then the Lord asked [Satan], 'What do you think of my servant Job? No one on earth is like him – he is a truly good person, who respects me and refuses to do evil'" (Job 1: 8–12 *CEV Bible*). A crucial further point is that Job meets God in the depth of his suffering, where he sees God with his own eyes (42: 5).

This is where Wiesel met God. Here we have two voices, Job's and Wiesel's, who listen to the deep intuitions from the core of their being, who draw from the deep well of their humanity formed by God when he created humans "in his own image" – an example of prevenient grace. Job and Wiesel both admit, to quote Job, "I *know* so little," (42:3) but choose the priority of faith over reason. They know they can't *know* the total situation (Vardy 1992: ch. 13). Like Job and Wiesel, Jesus of Nazareth, when facing death, knew he could not *know* the total situation, and chose the priority of faith over reason. Jesus on the cross forgave those who deserted him, those who denied they knew him, those who crucified him, and accepted those who were crucified with him, finally saying, "Father, I put myself in your hands" (Mark 15: 34; Matthew 27: 46; Luke 23: 34, 46).

Ivan Karamazov and Elie Wiesel both face the problem of belief in a loving God and the suffering and death of an innocent child. Ivan confidently thinks he has all the *knowledge* he needs (cp. Camus 1984: 174–8). Reason tells Ivan that, in facing such a contradiction, there can be no God. Elie has all the knowledge available, finds himself in that abyss of hell where, in terror, he looks into the eyes of God and faces the paradox of a loving God and the suffering and death of an innocent child.

While in that hell of seeing the child in Auschwitz, he finds an intuition welling up from deep within himself, partly including an aspect of the image of God amongst other unidentified internal resources, which is an intimation of prevenient grace. The eyes of God tell Elie that he is suffering there with the child and all of them. Faced with the choice these two men faced, do you trust in reason, or in something deeper, which is called faith for want of a better name?

In the twentieth century many have followed in the way of suffering love, like Gandhi, Martin Luther King, Steve Biko and Nelson Mandela. These are further examples where the role of "victim" is mysteriously and providentially changed to that of "victor". And the role "of being blotting paper for evil" is also accepted by many people around us, and is often unnoticed. Such people include therapists, psychoanalysts, parents, ministers, colleagues, friends or neighbours. They intuitively set out to absorb evil in an attempt to heal the person or persons possessed with hatred, anger or fear, though they may not be able to put into words what they are doing. It is an intuitive activity which Michael Polanyi has taught us to recognize as *tacit knowledge*. If it inspires others to such suffering love, it shares in the expanding network of 'active love'.

Other world faiths offer different ways of coping with suffering and evil. Hinduism suggests that one should avoid evil by renouncing all cravings and desire and following the moral order (Holm and Bowker 1994: 88). Buddhism suggests that one should avoid cravings which lead to quarrels, strife and conflict; without desire one will not suffer or promote evil, for evil is due to ignorance (30). Judaism suggests that God created evil, even though he is not evil, but evil can only be overcome by obedience to God's will (127–8). For Job, Christ, Wiesel, and the rest of us, that means absorbing it.

Evil: a Summary

People can be involved in evil in five ways.

First, they can be involved with evil as victims. For instance there is a widespread recognition that an abused child or adult, if they are spiritually destroyed by what happens to them, may themselves become an abuser, perpetrating the downward spiral of evil through social influence too great for them to transcend. Bitterness, hatred, anger, fear and aggression may be the results. They are primarily the destroyed or maimed victim of evil, either physically or spiritually or both. Our first conclusion is that evil can destroy a person physically, by maiming or killing, or psychologically, by disempowering them.

Secondly, most positively, many people recognize evil and its compelling power in the hopefully occasional enactments of an evil deed. Surrounded by the positive influences of relatives, colleagues and friends, and the positive influence of their moral, religious and social beliefs, their rituals and customs and their own moral and spiritual consciousness, they largely triumph over evil influences or opportunities.

Thirdly, some, through situational, social or personality influences, unwillingly fall into evil ways and want to be cured. Their expensive recovery depends on those around them having the loving ability and power to absorb their evil as the means of enabling them to return to moral and personal health. A small number of this group may feel that they are "possessed" by evil and need an exorcism.

Fourthly, there are wholly evil people. Strangely, they are rarely found in prison (Scott Peck 1990b: 77). Usually, those trapped by evil get sent to prison. Wholly evil persons work very hard to appear good. They deceive themselves by a tissue of lies, which they use in three ways: to defend the fictitious image that they are good; to continue to get their own way; and to deprive many others of their rightful freedoms.

This misuse of power by evil people spiritually damages their victims in different ways and to differing degrees. Evil people are enemies of love and friendship, offers of which are regarded as humiliating. They prefer icy control and separation.

Fifthly, people can be involved in evil through contact with psychopaths. These are persons who are amoral. Actually, they cannot be moral or immoral because they do not seem to know the difference between right and wrong, and in this respect they are not fully human. They are therefore not included in a discussion of evil, even though their actions may be called evil by the morally aware.

Knowledge of evil is an aspect of knowledge of persons, as well as an aspect of moral knowledge and of personal knowledge, and it completes our review of insights into the nature of personal knowledge.

SUMMARY

- From work in the three disciplines of philosophy, psychology and psychiatry in the United States during the twentieth century, there has been an extremely strong emphasis on the irreducibility of the person.[11]
- *Pragmatism* argued that "what worked" for humans was knowledge, so James's defence of religious experience was that if it fills you with delight and has beneficial effects on your life, it is knowledge, it works and it is true.
- *Personalism* viewed all of reality, the whole universe and its contents, as the natural order in the mind of God, and persons created in the image of God as manifestations of Him. The person was irreducible, underpinned by the Person of God, in whose image persons can evolve and develop.
- *Humanistic psychology* emphasized the wholeness and irreducibility of being a person.

    An intrinsic or humanitarian orientation towards life should be cultivated, whether humanistic or religious, internalizing and making one's beliefs and values relevant to life.

    The atheistic humanist group emphasized the need to create freedom from the constraints of the social and physical environment in order to give one freedom to develop and use one's abilities, focusing on the higher values, wholeness of being and moral maturity and responsibility.

    The religious humanist group was concerned to explore what it means to be a human being enjoying the fullness of life (John 10:10). Reason and a transhuman agent, the God within, draw us towards our goal: love of God and neighbour.
- *Psychiatry and psychotherapy* focus on self-conscious awareness of voices received through dreams and when awake, and from the realm of transcendence. Realistic discipline, a focus on altruistic love and the cultivation of the higher spiritual values will free us from materialistic attachments, and free us to accept acts of grace and develop relationships, human and divine, in order to focus on the needs of others.
- *The disease of evil*: Spiritual development can be retarded or destroyed by laziness, and/or selfishness, leading to rejection of relationships and moral values.

Evil is a downward slope, beginning with the occasional vindictive act, hopefully recognized, so no tendency develops and conscience triumphs over imagined self-interest.

Some are dragged down through deprivation and social circumstances into lifestyles focused on anti-social and self-destructive behaviours, from which many would gladly escape and return to personal health and self-respect.

Some are totally selfish, ruthless and genuinely evil, seeking to cloak their behaviour behind a respectable facade and concerned with control and power rather than relationships and community.

Some people are trapped in the evil downwards spiral of being abused, unjustly treated and marginalized and are left in a soup of hatred, bitterness and envy.

Evil tragically destroys the humanity in some persons. This disease of evil is the challenge which the atheistic humanist and religious humanist groups are concerned to overcome, aiming at a global commonwealth of equal rights, equal resources and equal opportunities, which can be described as the goal of the Kingdom of God on earth.

NOTES

1 This theory of knowledge was pioneered in America by Charles S. Pierce (1839–1914) and was taken up and developed by William James and John Dewey.

2 For a brief study of the range of meanings of "ontology" see Audi (1995: 490).

3 The movement began and developed under the influence of Borden P. Bowne (1847–1910) and three of his students at Boston University.

4 Piaget, an important educational psychologist, is discussed more fully at the end of Chapter 4.

5 This is the phenomenological method of studying experience described in Chapter 9. Behaviourism is rejected.

6 Catherine Longman (née Entwistle) (1943–93), *Woman's Own*, 10 October 1959. Her "picture of the girl" here models the "picture of God" held by process theologians.

7 M. Scott Peck, *The Road Less Travelled* (1990a). This important book was on the bestseller list for five years and has probably had more impact on the public than any other single book of this kind. It is subtitled *A New Psychology of Love, Traditional Values and Spiritual Growth*. Cf. W. W. Dyer's *You'll See It When You Believe It* (1990), another national bestseller.

8 An agnostic humanist could suggest that this is a case of the one-sided presentation of "how things are". If health is a miracle, what of illness? Cf. the last section of this chapter on 'Personal Knowledge of Human Evil'. E.g. *The Independent*, 17 April 1999, p. 3. Milosovic's ethnic cleansing in Kosovo.

9 Scott Peck (1990b: 50–1 footnotes). Peck mentions the work of Fromm, another major psychoanalyst, who escaped the Holocaust and studied the evil of Nazism.

10 Cited by Surin (1986: 116). This is related by Elie Wiesel, a survivor of Auschwitz, in his book *Night*.

11 Reductionist views of human nature, like scientism with its emphasis on materialism, and behaviourism in psychology with its emphasis on manipulating controlled responses, had some following.

REFERENCES

Allport, G. W. (1969), *The Individual and His Religion* (1950), Toronto: Macmillan.

Allport, G. W. and Ross, J. M. (1967), "Personal Religious Orientation and Prejudice", *J. Pers. Soc. Psychol.* 5, 432–43.

Audi, R (ed.), (1995), *The Cambridge Dictionary of Philosophy*, Cambridge: Cambridge University Press.

Bowker, J. (1987), *Licensed Insanities*, London: Darton, Longman & Todd.

Bowker, J. (1998), *The Complete Bible Handbook*, London: Dorling Kindersley.

Brown, R. E., Fitzmyer, J. A. and Murphy, R. E. (eds), (1995), *The New Jerome Biblical Commentary*, London and New York: Geoffrey Chapman.

Campbell, A. (ed.), (1987), *A Dictionary of Pastoral Care*, London: SPCK.

Camus, A. (1984), *The Plague* (1960), Harmondsworth: Penguin.

Chisholm, R. M. (1966), *Theory of Knowledge*, Englewood Cliffs, NJ: Prentice-Hall.

Delaney, C. F. (1995), "Personalism", in R. Audi (ed.), (1995), *The Cambridge Dictionary of Philosophy*, Cambridge: Cambridge University Press, p. 575.

Dostoevsky, F. (1982), *The Brothers Karamazov*, (1880), Harmondsworth: Penguin.

Dyer, W. W. (1990), *You'll See It When You Believe It*, New York: Avon Books.

Frankl, V. E. (1963), *Man's Search for Meaning*, New York: Washington Square Press.

Frankl, V. E. (1977), *The Unconscious God*, London: Hodder & Stoughton.

Fromm, E. (1971), *Psychoanalysis and Religion* (1950), New Haven and London: Yale University Press.

Fuller, A. R. (1994), *Psychology of Religion*, Lanham, MA, and London: Littlefield Adams.

Hick, J. (1968), *Evil and the God of Love*, London: Fontana.

Holm, J. and Bowker, J. (eds), (1994), *Human Nature and Destiny*, London and New York: Pinter.

James, W. (1968), *The Varieties of Religious Experience* (1902), London: Fontana.

Macmurray, J. (1972), *Reason and Emotion* (1935), London: Faber & Faber.

Maslow, A. H. (ed.), (1970), *New Knowledge in Human Values*, 3rd edn, New York: Harper.

Peterson, M., Hasker, W., Reichenbach, B. and Basinger, D. (1991), *Reason and Religious Belief*, Oxford: Oxford University Press.

Pittenger, N (1970), *Christology Reconsidered*, London: SCM Press.

Richardson, A and Bowden, J. (eds), (1991), *A New Dictionary of Christian Theology*, London: SCM Press.

Roberts, J. D. (1991), *A Philosophical Introduction to Theology*, London: SCM Press; Philadelphia: Trinity Press International, p. 156.

Robinson, J. A. T. (1963), *Honest to God*, London: SCM Press.

Rogers, C. R. (1980), *A Way of Being*, Boston: Houghton Mifflin.

Scott Peck, M. (1990a), *The Road Less Travelled: A New Psychology of Love, Traditional Values and Spiritual Growth* (1978), London: Arrow Books.

Scott Peck, M. (1990b) *People of the Lie* (1983), London: Arrow Books.

Speake, J. (ed.), (1981), *A Dictionary of Philosophy*, (1979) London: Pan Books.

Surin, K. (1986), *Theology and the Problem of Evil*, Oxford: Blackwell.

Vardy, P. (1992), *The Puzzle of Evil*, London: Fount, HarperCollins.

Wiles, M. (1994), *The Remaking of Christian Doctrine* (1974), London: SCM Press.

Wright, D. (1967), "A Review of Empirical Studies in the Psychology of Religion", *Catholic Psychology Group Bulletin*, pp. 11–37.

4 | Are There Different Kinds of Knowledge?

As teaching is concerned with the exploration of knowledge, it follows that educational philosophers will have engaged in the study of the nature of knowledge. Rationalists have made the most important contributions to exploring the nature of knowledge and how to teach it. The rationalist, as we saw earlier, takes the view that reason is the sole route to truth. The basic units of thought are concepts, and the rational mind brings order to the stream of sense experience by applying concepts and organizing them into coherent systems so that everything experienced is given a place and therefore a meaning in an individual's understanding of themselves, society and the world around. The individual achieves this by using publicly shared systems of classification so that communication can take place between individuals and communities.

WITTGENSTEIN: KINDS OF LANGUAGE USE INDICATE
FORMS OF KNOWLEDGE

Most significant educational philosophers after 1953 began their thinking about the nature of knowledge with Ludwig Wittgenstein's *Philosophical Investigations*. Wittgenstein was an "analytic rationalist". In his book he introduced an analogy between human activity and games. One form of human activity is understanding. "Wittgenstein suggests that understanding consists in *being able to follow a rule*. Even if a research scientist changes a rule, then understanding involves following the new rule" (Bonnett 1994: 50). We have already mentioned Wittgenstein's new rule or verification principle. It states that "the meaning of a sentence depends on *the use* to which it is put". In other words, you must listen carefully to the way a person is *using* words in order to determine their meaning. This was his new verification principle; it showed whether a sentence was meaningful or not. If it was meaningful, it could then be tested for truth. This approach opened up the whole of language use for re-evaluation in different quarters.

OAKSHOTT: KNOWLEDGE IS THAT WHICH IS INTERPRETED AS SIGNIFICANT

Michael Oakeshott has been called a "conservative rationalist" (Bonnett 1994: 42). The foundations of his approach lay in his understanding of the relationship of the individual to society. He did not think that the human world was mainly concerned with physical things, but with what he called *intelligibles*. By that he meant everything in the human world, things (cups and spades, and so on), but also ideas, understandings, significances, beliefs, as well as institutions like marriage, the family, the school, the church and parliament. An *intelligible*, then, is any thing, idea, understanding, belief, institution or relationship which the individual *interprets* as important or significant.

Any individual grows up in the context of a *shared way of living*. This is expressed through *shared procedures*, through which an individual learns conventions, values and patterns of significance. "Through time we have developed sets of social rules, conventions, standards for discriminating and evaluating things. These shared procedures constitute our culture, and structure aspects of our lives as various as buying cornflakes, writing a will, or offering up a prayer" (Bonnett 1994: 42). Oakeshott seems to assume that *shared procedures* lead to *truth*.

Culture is not static, but is, as Oakshott suggests, an "ongoing conversation":

> In this conversation, which began with the dawn of human awareness, the language of feelings, sentiments, desires, recognitions, moral and religious beliefs, intellectual and practical enterprises etc. interplay, constantly creating the parameters and possibilities for human being. It is only through engaging in this conversation, entering the interplay of these various languages, that "self-disclosure" and "self-enactment" of the individual can occur. (cited by Bonnett 1994: 42)

Education, for Oakshott, is not about the promotion of child-centred rampant individualism, or about socialization, making the child fit for jobs and roles in society. Oakshott suggests that in school, areas of the curriculum should not be "encountered" or "delivered" but *lived*, for then pupils will pass on "the living tradition of our culture". This involves an apprenticeship approach to learning, i.e. through emulation (Bonnett 1994: 45). A limitation of this form of "conservative rationalism", one concerned with the transmission of culture, is that it leads to the question of how the development of individuality in the child can be safeguarded (Oakeshott 1971).

For Oakshott, then, knowledge consists of *intelligibles*, i.e any thing, idea, understanding, belief, institution or relationship which the individual interprets as important or significant. He recognizes that this involves "various languages".

HIRST: SEVEN FORMS OF KNOWLEDGE

In the philosophy of education, Paul Hirst, who was Professor of Education at the University of Cambridge and a "hard rationalist" (Bonnett 1994: 65), sought to develop a philosophical analysis of the subjects in the school curriculum. He distinguished different forms of knowledge, each of which "involves the development of creative imagination, judgement, thinking, communication skills, etc., in ways that are peculiar to itself as a way of understanding experience".[1]

Hirst distinguished three main criteria to identify different forms of knowledge. These were:

I   "*Certain central concepts* that are peculiar in character to the form   (of knowledge). For example, those of gravity, acceleration, hydrogen, and photosynthesis are characteristic of the sciences; number, integral and matrix of mathematics; God, sin and predestination in religion; ought, good and wrong of moral knowledge" (Hirst 1974: 44).

2   "In a given form of knowledge . . . the form has *a distinctive logical structure. . . .* ". (Hirst 1974: 44; italics mine). This includes a distinct family of concepts, procedures and sets of rules.

3   "Each form, then, has distinctive expressions that are *testable against experience in accordance with particular criteria* that are peculiar to the form" (Hirst 1974: 44; my italics highlight Hirst's key points). These are appropriate truth tests.

Hirst makes a further qualification of his third criterion: "All knowledge involves the use of symbols and the making of judgements in ways that cannot be expressed in words and can only be learnt in a tradition . . . from a master on the job" (Hirst 1974: 45).

In summary, then, Hirst suggested in 1970 that there are seven forms of knowledge:

✧   Formal Logic and Mathematics,
✧   Physical Science,
✧   Knowledge of Our Own and Other People's Minds,
✧   Moral Judgements,
✧   Objective Aesthetic Experience,
✧   Religion,
✧   Philosophical Understanding. (Hirst and Peters 1970: 63)

Hirst has removed History and Social Science from this list, which appeared in earlier lists, because he realized that while some facts can be discussed in terms of physical causation, like the Physical Sciences, there are also explanations of human behaviour which involve "intentions, will, hopes and beliefs etc. . . . The concepts, logical structure and truth criteria of propositions of this kind, are, I would now argue different from, and not reducible to, those of the former kind" (Hirst 1974: 86).

What Hirst means is that History and Social Science include two forms of knowledge. Sometimes they are concerned with situations which involve causation, and are therefore in the realm of physical science; sometimes they are concerned with situations which involve other people's minds, which involve personal notions like intentions and hopes. This form of knowledge, the third in Hirst's list above, he calls "Knowledge of Our Own and Other People's Minds". I prefer to call this "personal knowledge" expressed through the use of personal discourse. So History and Social

Science are, for Hirst, appropriate subjects in the curriculum of universities and schools, but he regards them technically as fields of study, like geography, which use various forms of discourse appropriate to the precise matter under discussion.

Hirst's third form of knowledge, understanding our own and other people's minds, which was absent from the first formulation in 1965, shows his increasing recognition of the distinct nature of human feeling, imagination and direct personal knowledge of other persons. Hirst states that this form of knowledge includes "concepts like those of believing, deciding, intending, wanting, acting, hoping, enjoying, which are essential to inter-personal experience and knowledge" (Hirst and Peters 1970: 63). As I said in the last paragraph, I prefer to call this 'personal knowledge'

Hirst also ceased talking about Literature and the Fine Arts, and talked instead of Objective Aesthetic Experience (Hirst 1974: 87).

Hirst notes two other classifications of knowledge which concern us practically. First, there are the curriculum subjects: organizations of knowledge which are not forms of knowledge or subdivisions of these disciplines, but each include several forms of knowledge. These he calls Fields of Knowledge. We have already noted that History, Social Science and Geography are examples. Engineering and Medicine are examples from the applied sciences. Secondly, while moral knowledge is a distinct form concerned with what ought to be done in practical situations, in practice moral values and judgements arise alongside questions of fact and technique in, for example, political, legal and educational theory, which Hirst cites as the clearest examples of "fields where moral knowledge of a developed kind is to be found" (Hirst 1974: 46).

In discussing language and thought he affirms that intelligible thought involves symbols, usually language: "Concepts and propositions are units of meaning and not psychological entities . . . "; "Understanding a form of thought involves mastering the use of the appropriate language game"(Hirst 1974: 83).

In Hirst's 1974 reflections on his account of forms of knowledge, he accepts physics as a subsection of physical sciences, music as a subsection of literature and the fine arts, and Christian theology as a subsection of religion. All these subsections he calls disciplines. However, he does not accept that a discipline may have concerns wider than its basic form of knowledge. Importantly, he recognizes that "religious studies . . . is now usually cross-disciplinary, and so employs different modes of discourse on different topics or dimensions of an argument" (Hirst 1974: 97). Also there may be educational reasons for studying educational units which are multi-disciplinary.

If you, the reader, were happy sorting the words given to you earlier into families (see Chapter 2 above), then you largely understand the forms of knowledge argument outlined by Hirst.

Comment on Hirst

Bailey feels that Hirst's tight logic of the forms of knowledge leaves out many "practical and expressive activities thought by many to be important" (Bailey 1984: 74). Bailey is a "soft rationalist" (Bonnett 1994: 71) who, while very sympathetic to the work of Hirst, is well aware of its limitations. He noted that Hirst failed to take adequate account of the many language games listed by Wittgenstein, even though he was aware of them, such as:

> Giving orders and obeying them
> Describing the appearance of an object
> Constructing an object from a description
> Reporting an event
> Speculating about an event
> Forming and testing a hypothesis
> Presenting the results of an experiment in tables and diagrams
> Making up a story; reading it
> Play acting
> Singing catches
> Guessing riddles
> Making a joke; telling it
> Solving a problem in practical arithmetic
> Translating from one language to another
> Asking, thanking, cursing, greeting, praying. (Bailey 1984: 81; Wittgenstein 1967)

Wittgenstein said: "It is interesting to compare the multiplicity of the tools in language and of the ways they are used, the multiplicity of the kinds of words and sentences, with what the logicians have said about the structure of language" (Wittgenstein 1967: 11e–12e cited by Bailey 1984: 81). Bailey makes the point that "whatever Hirst may be doing, Wittgenstein was not reducing meaning to seven sets of characterizations!" (1984: 81).

Bailey made the more significant point that "*meaning* has a different sense when connected with notions like *significance* and *importance*. In that sense religion and art can *be both*. This sense is not metaphorical or parasitic on the 'meaning of statements' usage, but is rooted in language in its own right."

PHENIX: REALMS OF MEANING

While Hirst was exploring the implications for knowledge in the UK curriculum of "Wittgenstein's criterion of meaning residing in the way language is *used*", Philip Phenix was doing exactly the same in the United States. He suggested that a general education should incorporate six fundamental realms of meaning which comprise "the basic competencies that general education should develop in every person" (Phenix 1964: 8).

The six realms are listed as symbolics, empirics, esthetics, synnoetics, ethics and synoptics, and are seen as providing "the foundations for all the meanings that enter into human experience. They are the foundations in the sense that they cover the pure and archetypal kinds of meaning that determine the quality of every humanly significant experience" (Phenix 1964: 8; see table 4.1 below).

Meaning for Phenix includes not only logical thinking, but also feeling, conscience and imagination. Meaning has four dimensions:

> First, that of inner experience, including the qualities of reflectiveness, self-transcendence, which all varieties of meaning exemplify. Secondly, there is the dimension of rules, logic

and principle, each type of meaning being defined by a particular logic or structural principle. Thirdly, there is the dimension of selective elaboration . . . leading to the development
of those that are significant and have an inherent power of growth and elaboration . . .
Finally, there is the dimension of expression, for the meanings . . . are communicable
through symbols. Symbols are objects that stand for meaning. (Phenix 1964: 21–4 cited by
Hirst 1974: 55)

Phenix, after exploring the logical patterns of disciplines, postulates nine generic
classes (Phenix 1964: 25), which, in practice, he clustered into six realms of meaning.
Every cognitive meaning has two logical aspects, namely, quantity and quality.
Knowledge relates to a knower relating to the thing known. There are three degrees
of quality: the singular – one thing; the general – a selected plurality; the comprehensive – a totality. There are also three qualities of meaning he calls: fact – what exists;
form – imagined possibilities; and norm – what ought to be. By pairing the various
aspects of quality and quantity, the nine basic or generic classes of cognitive meaning
can be distinguished as outlined in table 4.1 below.

We are not concerned here with Hirst's detailed evaluation of this classification of
knowledge (Hirst 1974: 57–8) but while he finds empirics, esthetics and ethics acceptable forms of knowledge, Hirst does not regard symbolics, synnoetics and synoptics
as fundamental categories of meaning, since they are all complex in nature (66).

Hirst was concerned to defend the thesis that forms of meaning and forms of
knowledge are identical and are expressed in true propositions. "Meaning, truth and
knowledge go together as they are logically divisible by the same criteria" (Hirst 1974:
67). Phenix, however, has his existential categories, which include kinds of experience, feeling, attitude, symbols and skills which can be defended as aspects of personal
knowing.

Table 4.1   Phenix's logical classifiction of meanings

Generic classes		Realms of meaning	Disciplines
Quality	**Quantity**		
General	Form	Symbolics	Ordinary language, maths, non-discursive symbolic forms
General	Fact	Empirics	Physical sciences, life sciences, psychology and social science
Singular	Form	Esthetics	Music, visual arts, arts of movement, literature
Singular	Fact	Synnoetics	Philosophy, psychology, literature, existential religion
Singular	Norm	Ethics	The varied special areas of moral
General	Norm	Ethics	and ethical concern.
Comprehensive	Fact	Synoptics	History
Comprehensive	Norm	Synoptics	Religion
Comprehensive	Form	Synoptics	Philosophy

Phenix is quite clear:

> reason and mind tend to be narrowly construed as referring to the process of logical thinking. The life of feeling, conscience, imagination and other processes is not rational in the strict sense and is excluded by such a construction, and the idea of a man as a rational animal in the strict sense is, accordingly, rejected as being too one-sided. (Phenix 1964: 21)

Bailey points out that Phenix talks of *meaning* instead of *rationality*. Phenix says that "there are different meanings contained in activities of organic adjustment, in perception, in logical thinking, in social organization, in speech, in artistic creation, in self-awareness, in purposive decision, in moral judgement, in the consciousness of time, and in the activity of worship" (Phenix 1964: 21).

Bailey's criticism of Phenix is less narrowly focused on Hirst's demand for knowledge to be defined by his three criteria of "distinctive concepts", "logical structure" and "truth tests". While Hirst rightly indicated that the attempt to define nine logical classes collapses, it is nevertheless the case that, viewed from the perspective of pragmatism and personalism, it is a most successful classification of modes of knowledge, taking full account of the fact that personal and relational knowledge is direct and not propositional.

Hirst was explicitly seeking to reduce "knowing a person" to propositional knowledge about a person. Phenix implicitly assumes that "knowing William" is different from "knowing about William". On balance Phenix's account of the human search for knowledge and meanings, based on pragmatic assumptions, seems more inclusive than Hirst's.[2]

BAILEY: KNOWLEDGE OF THE HUMAN AND MATERIAL WORLDS

In his lucid discussion of Hirst's and Phenix's models of knowledge for the school curriculum, Charles Bailey points out that Hirst's "forms of knowledge" argument is not strictly valid for religion, music, visual arts, dance, drama and the imaginative aspects of literature (Bailey 1984: 76–7). He concludes that their inclusion is therefore justified rather on empirical grounds (92). On the other hand, Bailey accepts Hirst's refutation of the logical structure of Phenix's scheme, but appropriately defends it on pragmatic grounds (83–4). There are strengths in both models.

Bailey's model is based on a humanistic world-view: "The vision is that all human beings are born into – as it were – a kind of double world, a world of persons and a world of physical matter and structures" (1984: 107). Bailey draws inspiration from Oakshott (1975: 12–13). Oakshott makes two logically distinct kinds of enquiry or "goings on". One kind focuses on "exhibitions of mind or intelligence", i.e. on the "world of persons". This finds expression in subjects like ethics, history, literature, law and religion. The other kind of enquiry focuses on the physical world, like rock structure and the facial resemblances of children and parents, in subjects like physics, chemistry, biology and psychology. This logical distinction Bailey holds together by using Oakshott's "idea of practice", like practising language use, relating to other people and learning moral behaviour. Learning procedures in all modes of activity require *practice* (Bailey 1984: 109).

Bailey recognizes that "tests for truth", advanced by Hirst, limit the range of meanings to cognitive thinking, ignoring human feelings, hopes, significance and purposes. Education needs the serving competencies of literacy, numeracy, logical reasoning and physical education, but also the cultivation of dispositions, such as to attend, concentrate, co-operate, reason, imagine, inquire and organize time, materials, thought and action (Bailey 1984: 110–13). Bailey's soft rationalism is directly related to every person's need to relate to the human social world, on the one hand, and the physical world on the other. Bonnett finds this account "attractive when compared with the rather abstract feel of hard rationalism" (Bonnett 1994: 71).

BONNETT: KNOWLEDGE, REASON AND EMOTION

In his approach to children's thinking, Michael Bonnett reviews the rationalist accounts of thinking and knowledge and indicates that they fail to take account of the "subjective insight in a person's understanding" (1994: 99), for emotions, attitudes, dispositions and motives are central elements influencing a person's thinking. He turns to existentialism, noting that existence precedes essence, and that to become one's true self one needs to be self-aware, applying moral principles and making free choices, to become authentically one's true self and not abandon this responsibility by drifting along with the crowd, gang or classmates.

An "existentialist perspective on the development of thinking and understanding highlights the claim that freedom and responsibility give *subjective weight* to learning". Then "teaching the public forms of knowledge [rationalism] in the context of respecting and eliciting children's authenticity [existentialism] would help develop understanding which was rational, authentic and possessed *subjective weight*" (Bonnett 1994: 16).

SUMMARY OF EDUCATIONAL PHILOSOPHERS' VIEWS ON KNOWLEDGE

An exploration of language use has led to the identification of different forms of knowledge and ways of knowing, which include:

✧ Logic, Mathematics and Symbolics;
✧ Physical, Biological and Behavioural Sciences;
✧ Personal Knowledge (Phenix: Synnoetics; Hirst: Knowledge of Our Own and Other People's Minds): e.g. "I know William," "I know how George thinks," "I know Fred's arguments and his feelings"; which involves direct personal relationships involving intentions, will, feelings, hopes and beliefs as subjective knowing. Personal knowledge is a major aspect of history, economics, social sciences, and literature like novels, plays and biography;
✧ Ethics and Moral Judgement;
✧ Aesthetic Experience: Music, Visual Arts, Arts of Movement;
✧ Religion;
✧ Philosophical Understanding.

Phenix's important realm of meaning, which he calls synnoetics, relates to the existential aspects of philosophy, psychology, literature and religion. The form of knowledge alluded to, but not stated, is personal knowledge, which is explored further in the work of Polanyi, considered later (see Chapter 6; note that Polanyi is not mentioned in Hirst 1974).

All the last four "forms of knowledge" listed above involve both objective knowing and/or subjective knowing. Any of these forms of knowledge may find co-ordinated use in particular fields of knowledge, like geography, religious studies or medical ethics (Reiss and Straughan 1996).

As we have explored the meanings of "knowledge in the context of education", we need to explore next the work of Piaget, who studied the nature of knowledge through exploring the different stages in the development of thinking from childhood, through adolescence, to adulthood. Then, in the next chapter, we need to explore the contributions of scientists to our understanding of the nature of knowledge. First, Michael Polanyi, because of his exploration of the nature of personal knowledge and scientific knowledge, then the contributions of three scientist-theologians Barbour, Peacocke and Polkinghorne.

PIAGET: EXPLORATION OF KNOWLEDGE IN SCIENCE AND RELIGION

Piaget's work on child development focuses on how children develop their powers of discovery. This question especially attracted the attention of educational psychologists and teachers in the period 1960–85. The child slowly develops schemes of understanding in which details are co-ordinated. Piaget indicated that this takes place in stages. His interest in how knowledge was related to evolution was not of central interest to teachers who read his work; they were more concerned with how to promote children's learning. Here, we are interested in his underlying theory of knowledge.

The first stage that Piaget identified in the development of a child's thinking is called the *sensorimotor stage*. The child's earliest explorations are in co-ordinating perceptions and movements of objects.

Between birth and approximately two years of age the child identifies objects through its senses and will co-ordinate its mother or bottle with its mouth and perhaps its hands into the *scheme* of sucking. This may involve sucking the breast, a bottle, a thumb, a sheet, a toy or any other object. The *scheme* of sucking will involve all these activities. It will also develop and co-ordinate its hands touching and handling objects like his feet, toys or other people. The child *assimilates* new objects into existing *schemes*, but sometimes has to modify a *scheme* to *accommodate* new objects or information.

Initially a child thinks an object only exists when it can see it. A baby will watch a toy being hidden with a handkerchief, and then lose interest, on the grounds that it does not exist if it cannot be seen. A magical moment comes when the child triumphantly pulls the handkerchief away from the covered toy and gives a big grin, knowing now that the toy exists whether covered or seen. It has achieved the insight of the conservation of matter. The child learns to remember various co-ordinations of objects and to work out in advance similar acts. In the second year of life the child imitates and begins to make use of language to represent activities to itself (Lovell 1965).

The second stage is called the stage of *symbolic representation and pre-conceptual thinking* and lasts from approximately two years to five years of age. *Pre-conceptual thinking* takes place when a child is building up an understanding of the meaning of words which are factual concepts.

For example, Tom had a toy farm. It consisted of a green rectangular piece of plywood about 60 cm by 40 cm with a raised edge for a "hedge" and a gap in one side for a "gate". In it he placed his buildings, farm vehicles and implements, fences and animals. Travelling with a farmer around fields to inspect the sheep, he constantly called each field a farm. "Farm" is a *concept* which contains many elements. A *pre-concept* is formed when a child centres on one element of a concept, the green square, and thinks that that is the farm. Piaget defines a pre-concept as "absence of inclusion of the elements in a whole, and direct identification of the partial elements one with another, without the intermediary of the whole."[3]

Pre-conceptual thought is characterized by attention to one feature. For example, the child's thought may centre on himself. Coming into a harbour, looking over the front of a boat, Jim might say, "Look, Mummy, the lighthouse and the pier are bouncing up and down!" Jim is still but the lighthouse moves!

Intuitive Thought characterizes approximately the years from four to seven. Piaget defines thought as "internalized action". The child increasingly develops this skill during this period. Pre-conceptual thinking changes as concepts are formed and the child attempts to co-ordinate them into *schemes,* which are structures of understanding.

An example Piaget gives is of two rows of, say, five identical counters, but one row is 5 cm long and the other is 10 cm. If the child is asked if one row has more counters than the other, at this stage the child *centres* on length and often says the longer row has more counters than the shorter.

Another example of a failed co-ordination is when water from one vertical glass is poured into another narrower vertical glass, in which it reaches a higher level than in the first glass. The child says the amount of water has increased. Piaget calls this kind of thinking *pre-operational thought.*

Operation is a technical term Piaget uses to mean actions which can be performed and thought through and *reversed.* For example, if a child said that the quantity of water was the same in the glass and water experiment just described, how can one know whether or not the child has just guessed the answer, as children often do at this stage? So ask: "how do you know?" To which the answer must be: "If I pour it back into the first glass it will be the same amount" (Miles 1971: 25).

When that happens the child has reached what Piaget calls *the stage of Concrete Operational Thought.* "By concrete operations" Piaget means "actions which are not only internalized, but are also integrated with other actions to form general reversible systems" (cited by Miles 1971: 25). This means that the child can handle processes involving perhaps up to four factors which can be observed. So, in the counters experiment described above, the child will say they are the same in number, ignoring the length of the dispersal line.

Concrete operational thinking is normative approximately between the ages of seven to twelve or fourteen years. At this stage the child is able to handle operations with classes, e.g. white beads + brown beads = wooden beads, or operations with relations, e.g. the water and glasses experiment described above. However, at this stage the child has difficulty in handling figurative use of language. A young primary school child rushed into the lounge of his home on a winter evening after dark and told his father that he had seen God. The father was taken out into the hall, the light was switched off, and the child pointed triumphantly at a stream of diffused light in the centre of the window of frosted glass to the side of the solid wood front door. "Can I open the window please, George?" "Yes, Dad." "Look, the light comes from the street light, but you are quite right, George, to say that God is like light and can fill us with light and goodness."[4]

Formal Operational Thinking is the final stage of development. The most important property of this stage concerns the Real versus the Possible. At the concrete operational stage the child always begins with the real. At the level of formal operational thinking, a young person begins with the possible.

In general terms, there are four characteristics of formal operational thinking:

1 Thinking is usually propositional: verbal statements are substituted for objects, so a new type of thinking is used – propositional logic.
2 The method of thinking is basically making a hypothesis and then deducing possible consequences. Piaget and Inhelder called this hypothetico-deductive thinking (Piaget and Inhelder 1958: 253).
3 If the adolescent wants to test a hypothesis, they will attempt to isolate all the individual variables and analyze systematically all possible combinations. It is thus a method of combinatorial analysis.
4 Developing understanding of the use of figurative language, verbal symbols, metaphors and abstract concepts and conceptual schemes.

Two examples follow of answers by students to a question on the story of the Prodigal Son (Luke 15: 11–32), one at the concrete and one at the formal operational level. The question was: "Do you think the father was wise welcoming the younger son back?"

Concrete operational answer: "I'm not sure. In one respect it was nice to have his son back, but his son had also run away from him, and spent his money unwisely."

This simple answer combines three factors intelligently: "nice to have him back", "had run away", and "spent money unwisely".

Formal operational answer: "I think he was right in welcoming the son back, but 'wise' always implies one sort of expedient. He was probably drawing up a bit of trouble in that his other son was going to be cross with the younger son. But he was right to welcome him back, because when somebody has done something wrong and they admit it and try not to do it again, and be better, then it is the duty of the person whom they have wronged to accept them back – to help them. I think this applies to the father."

This sophisticated answer distinguishes the notions of "right", "wise" and "trouble", outlining the connotations of each. Then the student hypothetically examines the principle of forgiveness and acceptance as a general law of right behaviour, and then applies it to the father to indicate that he thought the father did the right thing.[5]

Piaget, in his work on child development, illustrated here by his own work and that of some of his followers, was concerned to observe the *psychological processes* involved in the development of increasingly powerful forms of reasoning in children. Michael Polanyi, however, in studying Piaget's work, recognized that Piaget was also illustrating the development of *kinds of knowing* at different ages and stages of development effected through biological maturation, ability, nurture and the internalized action of the growing child (thought = internalized action) (Chapman 1988: 381–2).

Piaget used psychology to answer questions about the nature of knowledge (Chapman 1988: 381–2). He underpinned his account of formal operational thinking with a logical model, which logical thinking would explore in solving a problem. This involved the use of a lattice structure and a group structure which could be integrated into one total system. The lattice structure begins with two propositions, e.g. "the church is round", "the church is Norman", and two more propositions that are the inverse of the propositions stated, "the church is not round" and "the church is not Norman". From these four propositions, it is logically possible to assemble a complete lattice structure of binary propositions, e.g. the church is round and it is Norman. The INRC group structure consists of four components: *identity*, which is an action, e.g. add weight to one side of scales; *negation*, e.g. remove the weight; *reciprocal*, e.g. put an equal weight on the other side of the scales; and the *correlative*, e.g. remove the reciprocal weight.[6]

There is much discussion about the relationship of the logical structures to the empirical work of Piaget. All we need note here is the direction of Piaget's interest in the nature of knowledge. As Chapman says, "For Piaget, psychology was primarily a means of answering epistemological questions" (1988: 381–2). He was concerned to show how "superior" forms of knowledge succeeded "inferior" forms of knowledge, and that was how he understood knowledge at each of the stages of thinking he describes. Further, he held that one form of knowing was "more advanced than another to the extent that it is more *de-centred*" (416–17). By that, Piaget means that a more advanced stage of understanding involved the co-ordination of a broader range of perspectives on that object, and, as such, more closely approached "the object as it really is", i.e. as it would appear under a co-ordination of all possible perspectives.

However, Chapman points out:

> because the number of possible perspectives on the objective is infinite, such "complete knowledge" is beyond finite comprehension. Nevertheless we can know that one model of the objective approaches the limit of "complete knowledge" more closely than another if we can determine that one of them has developed further beyond the initial centration on partial aspects and single perspectives. (Chapman 1988: 17)

As Chapman says, "the idea that more advanced forms of knowledge should be characterized by greater universality is hardly original with Piaget: what is new, is the attempt to make use of developmental psychology in seeking empirically based criteria for such increasing universality" (1988: 417). Piaget was clear that the emergence of a new stage did not eradicate but built on and included a former stage of knowing. This is paralleled in the understanding of science, where, for example, "Newtonian physics is recognized in relativity theory as being valid within a restricted field of application, although no such restrictions were originally part of Newtonian physics" (Chapman 1988: 418).

Piaget's use of mathematical models in understanding different stage levels of knowledge came close to structuralist thinkers of the 1970s concerned with the nature of knowledge, who also used mathematical models in a similar way. Both helped to explain different types of progress in science. "Normal" scientific progress results when a mathematical structure can be extended without changing its basic form; "revolutionary" progress involves replacing one mathematical core-structure by another (see Chapman 1988: 418 for details).

Piaget held that the operations of a person who is grounded on their sense experience formed the basis of the development of an elementary understanding of arithmetic. This was the foundation of the development of mathematical structures, which in turn aided the discovery of the nature of physical reality (Chapman 1988: 424).

Piaget and Garcia revised Piaget's earlier accounts of propositional logic in 1987. In practice this revision means that there can be no purely formal stage of development. The content to which formal operations apply are more abstract than the contents of concrete operations. Secondly, there is no final stage of formal operational thinking because "higher-order operations can be formed by reflective abstraction" (see Chapman 1988: 427 for details). Structures develop at different rates across different areas of content and contexts (428). This is well demonstrated by the different age levels of formal operational thought reached in different subject areas in research projects directed by Professor Lovell in the University of Leeds.[7]

Chapman concludes his reflections on the relation of Piaget's experimental work to his use of logic and mathematical structures by saying that they are an extension of the self-organizing functions found in the biological world into the development of human thought.

Finally, Piaget's thought has an underlying metaphysic. As a young man he was inspired by the problem of reconciling science and religion, but he wrote little on the topic after his last essay on "immanentism" in 1930. He certainly seems to have felt that the value and meaning of life were provided by the directionality of development and evolution. His constructive evolution is open towards the future, leading to the idea of an "ideal equilibrium" which has provided a reconciliation of the "is and the ought".[8]

The ideal equilibrium at the pinnacle of the evolution of biological forms and moral consciousness is the norm of reciprocity, "which is the equilibrium form of relationship of members of an ideal community of equals" (Chapman 1988: 435). His reconciliation of "being" and "value" led him to believe that higher values could be the subject of scientific investigation. Criticism of Piaget comes from two quarters: scientism and logical positivism recognize only material reality but not the tendency towards organization *within* reality; metaphysical religion tends to describe its values as real.

Chapman (1988: 435–6) sums up the situation:

Piaget's early conception of immanence was advanced as an explicit alternative to both of these one-sided views. Against mechanistic science, he argued that a tendency towards higher forms of organization exists at all levels of reality, that this tendency is consistent with natural laws, and that it imbues nature with a hierarchy of values corresponding to the respective levels of organization.

Against metaphysical religion, Chapman argued that this same tendency towards higher forms of organization provides human beings with "higher values" in the form of ideals. These ideals affect reality only in so far as they are embodied in human thought and action, and that, consequently, "God" cannot be identified with any supernatural force that affects the natural world from outside. Instead, God is identified with supreme values and universal norms of thought. "God is not a Being who imposes Himself upon us from without: His reality consists only in the intimate effort of the seeking mind" (435–6).

Chapman points out that Piaget's concern with acts of valuation is evident in its directionality. Piaget holds that identification of the spirit with supreme values and norms of thought implies that God is within both, but also transcends the individual, for "the spirit *in its fullness* transcends human knowledge, for like 'reality itself', it is known only through successive approximations, as a 'limit' that can be approached but never reached" (436).

In conclusion, Piaget finds science and religion not to be two distinct areas of knowledge, but rather an interpenetrating of science with values, for values evolve "upwards" towards "higher values". Piaget is also clear that the struggle for "good" is not purely intellectual, for "every act of love makes man a collaborator of God's" (Chapman 1988: 436).

NOTES

1   Note that in 1965 Hirst, as his title indicates, was exploring the nature of liberal education. Our interest here focuses only on the nature of knowledge (Hirst 1974: 38).
2   In the light of philosophical work by Neo-Aristotelian thinkers Hirst later came to reject his earlier "hard rationalism" in favour of an account of knowledge that stressed the distinctive character of practical reason and knowledge alongside that of propositional or theoretical reason and knowledge (see Hirst 1999).
3   Piaget (1951: 226); cf. Miles (1971: 22). The example comes from a friend.
4   From a church-going friend.
5   Miles (1971: 203, 209). See this text for a fuller discussion.
6   E.g. Conjunction – the church is round and it is Norman; Non-implication – the church is round and it is not Norman; Negation of reciprocal implication – the church is not round and it is not Norman. For a full account of Piaget's theory see Piaget and Inhelder (1958: 294–300), and for a full account of this example see Miles (1971: 27–8).
7   Consult the Faculty of Education Library, University of Leeds, 1960–83.
8   Chapman (1988: 434). This has been worked out theologically by Tillich (1969).

REFERENCES

Bailey, C. (1984), *Beyond the Present and the Particular*, London and Boston: Routledge & Kegan Paul.
Bonnett, M. (1994), *Children's Thinking*, London: Cassell.
Chapman, M. (1988), *Constructive Evolution*, Cambridge: Cambridge University Press.
Hirst, P. H. (1965), "Liberal Education and the Nature of Knowledge", in P. H. Hirst (ed.), (1974), *Knowledge and the Curriculum*, London: Routledge & Kegan Paul.

Hirst, P. H. (1999), "The Nature of Educational Aims", in R. Marples (ed.), (1999), *The Aims of Education*, London: Routledge.

Hirst, P. H. and Peters, R. S. (1970), *The Logic of Education*, London and Boston: Routledge & Kegan Paul.

Lovell, K. (1965), *Educational Psychology and Children*, London: University of London Press.

Miles, G. B. (1971), "A Study of Logical Thinking and Moral Judgements in G.C.E. Bible Knowledge Candidates", MEd dissertation, University of Leeds.

Oakeshott, M. (1971), "Education: the Engagement and Its Frustrations", in R. F. Dearden, P. H. Hirst and R. F. Peters (eds), (1972), *Education and the Development of Reason*, London and Boston: Routledge & Kegan Paul.

Oakshott, M. (1975), *On Human Conduct*, Oxford: Clarendon Press.

Phenix, P. (1964), *Realms of Meaning*, New York: McGraw-Hill.

Piaget, J. (1951), *Play, Dreams and Imitation in Childhood*, London and Boston: Routledge & Kegan Paul.

Piaget, J. and Inhelder, B. (1958), *The Growth of Logical Thinking from Childhood to Adolescence*, London and Boston: Routledge & Kegan Paul.

Reiss, M. J. and Straughan, R. (1996), *Improving Nature? The Science and Ethics of Genetic Engineering*, Cambridge: Cambridge University Press.

Tillich, P. (1969), *Morality and Beyond* (1964), London: Fontana.

Wittgenstein, L. (1967), *Philosophical Investigations* (1953), Oxford: Blackwell.

5 | Changing Views of Scientific Knowledge

The twentieth century began with Newton's understanding of the universe as a clear view of the way things are. This view is still held by the man or woman in the street and is still correct for large areas of everyday matters. But the Newtonian view has been superseded in advanced science, operating beyond the realm of normal sense perception, by what was discovered during the first half of the twentieth century. Four revolutions have taken place in scientific thinking in the last hundred years, three of which, relativity, quantum mechanics and chaos theory, have radically modified the way we now understand Newton's views. The fourth theory is a revision of Darwin's theory of evolution. We will summarize the Newtonian view before reviewing the three revolutionary views in physics and then attending to a revised Darwinian view.

CLASSICAL PHYSICS

The Newtonian world-view was *realistic,* i.e. it was thought to describe the world as it exists. Such a view is called *classical realism,* or sometimes *naïve realism,* for it largely pictures a one-to-one correspondence between what is out there in the world and the words or models used to describe it. The movements of, for example, planets or pendulums are held to be governed by the rules of clockwork. Planets have been created, "wound up", set in motion, and then left alone. Such a view is *determinist.* It held that the future of systems of matter or bodies in motion could be predicted accurately if one had precise knowledge of their present position and movement. Such a view was also *reductionist.* Large bodies, like planets or pendulums, were studied to discover the constituent units of which they were composed. So attention was focused on molecules and atoms. Some thought that the world consisted of 'nothing but' atoms and molecules in discrete combinations.[1] Finally, knowledge of the world was *objective.* That is, the scientist thought that he was describing what he saw as if he were "an external observer" viewing the world, or his objects of study, from "outside".

Up to the beginning of the twentieth century Newton's understanding of physics dominated the subject, and is known as classical physics. Space and time were understood as being separate. Space was an empty container in which objects had their place. Time passed, as the clock seems to indicate, mechanically and uniformly everywhere and from past to future. Matter existed in discrete units that were changeable through physical manipulation or chemical interaction, which can be understood. Matter was neither created nor destroyed, only transformed.

RELATIVITY

Relativity challenges all of this. In 1905 Einstein proposed the theory of *special relativity*. This was based on the claim that the speed at which light travelled was an absolute constant of nature. He also demonstrated that *space* and *time* constitute a continuum. In other words space and time are intimately interrelated. In 1915 Einstein outlined his theory of *general relativity*, including gravity in his earlier ideas: mass and energy are essentially the same and interchangeable ($E = mc^2$); gravity and acceleration cannot be distinguished, they are interrelated. For example, a clock that is moving runs more slowly than one at rest.

If you left the earth in a spaceship in the year 2200, travelled at a very high velocity for five years, and returned, you could find yourself in the year 3000 (cf. Barbour 1990: 109)! "Matter is, if you will, in the elastic matrix of space-time. Instead of separate enduring things externally related to each other, we have a unified flux of interacting events" (111).

A major effect of relativity theory was to end the Newtonian belief in the total separation of space and time when considering the structure and organization of the universe. "It is important to understand that although relativity was a very great revolution *within* physics, in many ways it was a natural development of Newton's physics, of no great significance to the man in the street. We still have laws, we still have cause and effect."[2] Of course, it remains the case that matter, space, time and gravity appear separate in our everyday experience, and is a key reason why the Newtonian view still describes the world of everyday experience very adequately when speeds are very much less than the velocity of light. Nevertheless, this picture should be expanded by seeing the universe as an interacting flux of integrated events. This hints at a worldview in which the focus is no longer on the elements but on the interrelated whole. "Perhaps the most significant implication for religion comes from cosmology – the idea that time might have singularities where it begins and ends."[3]

QUANTUM MECHANICS

If science teaches you anything, it is that the world is full of surprises. Common sense is not the measure of everything. Quantum theory tells us that the way things behave on the scale of atoms or smaller particles is totally different to the way that large objects behave in our everyday world. (Polkinghorne 1994: 16)

This insight arose from the work of Planck and Einstein, demonstrating that light not only shows wavelike properties, but also sometimes behaves as if it were made up of tiny particles, i.e. quanta.

The strange behaviour of electrons is demonstrated in a famous experiment involving four stages:

1   A barrier with two slits in it is placed between the source of a beam of electrons and a screen which records hits by them. The electrons make an undulating pattern on the screen that physicists call a diffraction pattern, which is characteristic of wavelike behaviour. Most of the electrons arrive at a point on the screen that is equidistant from the two slits.

2   If slit two is closed, the beam of electrons forms a one-slit cluster pattern on the screen opposite slit one. If slit one is closed and slit two opened, there is a one slit cluster pattern opposite slit two.

3   If only one electron is fired and both slits are open, the electron may hit anywhere within the continuous unified undulating two-slit symmetrical wave pattern on the screen.

4   If one electron is fired and only one slit is open, it hits the screen within a one-slit cluster pattern, either opposite slit one or slit two.

How then can one electron, fired with two slits open, end up *anywhere* on the screen in the two-slit symmetrical wave pattern area centred on the midpoint? Physicists say that the individual electron has gone through both slits!

A number of insights emerge from quantum theory which we need to understand:

1   Heisenberg formulated his *uncertainty principle*, which stated that one could know where an electron was (position), in which case one cannot know what it was doing (its momentum), or one can know its momentum but not its position.

2   Dirac formulated the "superposition principle", which stated that these two physical states, going through slit one and going through slit two, could be added together. In other words electrons travelled in both ways at once. This "explains" the third stage of the experiment, i.e. that the electron passed through both slits in order to end up *anywhere* in "the continuous unified undulating two slit symmetrical wave pattern" area (Wilkinson 1993: 53). This is not "common sense", and it is not possible to "picture it" in any meaningful way.

3   The wave theory and particle theory are *complementary*, i.e. either way of looking at the behaviour of electrons is possible, but one cannot use both at the same time.

4   It turns out that "once two quantum entities have inter-reacted with each other, they retain the power to influence each other no matter how widely they subsequently may separate" (Polkinghorne 1998: 31). This is called *nonlocality*, or togetherness-in-separation.

5   Although the electron 'goes through both slits', if we actually look to see where it is, i.e. make a measurement of its position, we shall always find it at one slit or the other on any particular occasion of doing so. In other words, measurement makes definite what had previously been indefinite. How this comes about is not fully understood. There is continuing debate about the 'measurement problem' in quantum theory.

How does quantum theory affect our understanding of the nature of scientific knowledge? The coming of quantum mechanics has shown that the *world is full of surprises*, because of the differences already mentioned and now summarized.

Objective results can be obtained using Newtonian principles, and that is one kind of *knowledge*.

Quantum measurement interferes with the behaviour of what is measured in the world of quantum mechanics, so *quantum reality* is different from *objectivity*. In this case, what is real cannot be known objectively, but it can nevertheless be known. The understanding of quantum measurement requires theoretical interpretation, but the judgement of what interpretation to choose is based on non-empirical criteria (Polkinghorne 1991: 89) and that means it is a subjective judgement based on preferred theory and argument. The observer probably influences and thereby modifies reality; here knowledge of reality is not objective, but it is *intelligible*.

Before the advent of quantum theory, physicists were already concerned with the smallest constituents of matter and their nature. This search concluded by the end of the 1970s that protons and neutrons were made up of quarks and gluons. Quarks are invisible, but they are believed to exist because they help to make sense of the strange patterns of the interactive behaviour of protons and neutrons. Knowledge of quarks is not *objective*, but it is about *reality* and is *intelligible* (Polkinghorne 1994: 98).

Finally, electrons which have interacted continue to influence each other no matter how far apart they travel. If one is measured, it affects the other. This discovery has an effect on how we understand the nature of knowledge. The concern of the physicist with the smallest elements in matter, like quarks, has been forced to take on board the fact that a holistic principle of togetherness operates even amongst the smallest particles in certain situations.

The togetherness, or the whole, controls the parts, which is an example of what is called "top-down" causality. In the past science has often concentrated on "bottom-up" causality. For example, my fitness is partly the result of my biochemical make-up, my genes and the exercise taken. Recently there has developed a concern for "top-down" causality, of which a human example, now, would be me willing my hands to move precisely as my mind dictates in order to enter these words into the word-processor, or deciding for myself to take exercise. There will be more about both "bottom-up" and "top-down" causality later.

Quantum theory is more important than relativity in the sense that (a) it makes the picture of the physical model of the universe more abstract, and (b) it removes the notion of *total determinism*.

CHAOS THEORY

This theory is exemplified by what is picturesquely referred to as the "butterfly effect". "The earth's weather systems are so sensitive to small disturbances that a butterfly stirring the air with its wings in the African jungle today, could have consequences that grew so rapidly [exponentially] that they would bring about storms over London or New York in about three weeks' time" (Polkinghorne 1998: 41).

Chaos theory is the name given to hypersensitive systems, and its discovery began with a meteorologist called Lorenz in 1961. Weather systems never quite repeat themselves, i.e. they are *aperiodic*, with imperceptible differences causing large changes in behaviour. Weather forecasters found it very difficult to give a long-term forecast for them. It was possible to produce a forecast for perhaps two or three days, but beyond six or seven days they were worthless: weather was long-term *unpredictable*.[4]

Lorenz turned towards the mathematics of systems and became more successful when he put in a variable for the differential effect of warming the Atlantic Ocean and the east coast of America. He mimicked both the *aperiodicity* and sensitive dependence on initial conditions – *unpredictability*. He reduced his initial twelve equations down to three, which represented non-linear relationships, which means they were not proportional. The equations are *reflexive*, i.e. they feed back upon themselves, and *non-linear*, which means that, for example, if the cause is doubled it does not double the effect.[5] However, while the equations give a model of the kinds of effects of chaos theory, it can never be predictive because it is never possible to predetermine all the influences which may contribute to the ultimate effect.

Chaotic systems exhibit fractal-like behaviour. A fractal is a way of realizing an infinite degree of structure. For example, take a triangle and add a triangle one-third of its size to each side so that it looks like a star of David. Then take each of the twelve sides and add a smaller triangle on to the middle third, and so on to infinity (Gleick 1988: 98).

Chaotic behaviour is limited, on the one hand, by its "attractor", the set of possible motions towards which it tends and, on the other, by the boundary or "watershed" between one kind of fractal basin and another which would provide a competing "attractor" (Gleick 1988: 235). The influences which prevent chaos theory being deterministic can include those of animate creatures, including human beings, who have a measure of freedom of action.

Professor Polkinghorne makes the point that "just as Heisenberg's uncertainty principle led most physicists to believe in quantum indeterminacy, so it is suggested that chaos theory should encourage belief in a more subtle and supple physical reality than the clockwork world of Newton" (1988: 42). Six insights emerge.

First, it is a relief that the pattern of interaction found in chaos theory can be known in everyday life and is both pictured and experienced by people! Here is an example from folklore which illustrates this brief introduction to the theory:

> For want of a nail, the shoe was lost;
> For want of a shoe, the horse was lost;
> For want of a horse, the rider was lost;
> For want of a rider, the battle was lost,
> For want of a battle the kingdom was lost! (cited by Gleick 1988: 23)

Clearly, such a system is not predetermined, but "open" to influence from a myriad range of sources, inanimate, animate and Divine.

Secondly, there is another characteristic of chaos theory which is important. A study of large-scale patterns shows up, for example, the repetitions of large-scale geometrical patterns which repeat themselves on a sequence of increasingly smaller scales, known as *fractals*. Examples can be found in natural forms, such as the shape of a fern leaf (Gleick 1988: 238). This shows up relationships between the parts of these patterns, which indicates a larger "whole".

This *holistic and anti-reductionist* pattern is the focus of attention rather than the constituent parts. James Gleick describes it thus: "Chaos is anti-reductionist. This new science makes a strong claim about the world . . . the whole cannot be explained in terms of the parts. There are fundamental laws about complex systems . . . they are laws of structure and organization and scale . . . " (cited by Barbour 1998: 183).

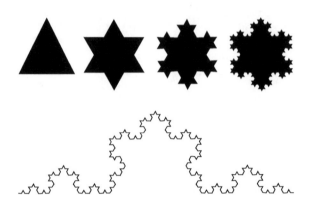

Figure 5.1   The Koch Snowflake

A third significant feature of chaos theory is the discovery of the emergence of order out of chaos. The second law of thermodynamics states that in every closed system, there is an increasing tendency to disorder expressed through increasing decay. Animals do not violate this system, for they receive energy and materials from the sun and the material environment and so maintain their order while living.

Prigogine discovered how ordered systems could emerge from disordered inanimate nature. For example, complex patterns of convection are created when air molecules are heated from below and cooled from above. The air molecules were in a state of random movement (entropy) prior to the introduction of heat: order emerges from disorder.

A fourth characteristic of chaos theory is that it applies to many areas of science and social science, e.g. meteorology, chemistry, physics, biology, ecology, fluid dynamics and population changes. All these areas use mathematical models related to the elegant and beautiful equations of Mandelbrot, which model the irregular and fragmented shapes, which are jagged and broken. These shapes are called *fractals* and are described above. One of the most celebrated examples of them is the Mandelbrot set, which has been the subject of so many striking posters.

Examples from nature range from snowflakes to the discontinuous dust of galaxies. So now, instead of disciplines remaining separate areas of isolated study, conversations take place between them on chaos theory. One mathematician had a dream about Mandelbrot's equations. In it he was dead and suddenly heard the unmistakable voice of God remarking "You know, there really was something to that Mandelbrot" (Gleick 1988: 114).

Another characteristic worth mentioning about chaos theory is its ability to model networks of patterns which had seemed independent but were shown to be together with "invisible" lines (Gleick 1988: 228). Again, the whole is greater than the sum of the parts, and needs a holistic rather than a reductionist explanation. This theory could well model the apparently isolated centres of galaxies or the secondary cancer growth areas in a human body.

The final comments on chaos theory I leave to Prigogine, who was awarded the Nobel Prize in 1997 for his work on the thermodynamics of non-equilibrium systems. At the end of *Order out of Chaos* (Prigogine and Stengers 1988), he supports the view of Weyl that scientists should embrace not only the scientific approach to life with its *theoretical knowledge* based on theoretical constructions to model the world, but also *direct knowledge* (italics mine) based on perception, thought, volition, feeling and doing. He not only supports the "coming together" of these two kinds of knowledge, but also accepts that we need a new relationship between man and nature and man and man. We can no longer accept the separating distinction between scientific and ethical values.

Ideas of instability, aperiodicity and unpredictability apply not only to the material world but also to the social world: "Societies are immensely complex systems involving a potentially enormous number of bifurcations exemplified by the variety of cultures that have evolved in their relatively short span of human history" (Prigogine and Stengers 1988: 313). In this dangerous and uncertain world Prigogine and Stengers end on a note of

> qualified hope that some Talmudic texts appear to have attributed to the God of Genesis . . . Twenty-six attempts preceded the present genesis, all of which were destined to fail. The world of man has arisen out of the chaotic heart of the preceding debris; he too is exposed to the risk of failure, and the return to nothing. "Let's hope it works" [Halway Sheyaamod] exclaimed God as he created the World, and this hope, which has accompanied all the subsequent history of the world and mankind, has emphasized right from the outset that this history is branded with the mark of radical uncertainty.[6]

Prigogine and Stengers, like the nameless mathematician mentioned above, are here assuming the complementary nature of scientific, personal and religious forms of knowledge and their attendant insights.

DARWIN'S THEORY OF EVOLUTION AND ITS DEVELOPMENT

Darwin's Theory

The *Origin of Species* was published in 1859. The theory states that the natural selection of variations in a species is the main source of evolution. It contains three main points: random variation – small differences amongst members of a species; the struggle for survival – more are born than survive; the survival of the fittest – this leads to natural selection. That is, it is obvious that the fittest have the properties most likely to survive, and so a species will evolve by using its most appropriate genes for a particular ecological environment. A species will not survive unless it knows how to "relate" successfully to the forms of life, vegetation and climate in the neighbourhood. In the *Descent of Man* (1871) Darwin extended evolutionary theory to include humans. He thought our moral and mental faculties differed from those of animals in degree rather than in kind.

Darwin's theory is sometimes outlined as three "assumptions" from which three deductions can be made:

Assumptions:

1   Like creatures breed more like creatures, with variations.
2   Some variations are more inclined to survive than others.
3   All creatures produce more offspring than can survive.

Deductions:

1   The more favoured varieties are more likely to survive and breed.
2   There will be more favoured varieties in the next generation.
3   The species will tend to evolve over time (Rose 1997: 181, adapted).

However, there are three problems. First the mechanism for the transmission of similarities and differences was not understood. Darwin thought it was due to "gradualism", the slow change in succeeding variations. Secondly, the argument of design posed a problem: how could there be half an eye? Thirdly, how were new species formed through evolution?

The first problem in the development of the theory of evolution was solved in 1930 when an acceptable synthesis of Darwin's insight into mutation and Mendelian genetics became available. This was called Neo-Darwinism, or the modern synthesis, and provided a way of testing the inheritance of variations that was given mathematical expression (Rose 1997: 187).

Professor Richard Dawkins offers an answer to the second problem. He suggests that 1 per cent of an eye is better than no eye. Also image-forming eyes have evolved at least forty times in different groups from those which share the antecedence of the human eye. The argument is accepted in principle by Rose (1997: 194).

Concerning the third problem, that of natural selection, we must first note that the six-point summary outlined earlier does not include a mechanism for the formation of new species (Rose 1997: 195). However, Darwin had in fact found a major reason for the development of separate species: the geographical division of breeding groups from the same original species, isolated in different areas or regions (Darwin's Galapagos finches, Rose 1997: 205).

The Structure of DNA

The discovery of the structure of DNA in the 1950s led to a further development of evolutionary theory. Dawkins offers an attractive picture of DNA. Seeds from a willow tree rain down on the Oxford canal, making it white. Inside each seed is "the tiny capsule that contains the DNA, the genetic information" (Dawkins 1991: 111). "There is enough storage capacity in the DNA of a single lily seed . . . to store the *Encyclopaedia Brittannica* 60 times over."

"Amazingly only about 1 per cent of the genetic information in, for example, human cells seems to be actually used . . . nobody knows why the other 99 per cent is there" (Dawkins 1991: 116). "The DNA is arranged along stringy chromosomes, like long computer tapes" (117): Each person contains forty-six chromosomes, twenty-three from each parent. DNA is the master substance of growth and heredity. It is estimated that the molecules found in the forty-six chromosomes of humans require a molecular thread about five feet long.

The impact of this developed Darwinian theory is that it reaffirms insights discovered in geology, before Darwin, that the world is not static but changing according to discernible rule-bound processes, though Monod and other biologists like Dawkins talk about rule or necessity and random variation or chance. They argue that "cumulative selection, by slow and gradual degrees, is the explanation, and the only workable explanation that has ever been proposed, for the existence of life's complex design" (Dawkins 1991: 317).

Reductionism

But Monod and Dawkins confuse biological discourse with philosophical discourse and assert that purposeless, blind chance is the key link in the process. This insertion of the concept of purposeless blind chance, a metaphysical concept, into a scientific theory transforms the mode of discourse from scientific into metaphysical discourse, affirming a materialist world view. One can say that, but let us not pretend it is science. It is *scientism*, the metaphysical view that science provides an all-inclusive outlook on life and philosophy for living.

Ultra-Darwinism

More trenchant criticism of the "selfish gene" comes from Professor Steven Rose (1997). He argues that Ultra-Darwinism – the views advanced by Dawkins and others – is simplistic in at least four respects. First, the individual gene is not the only level at which selection occurs. Secondly, natural selection is not the only force driving evolutionary change. Thirdly, organisms are not always flexible to change. Fourthly, organisms are sometimes passive, but sometimes *active* in effecting change. This makes possible downward causation as a factor in genetic evolution, in sexual pairing, for example. The heart of modern biology is not the gene, but the nature of the individual living unit, the organism, which has a lifeline existing in four dimensions (Rose 1997: ch. 6).

Ultra-Darwinism is criticized for its reductionist metaphysical assumption, which is that the purpose of life is reproduction – the mind is designed to get genes into the next generation. Two further assumptions are made, one concerning an object and one concerning a process. The object is the gene, pictured as the minimal life form, like a billiard ball atom, concerned only with replicating itself. The process is the assertion that adaptation "eliminates" inferior variations. Rose discusses the inadequacies of these assumptions (Rose 1997: 209–10). In so doing, he outlines the framework of what is called "Revised Darwinism".

Revised Darwinism

We need only note here four comments that Rose makes, as we are only interested in the forms of knowledge, not the detailed arguments:

First, Rose notes that molecular biologists, like Dawkins, suggest that nucleic acid script in the gene was the beginning of evolution. Rose disagrees, saying there were already cells, organisms in existence, ready to receive the DNA code.

Second, the script of genetic evolution is a record, but it does not comprise the history of life, which is about *organism* not just molecules.(Rose 1997: 270f.).

Third, Rose embraces multiple forms of explanation (1997: 9–13). He accepts reductionist methodologies when they unravel, often successfully, the secrets of mechanisms in all fields of science, but rejects them when they "tip over into reductionist philosophy" (295). He offers a biologist's decalogue, which is materialistic, to make "biology whole again" (302f.).

Here is a brief summary of Rose's decalogue. Our history shapes our knowledge. There is one world but many ways of knowing. It has levels of organization which are fundamentally irreducible. In living systems, causes are multiple and can be described at many different levels and in many different languages. Being and becoming, living organisms exist in four dimensions, the three of space and one of time, and cannot be "read off" from the single dimension that constitutes the strand of DNA.

The stability of organisms is achieved through dynamics: organisms are unstable open systems in which continuity is provided by a constant flow of energy through them. Organisms are in constant interaction with their environments which interpenetrate. The structure of the interaction of our lifeline trajectories and our changing environment, which can occur at many levels, from the molecular to the species, constrains evolution. The past is the key to the present, for organisms can only respond to present opportunities and limitations. Human beings and other living organisms have the ability to construct their own futures in circumstances they did not choose. "And therefore our biology makes us free" (summary of Rose 1997: 302–9).

Fourth, and most important, is Rose's philosophical clarity in drawing Dawkins's attention to the way that he, Dawkins, *uses* metaphor, analogy and homology (Rose 1997: 32–4). A metaphor is a name or descriptive term applied imaginatively, but not literally, to illustrate an aspect of another object or subject, e.g. "food for thought". An analogy is a resemblance of a function between two things which are essentially different, e.g. the heart can be regarded as a pump (33). "Homology implies a deeper identity, derived from an assumed common evolutionary origin" (34). Rose quotes examples of homology that include "the bones of the front feet of a horse are homologous to those in the human hand" and "chick memory is homologous to human memory" (34). However, he says that the claim that "animals biting and clawing at each other is behaviour homologous to human aggression" is a delusion (68).

Rose (1997: 87) shows, by example, how Dawkins misunderstands the use of analogy. He mentions how discussion of altruistic behaviour, a supposed characteristic of the selfish gene, allows a metaphorical relationship to be given homological status. That is, that Dawkins says that altruism has a genetic origin.

This is philosophical confusion, for altruism and selfishness are human characteristics and belong in the mode of personal discourse. Altruism and selfishness have nothing to do with biological discourse. It is quite a different matter to say that gene *x* often permits altruistic behaviour, and gene *y* often permits selfish behaviour. This has scientific plausibility without philosophical confusion (Rose 1997: 202, 294). But in each case of selfish or altruistic behaviour, that behaviour is the result of *humanly chosen purposes*, not *biological causes*. The genes may provide constraints on human choice, but they are not the causes of human behaviour. However, those genes may have other characteristics, and delimit different behaviours in changed environmental circumstances (cf. McGrath, 2005: 467).

Dawkins's fellow biologist, Professor Steven Rose, is better qualified than I am to pass professional scientific judgement on his use of language. Rose's charge (1997: 202): Dawkins is guilty of homology – he asserts that selfishness and altruism are chemical or biological properties of a gene. My defence of Dawkins: He overlooks philosophical distinctions. He is enthusiastically popularizing biology. This enthusiasm I defend. Dawkins means to illustrate, not to mislead, as I indicate in my comments in the paragraph above. But I must admit in the end that Dawkins's language use is misleading.

However, Rose is guilty of the same charge, i.e. confusing modes of discourse, he makes against Dawkins. The evidence: the book's last sentence, which I must admit I like: " . . . it is therefore our biology that makes us free." The charge: the confusion of biological discourse with personal discourse.

Argument against Rose: "Freedom" is only *exercised* by human beings. *Homo sapiens* is a species. A "human being" is a member of the species *Homo sapiens* who has been subjected to an inspection involving *an interpretation* and *an evaluation* AND who has been awarded the diploma of being a human being. That is, the person is seen to be "educated" or "sociologized" or a "nice person" or "easy to relate to" – any cluster of a thousand characteristics of "what it means to be a human being".

Are you, the reader, not convinced? All right, I say that a person who is amoral (i.e. without a sense of right and wrong) is less than fully human. A person who tells the truth, for example, is moral, and is fully human in that respect. A person who tells lies, but who knows that that is wrong, knows they are immoral, but is also fully human because they know the difference between right and wrong. Lee Harvey Oswald, the alleged assassin of President Kennedy, is supposed to have shot people on a Sunday afternoon as a enjoyable recreation. If he was without moral awareness, amoral, I say he was "less than a human being" in respect of not knowing the difference between right and wrong. However, he was still a member of the species of *Homo sapiens* – a biological fact.

Becoming a human being means being at the centre of an active network of personal relationships and having the human sensibilities to respond *responsibly*. This involves moral judgement, which Oswald allegedly lacked.

Biology does not set us free from psychological prisons created by, for example, a "friend", parent, teacher or partner who is oppressive. Psychological prisons are also created by guilt, sorrow, heartbreak, incompetence or laziness.

Biology does not set us free. Our biology can provide the genetic endowment for living fully, but we are all limited by metaphysical or philosophical boundaries and bound by psychological and/or social constraints. Rose's self-imposed boundary is materialism. The source of freedom lies elsewhere, in the spiritual resources given to human beings and developed carefully within traditions, like those of the world religions and ethical humanism.

DARWINISM AND COMPLEXITY THEORY

There is a concern among theorists to return to the principle of rule-bound processes without the loose cannon of chance. This has been pioneered by reflections on the part that complexity theory may have in throwing light on the process of evolution. Some discern a pattern of a self-organizing universe that develops higher levels of complexity which emerge from, and are different from, lower levels of complexity. Quantum theory may also contribute to this. Davies suggests that "a blend of molecular Darwinism and laws of organization complexity", which *create* information to allow for sudden jumps in complexity in the evolutionary process, would "amplify greatly the selectivity in the evolutionary process" (Davies 1998: 242–3).

Atoms and molecules often have twin characteristics. For example, consider the wave/particle duality of light. The wave can be seen as an information programme, like software in a computer, and the particle corresponds to the computer's hardware. It seems as though some quantum organizational process may explain the origin of big molecules (macromolecules) which contain a large amount of information.

The DNA molecule is a "crystal" with a stable structure of ten faces and is the central unit of reproduction. This is the "hardware". Chemists were surprised to discover a "quasi-crystal" with five faces. A normal crystal grows regularly, atom by atom, whereas a quasi-crystal requires "some sort of long range organization to make sure the right bits fit into the right places". Penrose thinks that quantum mechanics may be the source of the organizing system, for the quasi-crystal has a high information storage capacity (Davies 1998: 244f.).

The study of quantum computation also offers suggestions about the process of evolution. These contributions of quantum theory, if they are accepted by the scientific community at the end of the present exploration and discussion period, would certainly need a reconceptualization of the notion of "chance" in evolutionary theory.

Davies himself favours the suggestion that life began in the hot underworld, deep inside the earth's crust. He supports scientists who see "chance" playing a role, but within the constraints of an overall sense of direction towards the production of life (Davies 1998: 240). Davies recognizes that "for 300 years science has based itself on reductionism and materialism, leading inevitably to atheism and a belief in the meaninglessness of the universe" (246). Now an increasing number of scientists view the universe as a meaningful place, made so that it can "generate life and mind, bound to give birth to thinking beings able to discern truth, apprehend beauty, feel love, yearn after goodness, define evil, experience mystery" (De Duve 1995; Davies 1998: 246).

Is there about to be a paradigm[7] shift amongst scientist about evolution? The issue is not about evolutionary history itself, which is accepted by society, with the exception of creationists. We have noted that many professional biologists are committed to the ideas of chance and purposelessness. "However the trend of increasing complexity would provide evidence of purpose in the universe" (Davies 1998:

249–53). This could include a role for chance, or it might be that "the essential architecture of multicellular organisms might well be the product of certain mathematical principles of organization" (253).

It may be that we are beginning to catch a glimpse of "a self organizing and self-complexifying universe, governed by ingenious laws that encourage matter to evolve towards life and consciousness [in a] universe in which the emergence of thinking is a fundamental and integral part of the overall scheme of things" (Davies 1998: 256). This is a highly controversial issue which the scientific community is considering, and no doubt a consensus of opinion will emerge in the coming years, which will then establish any new insights as received knowledge and truth in the mode of scientific discourse.

At the end of his book, Davies adds one more sentence, which follows the previous quotation: "A universe in which we are not alone." This is a metaphysical statement. I share the view of Davies expressed in it, which is shared by Paley and denied by Dawkins (1991: 4f.), that behind the universe, God is the creative primary cause. For religious people, God is also the focus of the answer to the question of why the universe was created. Science deals with secondary causes as to how the universe is created, and we all await its judgement on the present continuing discussion.

Our interest in the study of knowledge, in this section concerning Darwinism and complexity theory, is to note two things. First, it is controversial, exploratory and undecided. Secondly, in contrast to the drift of arguments at the end of the nineteenth century in the direction of materialism and scientism, we can note at the end of the twentieth century significant scientific explorations of holistic explanations. Some of these look towards a religious focus.

SOCIOBIOLOGY

Discussion of evolution must also include reference to sociobiology, which is the study of social behaviour and cultural evolution. The subject has developed in the last thirty years, and Wilson (1978) is a key figure. His work defends genetic determinism, with less room for human freedom than many find comfortable. He thinks that all religion and ethics will be explained by biological knowledge as products "of the brain's evolution", and then "its power as an external source of morality will be gone for ever" (Wilson 1978: 201 cited by Barbour 1998: 256). Barbour makes the terse comment: "[It] seems to me inconsistent that Wilson never said that science is similarly discredited by its evolutionary origins, although it, too, is obviously 'a product of the brain's evolution'" (Barbour 1998: 256). He continues: "Wilson embraces a sweeping . . . reductionism that makes all the academic disciplines into branches of biology [which is] occasionally an explicit advocacy of what he calls '*scientific materialism*'" (257).

Barbour easily demonstrates that Wilson overstates his case and shows how cultural evolution is more significant today than biological evolution. Now "cultural innovation replaces mutations and genetic recombinations as the source of *variability* . . . [and] *the transmission of information* occurs through memory, language, tradition, education, and social institutions rather than through genes" (Barbour 1998: 257).

It must be noted that there has been a stage leap between the transmission of information through *genetic evolution,* which contains the sedimented memory of survival techniques of trial and error learning, and *evolutionary science* in which scientists *select* ideas for exploration which *intuitively* appear fruitful. "The number of possible theories is too large to test them at random" (Barbour 1998: 258).

HUMAN RESPONSES TO THE THEORY OF EVOLUTION

When Darwin's theory of evolution reached the general public, there was a range of responses, but in the popular mind these were polarized by the debate on evolution between Bishop Wilberforce of Oxford and Thomas Huxley in which "science defeated religion". This, of course, was in fact inaccurate because Wilberforce was guided by a distinguished conservative *scientist,* Richard Owen, who rejected evolution. Asa Gray, friend of Darwin and distinguished Professor of Natural History at Harvard, accepted evolution as the way in which God created, as did many senior Cambridge theologians and clergymen throughout the country. In Britain, Aubrey Moore was the "clergyman who more than any other man was responsible for breaking down the antagonisms towards evolution then widely felt in the English Church".[8]

A distinction has long been made between God as primary cause of creation and the secondary causes, which are the province of scientific investigation and discovery. Thus the Huxley/Wilberforce debate confused two quite separate debates. The first debate was scientific, and was between conservative scientists, like Owen, and scientists who supported evolution. The second debate was religious, between Christians who took the Bible literally and rejected evolution and Christians who accepted both the Bible in a non-literal way, in the light of the results of biblical criticism, and evolution.

Regarding biblical criticism:

> many people only half understood what the scholars were saying, and feared they were undermining the faith by denying the "truth" of Scripture. They found it disturbing that the Bible, a sacred book, should be examined in this way, and that the writers of Scripture should be treated as fallible human beings rather than as channels of divine truth. (Pirouet 1989: 11)

Add to that the feeling that the public got from the Oxford debate that "the theory of evolution was presented as an alternative to the biblical account" (Burke 1998: 44), and one obviously opts for the scientific view, in the light of the success of science in answering questions in the nineteenth century. There is little doubt that Professor Colin Russell and others are right in thinking that the "conflict was manufactured and unnecessary", but the damage was done (44).

What were the precise effects of the theory of evolution? This has been well summarized (based on Barbour 1998), so only the briefest account is necessary here.

The theory of evolution seemed to be a challenge to Scripture for those, such as creationists, who understood it literally. They read Scripture as a science text, so they thought the world was created in six days, with separate species. In the fifth century, Augustine warned against taking the Genesis accounts literally, as Galileo knew (Watts 1998: 4). Some contemporary atheistic evolutionists, like Dawkins, also still read

Scripture literally as science and so misunderstand and reject religion (Burke 1998: 54). Contemporary creationists are still with us (51–4).

It seemed to be a challenge to the idea that God was the primary designer and evolution was his method of creation. This was Darwin's view, which he reported in his *Autobiography* (1879), where he describes himself as a doubting Theist (Barbour 1998: 59). We have mentioned that his friend Asa Gray and others held this view. But those who saw chance as a central mechanism adopted a mechanistic philosophical position, which was nearly always atheistic. Today many still accept the distinction between primary and secondary causation, or centre on chance and adopt the atheistic mechanistic stance already mentioned.

The theory also challenged the dignity of humans, who are seen through Scripture, if taken literally, as in charge of all species (Genesis 1: 27–30). Evolution shows humans as the most developed creature in one of numerous parallel lines of evolution on earth. One line is neither higher or lower than the others (Rose 1997: 185).

In the evolutionary thesis, Darwin also claimed that the human moral sense evolved through natural selection. For some people this generated a hope which translated into social Darwinism and belief in progress, social and moral. "The Harvard philosopher John Fiske wrote that human dignity is reinstated because 'evolution shows us distinctly for the first time, how the creation and perfection of man is the goal towards which nature's work has been tending from the first'." (Barbour 1998: 61; see 72–4 for more detail)

Now many recognize that we live in a self-organizing and self-complexifying universe that is governed by laws, involving bottom-up and top-down causation, which lead to the emergence of life and consciousness. This can quietly restore human self-confidence and a belief that one is able to discern that life can be part of a purposeful providential plan.

A COMMENT ON SCIENCE AS THE ONLY FORM OF KNOWLEDGE

"Analytic philosophy has become increasingly dominated by the idea that science, and only science, describes the world as it is in itself . . ." This comment from Professor Putnam (1998: ix), one of the leading philosophers in the United States, precedes his treatment of the topic. He finds the idea that only science can describe the world as unsatisfactory. Later in his book, he explores Wittgenstein's "Lectures on Religious Belief ", which can "lead us to see our various forms of life differently without being . . . scientific". He indicates how an empirical belief and a religious belief differ. Putnam and Wittgenstein accept that "religious belief 'regulates for all' in the believer's life" (Putnam 1998: 145). The details of this philosophical argument do not concern us here. We just need to recognize that these two great philosophers recognize religious discourse as a form of knowledge which describes a form of life. Putnam makes it quite clear that "religious discourse can be understood in any depth only by understanding the form of life to which it belongs". To "understand the words of a religious person properly . . . is inseperable from understanding a religious form of life, and this is . . . a matter of understanding a human being" (154).

NOTES

1   This summary is based on Barbour's account (see Barbour 1990: 95f.).
2   Personal communication from Dr Waldram.
3   Personal communication from Dr Waldram.
4   The *Independent*, 4 November 1998, reported the loss of the schooner *Fantôme* and her crew of thirty-one, who tried to sail out of the path of hurricane Mitch, which then changed route and "followed the boat". Weather forecasting totally failed to predict the route of this hurricane even twelve hours ahead.
5   A mathematical formulation of this is the "Feigenbaum sequence". See Prigogine and Stengers (1988: 169).
6   The Talmud (500 CE) was the written commentary on the Oral Torah, the oral teachings written down in the Mishnah (200 CE), which interpreted the Torah.
7   A paradigm shift is when we completely change the way we look at something. For example, once the earth was viewed as flat, but now we know it is round, which is a paradigm shift of outlook; once it was thought that humans and animals were created as separate species, but now we all agree, apart from creationists, that species have emerged through evolution.
8   Moore (1979: 260). Moore is the main authority on these issues. See also Burke (1998: 44).

REFERENCES

Barbour, I. G. (1990), *Religion in an Age of Science*, London: SCM Press.
Barbour, I. G. (1998), *Religion and Science*, London: SCM Press.
Burke, C. D. (1998), "Evolution and Creation", in F. Watts (ed.), (1998), *Science Meets Faith*, London: SPCK.
Davies, P. (1998), *The Fifth Miracle*, Harmondsworth: Penguin.
Dawkins, R. (1991), *The Blind Watchmaker* (1986), Harmondsworth: Penguin.
De Duve, C. (1995), *Vital Dust*, New York: Basic Books.
Gleick, J. (1988), *Chaos*, London: Sphere Books, Cardinal.
McGrath, A. (2005), *Dawkin's God*, Oxford: Blackwell.
Moore, J. R. (1979), *The Post Darwinian Controversies*, Cambridge: Cambridge University Press.
Pirouet, L. (1989), *Christianity World Wide: AD 1800 Onwards*, London: SPCK.
Polkinghorne, J. (1991), *Reason and Reality*, London: SPCK.
Polkinghorne, J. (1994), *Quarks, Chaos and Christianity*, London: SPCK.
Polkinghorne, J. (1998), *Science and Theology*, London: SPCK.
Prigogine, I. and Stengers, I. (1988), *Order out of Chaos*, London: Fontana Books, Flamingo.
Putnam, H. (1998), *Renewing Philosophy*, (1992), Cambridge, MA: Harvard University Press.
Rose, S. (1997), *Lifelines: Biology, Freedom, Determinism*, Harmondsworth: Penguin.
Watts, F. (ed.), (1998), *Science Meets Faith*, London: SPCK.
Wilkinson, D. (1993), *God, the Big Bang and Stephen Hawking*, Tunbridge Wells: Monarch.
Wilson, E. O. (1978), *On Human Nature*, Cambridge, MA: Harvard University Press.

6 | The Integrity of Science

PERSONAL KNOWLEDGE

The Nature of Scientific Knowledge

The nature of science is explored from 1958, the year that Polanyi's book *Personal Knowledge* appeared (1958; Gelwick 1977). It begins by questioning the way that science was portrayed. For example, Bertrand Russell in his book, *The Scientific Outlook* (1937) adopted an objective view of science which was mechanistic, denying the use of freewill in the process of discovery. Today a more holistic view of science is presented in school science textbooks. The scientific method is described as involving the following:

- making observations
- constructing scientific models
- generating hypotheses
- making predictions
- designing and carrying out experiments.

In general the scientific method starts with *making extensive and well planned observations*, being very careful and unbiased about the *facts* observed. Sometimes they are *counterintuitive*. The next step is to construct *effective models* and *hypotheses*, which make a minimum of special assumptions to explain the observations, apply them to a wide range of phenomena, and which lead to exact numerical *predictions* that can be checked. We must all realize the immense labour and logical analysis which goes into working out the implications of a new model. This stage has to be undertaken rigorously, with little room for intuition. Emotional detachment is essential, an idea jealously guarded by the scientific community. "From the hypothesis various *predictions* can be made, and these may be tested by carrying out appropriate *experiments*" (Roberts *et al.* 1994, ch. 1). "This is not to say that science proceeds logically and impersonally. It never does! It is quite true that it proceeds by *intuition, serendipity, looking for the beautiful* and having *hunches*."[1]

However, some philosophers, including logical positivists, and many scientists who are reductionists regard scientific findings as impersonal objective knowledge, because they view scientific findings from results backwards to beginnings – a top-down perspective, quite different from the start of an exploration, viewed bottom-up, which was Polanyi's perspective. So there is the need for a balance between the personal approach to an investigation and the need to be impersonal in establishing experimental evidence, checking facts and exploring the implications and applications of any new model quantitatively.

Polanyi, viewing from the bottom-up, regarded scientific knowledge as the fruit of a scientist's *intellectual passions*. Emotion is the primary source of motivation towards scientific discovery. But why is emotion regarded with suspicion? Dr R. T. Allen describes the climate of suspicion and its sources. Terms like *pathos, passio* and *affectus* associate emotion with being affected and being passive. To be moved by an emotional experience implies "that one is controlled, not by one's own reason and will, but by essentially non-rational or alien causes". Emotions can be regarded as irrational, because they overwhelm us and "distort our judgement and disrupt our actions". Emotion is also thought to be irrational because it is connected with values, and so is linked with the "powerful 'emotive theory' of values" (Allen 2005: 42).

Polanyi regarded emotion as "an intelligent active response to something apprehended". So emotion has a key role in the life of science. Polanyi argues against the idea of science as "exact, critically tested, detached and impersonal knowledge" (Allen 2005: 43). He seeks to show that the scientist must passionately engage and actively care about his science. Scientific passions or emotions have three functions, *selective, heuristic*, and *persuasive*:

- ✦ *Selectivity*: involves the value of choosing a scientific issue worth exploring, the pursuit of which "relies on a sense of intellectual beauty".
- ✦ *Heuristic*: the heuristic passion, the desire to discover "requires a passionate motive to accomplish it. Originality must be passionate." (Polanyi 1958: 143; Allen 2005: 45)
- ✦ *Persuasive*: the third function of emotion is to persuade the scientific community that one's discoveries are correct: the scientific community needs convincing, especially if a scientific pioneer has had to use both new concepts and framework, and that involves emotional commitment.

During Polanyi's research in physical chemistry in the period 1913–20, in one of his projects he proposed a theory concerning the absorption of gases by a solid non-volatile adsorbent. Many experiments seemed to disprove this theory. He was asked to give an explanation of his theory before the Kaiser Wilhelm Institute for Physical Chemistry in Berlin, in front of Fritz Haber and Albert Einstein. They rejected his theory on the grounds that he had not taken account of new knowledge of the electrical concept of interatomic forces. Polanyi persisted with further evidence at a later meeting but his theory was rejected. Only after 1930, from the work of others, was Polanyi's theory vindicated, and it is still in use today. Now, from this and the experiences of others who make discoveries at the frontiers of knowledge, he recognized the four key features of scientific discovery to involve the following:

I   The recognition of a problem that is ripe for solution by one's own powers, one's personal conviction – there are no rules which suggest this.

2   The ability to stick to a theory, with supporting evidence, despite contrary evidence advanced by the scientific community, which may challenge it.

3   The recognition that all new theories must be subject to close scrutiny by the scientific community and that apparently invalid theories must be rejected. This is sometimes called the "rule of orthodoxy", but is better referred to as the "role of orthodoxy".

4   The recognition that some facts advanced by the scientific community in rejecting a theory are, in time, shown to lack the power to overturn a new theory.

This process, as Polanyi knew from his own experience as a research scientist of great distinction, shows how science advances. It is not through impersonal observation, hypothesis construction and objective experimentation, but through personal, subjective, individual vision, experimental commitment and dialogue with the scientific community, which is sometimes referred to as "conviviality".

Polanyi knew that the process of scientific discovery was not "objective". The false view of science can be called "the idea of objective scientific knowledge" Gelwick (1977: xvii). Polanyi was quite clear that what came first in scientific discovery was the sureness of the scientist, his or her faith "in the independent reality of truth – scientific, moral, aesthetic or religious" (Scott 1996: 4).

In science pioneering faith and observation, theory, deduction and verification are all needed and kept in balance.[2] Polanyi was taking the "bottom-up" view, while many scientists and philosophers in particular take a "top-down" view, looking back on the assured results of science, like Bertrand Russell and Karl Popper who tried to make science an objective reality (Scott 1996: 31). They suggest that the rules of science are more important than the activity of scientific discovery, and if the latter does not fit the rules, it is distorted or omitted. Russell says rules and induction identify a problem, while Popper says there are no rules. Polanyi says "yes" and "no". For example, there are rules for painting a picture, but art comes first and rules are worked out by analyzing how the best painters work.

Polanyi said: "The capacity for making discoveries depends on natural ability, fostered by training and guided by intellectual effort. It is akin to artistic achievement . . . unspecifiable but far from accidental" (Scott 1996: 31). Einstein recognized that there was no logical path leading to the discovery of the laws of physics. "They are only to be reached by intuition based on something like intellectual love" (Scott 1996: 32; Polanyi 1958: 104).

Russell's account collapses because he says the first stage of scientific discovery consists in observing the significant facts. But that conceals the very act of discovery. The whole process of discovery begins by asking significant questions about a problem. What are the key questions? The search is passionate, driven by intellectual love and care and involves activity, intuition and imagination, which all circle the questions and lead to the discerning of the significant facts and then theories.

The work of the creative imagination has been demonstrated by many, like Poincaré, who might be obsessed by a problem that is initially insoluble. After a period of concentrated effort to solve the problem ends in failure, the problem is relegated to the back of the mind while one attends to other matters. Then, later, suddenly, a moment of illumination may occur. The penny drops. The search will have been guided by notions of quality, beauty and economy or simplicity. But judgement is central to discovery. Which lines of exploration should be explored? One only knows the significant facts retrospectively, *after* the discovery is made, not before.

How could philosophers of science like Russell get their accounts of science so wrong? Scott identifies the reasons. First, they wanted to see discovery their way, so that the rules of science could be kept strict. Secondly, they saw it the way they did because they looked at discoveries from the successful results *backwards*, from the end to the beginning of the research. Polanyi noted that the results appear to be predetermined when viewed in that way. Yet in *looking forward* before the event, the act of discovery appears personal and indeterminate – a vision which becomes an obsession. Russell and Popper were not looking at the process of discovery, but at the tidied up results. Polanyi gives many examples of the dynamics of discovery.

Scott compares an objective view of Madame Curie's discovery of radium with her daughter's account. The former account suggests that Madame Curie followed "a carefully planned method of separation and analysis", but it actually took her forty-five months from her announcement of the probable existence of radium to the preparation of a decigram of pure radium. That evening, after her child was asleep, she said to her husband Pierre, "Suppose we go down there for a moment." They stole into the shed and stood looking at the radium, glowing blue in the dark. "Look! Look!" she whispered, like a child. Here was intellectual love, judgement and beauty.

But Popper tried to construct a house of science which excluded passionate personal judgement, the obsession with a problem, the intellectual leap or the moment of illumination. He argued that there was no proof of the laws of nature, but there was disproof. The only knowledge of which you are certain is that you are wrong. So, for example, "All swans are white." No! You have seen one black swan, so the theory is wrong.

However, a theory is only abandoned if a fact is *accepted* by the scientific community as evidence against a theory. For example, the quantum theory of light was for a while contradicted by experiments, but scientists assumed that the contradictory evidence would be explained sometime, somehow.

In the end Popper admitted that this intuitive, passionate searching was involved in discovery when he said, "Science does not rest upon rock bottom. The bold structure of its theories rises as it were above a swamp." Piles can be driven into this swamp, not to rock bottom but deep enough "to carry the structure at least for the time being" (Scott 1996: 43). Science is a house built on piles in a swamp.

So after all, scientific knowledge is not wholly objective and guaranteed by impersonal tests. However, we can distinguish between scientific knowledge and the process of scientific enquiry. Scientific knowledge is the accumulated knowledge of the physical world that we accept, knowing it is trustworthy, but also provisional and open to correction. So we can safely build bridges, make cars, weave cloth, grow plants, produce medicines, make telescopes and other things. The process of scientific enquiry involves the matter or process under investigation, subjective judgements, and a consideration for beauty and simplicity, the results of which are tested.

Tacit Knowing

However, there is another key element involved not only in discovery but in all understanding that Polanyi called "tacit knowing". He traced the origins of tacit knowledge back to the earliest forms of life where sense perception, the need for judgement and the need to learn from experience can be found.

The human mind continues this striving to make sense of the world. These tacit powers of mind indicate, first, as Polanyi said, that "we know more than we can tell". For instance, we can ride a bicycle or recognize the face of a friend, but cannot explain exactly how we do it. Tacit knowledge cannot necessarily be put into words, but it is not second-rate knowledge: we *know* how to ride a bicycle or recognize a friend's face.

Secondly, such knowledge depends upon caring enough to concentrate on the focus of our attention.

Thirdly, knowledge depends upon relating a collection of facts together in a particular way – as in riding a bicycle. However, attention to a collection of facts is subservient to the focus of the problem – keeping our balance.

Polanyi talks about "focal awareness", i.e. keeping our balance, and "subsidiary awareness", i.e. awareness of the many relevant factors, which would include the size of bicycle, the road surface and the steep curve in the road.

It is of course possible to write out the rule for keeping our balance: the rule is that the angle of imbalance (a), is compensated by measuring the radius of the curve of imbalance (r), which must be proportional to the square of the velocity (v) over the angle of imbalance: $r \sim v^2/a$ (Scott 1996: 49). That is knowledge, but it does not help a child to learn to ride a bicycle.

The child learns to ride a bicycle by processing non-verbal visual and bodily clues – it is tacit knowledge. Here, then, is real knowledge, based on details we often cannot name or check. Further, Polanyi claims that tacit knowledge is the major form of knowledge, and is based on intuition and the integration of perceived information and skills. It is also prior to the verbal expression of data, hypotheses and processes of testing and expressing rules which govern processes.

We have seen how knowing the rule of keeping our balance does not help us learn how to cycle. This tacit dimension, the key to Polanyi's claim that "we know more than we can tell", can be described in another way. The explanation is the double intentionality of consciousness. Consciousness is always a double act, not two separate acts. One act integrates two modes of consciousness, subsidiary or proximate attention and focal attention. We attend from one set of things (subsidiary attention) to attend to another thing or objective (focal attention) (Allen 1992: 6).

For example, in driving a car, a driver's focal attention is to keep to the left-hand side of the road, but subsidiary attention picks up the interval between telegraph poles and the line of curvature of the road to indicate the curvature of bends ahead, the cars in front approaching the driver, and the cars which the driver is approaching. Indeed, the driver's focal attention moves from hazard to hazard, a slow car in front, a difficult bend, an approaching car, in order to decide whether to overtake before or after it passes. The integration of these two modes of awareness is always the responsibility of the driver, who is always involved and committed to the task of driving carefully. It is a tacit integration, because, in court after an accident, one cannot recall all the tacit data which helped one driver avoid an accident which befell another driver.

Polanyi's Philosophical Aims

Polanyi says: "I believe that the function of philosophic reflection consists in bringing to light, and in affirming as my own, the beliefs implied in such of my thought and practices as I believe to be valid . . . " (Allen 1992: 8; Polanyi 1958: 267)

His aim is "to restore the capacity [of a person] for explicitly holding unproven beliefs which we might conceivably doubt and which might be false" (Allen 1992: 8). His basis for this is to state his ultimate beliefs which are based on his proximate beliefs and practices. This is a parallel process to moving from subsidiary awareness to focal awareness, with the reservation that in sense experience the focal experience is supported by subsidiary experience, but ultimate belief may be intuitively inferred from proximate beliefs and practices.

Gestalt Knowledge

Polanyi found that two other areas gave particular insight into the nature of human discovery: Gestalt psychology and the work of Piaget.

The Gestalt view is that you first see "the whole", whether it be a face, a landscape or a building, and then you may go on to see the specific items which make up the whole. You don't first see two eyes, two lips and a nose and "build up" a face you then recognize! Imagine making a portrait of a man made by typing letters of the alphabet in a particular sequence. If you were to attend to the letters as if they were words and sentences, it would be nonsense, but if you "stand back" and see the whole, not as lines but as a picture, you will see the portrait – and recognize it is Winston Churchill! The focal awareness in this case is the whole, while the letters on the page provide points of subsidiary awareness. Recognizing a face may involve psychological processes but the result of the process is tacit knowledge. You can change your focus from the details of a picture to the whole person pictured, for it might have been the eyes that convinced you that the picture was Winston Churchill.

Perception is similar to scientific intuition. Keynes said of Newton, "His peculiar gift is the power of holding continuously in his mind a purely mental problem until he has seen right through it. I fancy his pre-eminence is due to his *muscles of intuition* being the strongest and most enduring with which a man has ever been gifted" (Scott 1996: 56). Polanyi transformed Gestalt psychology into a theory of knowledge.

The Development of "Higher" Ways of Knowing

Polanyi's theory of knowledge was built up from watching skills which have been developed through evolution. Piaget's work on child development focuses on how children develop their powers of discovery. The child slowly develops *schemes*[3] of understanding in which details are co-ordinated.

An early example is the co-ordination of action involved in sucking. This begins as a reflex action focused on nipple or bottle, but develops an exploratory role, *assimilating* into the sucking *scheme* new objects, such as a thumb, toys, sheets, and any other object which can be explored by the mouth.

Clearly, the teddy's squeak cannot be assimilated into the sucking *scheme*, so this requires accommodation, which involves the development of a *manual scheme*. The hand rather than the mouth is the instrument now used in the tactile explorations of the child. It is through natural maturation and the child's exploratory action and thought that more sophisticated ways of knowing are developed.

Piaget's studies led him to formulate his theory that development takes place in stages, as we noted in the previous account of his work (see Chapter 4). Polanyi recognized that the child had a dim foreknowledge of what he was about to discover. "We can account for this capacity of ours to know more than we can tell if we believe in the presence of an external reality with which we can make contact. This I do" (Scott 1996: 59; Polanyi 1960: 133).

Polanyi accepted Augustine's view that faith precedes the quest for understanding, which is a search for knowledge.[4] His approach is different from that of Popper, who says, "Science does not rest upon rock bottom. The bold structures of its theories rises as it were above a swamp." Into this swamp we can drive down piles, not to the rock, but deep enough "to carry the structure for at least the time being" (Popper 1959: 11; Scott 1996: 43).

Scott pictures Polanyi as being "more like Brer Rabbit 'born and bred in a briar patch' and unafraid of being thrown back into one" (1996: 59). She goes on, "There is then no finished certainty to our knowledge, but there is no sceptical despair either. Through all our different kinds of knowledge there is a reasonable faith, personal responsibility and continuing hope."

A Hierarchy of Forms of Knowledge

Polanyi found that the work of both the Gestalt psychologists and Piaget gave support to his most profound revolutionary idea, *tacit knowing*, which lies behind our curiosity and drive to know, which is towards discovering the real world, for Polanyi was a realist: "I declare myself committed to the belief in an external reality gradually accessible to knowing" (Scott 1996: 65; Polanyi 1960: 133). He also accepted that we live in a "many level world".

The tendency to explain life in terms of physics and chemistry has been a deeply ingrained legacy in British minds from the days of the dominance of logical positivism. Then it was assumed that real knowledge was empirical, i.e. it could be tested for factual accuracy through the five senses, or scientific, i.e. tested through observation, hypothesis and evidence.

For example, many have suggested that the discovery by Watson and Crick of the "double helix", the DNA chain that contains the information which transmits the data which models an offspring on the parents, was proof that living forms were physically and chemically predetermined. But the formula is not just a biochemical blueprint, says Polanyi; it also contains the programme of a "builder". He illustrated his point by reference to a sign made of pebbles at a railway station on the Welsh border, which said, "Welcome to Wales by British Rail." He said the arrangement was not due to the chemical composition of the stones, or the physical distances between them, but to the person who arranged them in that way to convey the chosen greeting. They were arranged for a purpose. So was the DNA code, says Polanyi (Scott 1996: 116–17; Polyanyi 1958: 33).

Bertrand Russell fell into the trap of thinking that he was nothing but a collection of electrons and protons. As Scott says, "Russell, with gloomy fortitude, submits to be explained, and explained away as a person, but he takes comfort in thinking *our values* are somehow quite different. We ourselves are the ultimate and irrefutable arbiters of value" (Scott 1996: 118).

This leaves values without a context, in the way they can be found in some existentialist novels, like those of Sartre or Camus, where characters may drift from one situation to another with little purpose. Polanyi's explanation of life is more inclusive and adequate. He felt that life operated "by principles *made possible* and *limited* by physical and chemical laws, but not *determined* by them" (Scott 1996: 118).

To make sense of his argument, Polanyi introduced an important original idea which he calls *boundary conditions*. For example, machines and living things do obey chemical and physical laws which limit what they can do. But without violating those laws they are also controlled by another open set of laws. These are the freewill of the creator or operator of a machine deciding when and how to use the machine, and the freewill of a person deciding where they wish to go.

These laws governing human activity impose *boundary conditions* on the laws of physics and chemistry and all other "lower level" laws, psychological, sociological or biological, while the laws governing human activity he calls "the principle of marginal control". This allows for human freedom within the limits of the physical constraints imposed by the laws of "lower levels".

In the last chapter of *Personal Knowledge*, Polanyi explores how the hierarchy of levels could have emerged, for Darwin did not allow for them with his reduction of life to "chance" and "necessity". Moreover, Darwin was looking at the process of evolution from behind, after the event, rather than from before the event. However, we will return to this topic of a hierarchy of levels when we examine the work of Arthur Peacocke. We need to look at three other aspects of Polanyi's work: his views on mind and body, on the nature of being a person and on the nature of aesthetics, literature and religion.

Mind and Body: Mind and Brain

The modern discussion of mind and body began with Descartes' view that a person consisted of mind and body, mind and matter, mind and brain. In one sense Polanyi agreed with this but in a limited sense, that mind transcended matter and in this sense was different. Mind and body were not two aspects of the same thing, otherwise bodily mechanisms would determine human thought – the position of behaviourists, which Polanyi rejected. However, a person exists on all levels – chemical, physical, biological, psychological and personal, to name some of them. A person is therefore a unity, for the mind uses all aspects of a person and all data collected through the senses in order to feel, think and act.

Tacit knowing forms a major mode of knowing, with its changing concern from subsidiary awareness to focal awareness, or sometimes the changing focal concern in a changing situation, e.g. when driving a car, or helping a child learn to ride a bicycle.

The Nature of a Person

Polanyi regarded the person as the centre of responsible and skilled judgement. He finds a person more real than atoms. His ideas of tacit knowledge and a many-level world help us see persons as fully human. The picture of a human being must include a recognition of the abilities of subsidiary and focal awareness, and the abilities to feel, sense, think, imagine, explore and to know at many levels of knowing.

All this was consistent with the developing *schemes* outlined by Piaget which indicated how a human built up a sense of space, time, co-ordination, and ways of knowing at a physical, empirical, aesthetic, scientific, moral or personal level.

Polanyi used "personal knowledge" to assert that all levels of knowing were personal, i.e. subjective not objective, for so-called objective knowledge is in fact based on personal judgements. Like Piaget, he related biology to psychology, showing how species build up a structure of knowing and adapt it. But knowing also changes a person, encouraging the adoption of a new vision of life as *schemes* are changed.

Polanyi developed Piaget's two patterns of learning: that of building up a *scheme* by assimilating new material into it, or extending the *scheme* by the addition of new material to it, or of *adapting* the *scheme* and transforming it into a new *scheme* because the old scheme could not accommodate the new material. Polanyi called these two tendencies *dwelling in* and *breaking out*.

In summarizing Polanyi's insights into his view of the personal, Scott made four points.

1   Persons are individuals who can be recognized accurately by those who know them, yet we cannot *define* the human shape.
2   Persons recognize each other as individuals through empathy or in dwelling, for only a person *knows* another person.
3   We know a person in a particular way from our own perspective, so a son's view, a wife's view and an employer's view will be very different.
4   Polanyi recognizes the importance of talking about the human soul, but uses rather the language of *emergence* and *achievement*.

Polanyi is clear that his theory of knowledge implies that a *mind exists*: being is central to personhood. A person is free to fashion one's own knowledge according to one's own judgement, but also, as a member of society, loyalty and the respect for truth are both respected and cultivated by the citizen (Scott 1996: 155–9).

It is important to recognize that duality exists at the interface of two levels of knowing, for the duality of brain and mind exists when the brain of the individual co-ordinates physical things. These include the features of a face, the sound of a voice, the style of walk and other physical factors, and the mind of the individual fuses them into a recognition of a person they know well, and they say, "Hello Dad!"

There is a mutual "indwelling" and sympathy as one recognizes the other. The individual attends from the factors at the level of subsidiary awareness to a focal awareness of the person they are looking at. Dualism only exists at the interface between different levels of knowing, at which time the individual is engaged in an activity of dual control, from one level to the other.

Kinds of Knowledge: Aesthetics, Literature, Religion

Polanyi extends his ideas on tacit knowledge into his exploration of the nature of aesthetic experience "first applying them to the understanding of metaphors, symbols, painting and poetry, then to myth and ceremony and lastly to religion, ritual and belief" (Scott 1996: 165). Drusilla Scott first met Polanyi in 1960 and carried on a close friendship with him until his death in 1978. She says, "Wordsworth, like Polanyi, grew up and developed his great imaginative powers and his idealistic spirit in a time of comparative political quiet." However, political upheaval, turmoil of ideas, violence and tyranny shattered both men for a time. Polanyi escaped from Germany to England, from where he viewed the fruits of Nazism. Wordsworth viewed the fruits of the French Revolution. Scepticism, which assumed "that nothing, no moral faith or transcendent belief can stand which cannot be proved" (Scott 1996: 171), took hold of both for a while.

Wordsworth found a poet's answer to the state he was in. He returned to the intense sense experience of Nature and of human beings that he had had in his childhood and youth. Scott suggests Wordsworth reclaimed his past and admiration, hope and love, through tacit knowledge. Polanyi went back to his convivial knowledge of the men he had worked with. Neither of them saw the situation as Russell did. Russell chose honest scepticism rather than dishonest faith (Russell 1937: 56; Scott 1996: 8).

Wordsworth and Polanyi both discover an underlying religious dimension to their experience. "For Wordsworth's rediscovered vision was religious in the sense that it was a direct apprehension of the divine through the world of the senses and the feelings, a passionate intuition of

"a motion and a spirit that impels
All things, all objects of all thought
And rolls through all things."

It was God revealed through the majesty of Nature and the greatness of the human mind.[5]

Polanyi expresses his own religious convictions as the metaphysical assumptions which undergirded his understanding of the evolutionary process up through the emergence of the higher values of human beings. Man "is strong, noble and wonderful so long as he fears the voices of his firmament" (Scott 1996: 178). But Polanyi warns that if man dissolves through critical examination what he respects – law, art, morality, tradition, God – then he empties himself of meaning as well.

Polanyi wrote to Drusilla Scott on one occasion making it clear that "I am of course aiming at the foundations of religious faith. I have been doing so since I started thinking about matters in general terms twenty-five years ago. But I became increasingly reticent about this as time went on" (Scott 1996: 182). Polanyi was of Jewish origin, but without a specific religious upbringing. His religious interest was awakened by reading *The Brothers Karamazov* in 1913, and he became a "completely converted Christian". He was baptized into the Catholic faith in 1919, but developed a great admiration for Protestant Christianity. Polanyi wrote:

> The Christian enquiry is worship, it resembles, not the dwelling within a great theory of which we enjoy complete understanding, nor an immersion in a musical masterpiece, but the heuristic [trial and error learning] upsurge which strives to break through the accepted framework of thought, guided by the intimations of discoveries still beyond our horizon. Christian worship sustains, as it were an eternal, never to be consummated hunch, a heuristic vision which is accepted for the sake of its unresolved tensions. (Scott 1996: 184; Polanyi 1958: 199, 281)

Polanyi accepts tradition and new discovery as elements in science, religion and other forms of knowledge. He was quite clear that "Divine service can mean nothing to a person completely lacking the skill of religious knowing" (Scott 1996: 184; Polanyi 1958: 282). For him, religion involved a tacit integration of clues to a higher level of meaning, in a universe which consists of a hierarchy of levels of meaning. One should talk of a supra-natural level of meaning, rather than supernatural.

"Supra-natural" implies the complementary levels of meaning of the one reality – the wholeness of knowledge, while "supernatural" implies a dualism, and Polanyi consistently repudiated the reality of higher intangible things. The real miracle is a physically unchanged situation which points towards a changed and heightened spiritual significance. That was the way Jesus used parables. He told a story of a physical activity to illustrate some spiritual truth: he did not confuse the levels of understanding.

Scott quotes Dillistone in support of Polanyi's contention that there are no such things as *bare facts*, for interpretation enters all knowledge. A key question is the relation of symbol to reality.

> The word "incarnation" is a symbol, so is "resurrection" . . . A symbol is that which holds or draws two things together . . . It always serves to maintain a two way relationship, between the seen and the unseen . . . The slippery slope is downward to a view that would regard everything as no more than lifeless mechanism. The upward leap is towards an exhilarating ascent of symbolic interpretation opening out new vistas. (Scott 1996: 191)

She leaves a verdict on Polanyi's religious position to Professor W. T. Scott, who said, "I am convinced both that he [Polanyi] considered the Christian religion at its best to involve an encounter with and surrender to a pre-existing reality, and that he must have some vision himself, however ineffable, of this reality" (Scott 1996: 192).

LEVELS IN THE HIERARCHY OF KNOWLEDGE

In his search for an answer to the question: 'What is a human being?' Peacocke begins with the fact that human beings have evolved from star dust, the material constitution of nature. The sciences set out to examine and describe the different levels of organization at each increasing level of complexity, which are described by different

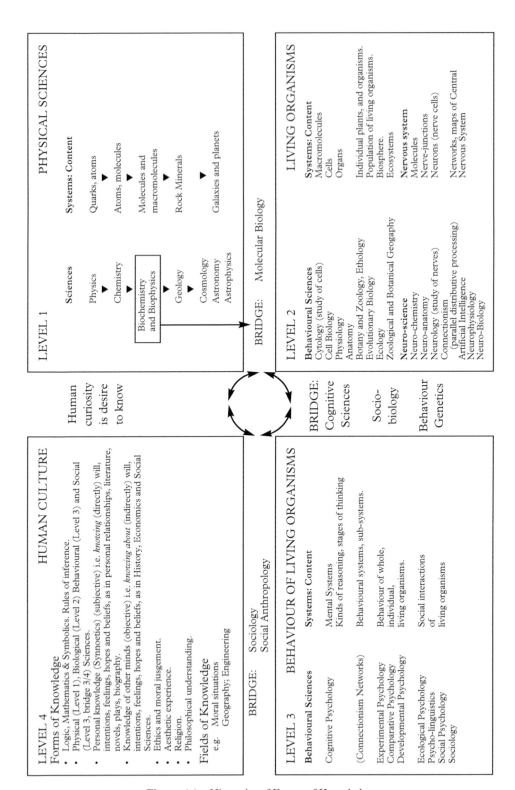

Figure 6.1 Hierarchy of Forms of Knowledge
(compiled from Hirst (1974), Phenix 1964; see chapter 4, above), and Peacocke, 1993, 1996;
see note 6, page 124).

disciplines within different forms of knowledge (Barbour 1998: 230–1; Peacock 1993: ch. 12; 1996: ch. 3; Polkinghorne 1986: ch. 6).

Peacocke suggests that four focal levels of complexity can be distinguished:

1   *The physical world* constituted by matter, which is the domain of the physical sciences.
2   *Living organisms*, which are the focus of biological sciences.
3   *The behaviour of living organisms*, which is the concern of the behavioural sciences, sometimes called the social sciences.
4   *Human culture.*[6]

While the outline on the opposite page (figure 6.1) is largely that of Peacocke, I have modified it to incorporate the insights of the educational philosophers and Polanyi discussed earlier. Clearly Level 4 is formed through human insight and endeavour. But also, all the insights and knowledge discovered in Level 1 have been discovered by humans, so the levels are therefore cyclical, and are all personal knowledge in the sense used by Polanyi. However, hereafter the term "personal knowledge" will be used in a limited sense to refer to direct acquaintance of, and interrelationships with, persons indicating their feelings, moods, intentions, wishes, hopes, fears, frustrations and loves of all kinds – personal, vocational, creational, recreational, material and spiritual. It is included in Synnoetics by Phenix.

In the diagram showing Levels 1 and 2, solid arrows pointing downwards refer to part-whole hierarchies of structural and/or functional organization. For example, molecules and macromolecules in Level 1 are "parts" of the "wholes" in Level 2. In Levels 1–3, examples are given of *systems* studied within these levels, together with their related discipline.

Level 2 outlines the part-whole hierarchy of levels of organization in the nervous system, all of which are included in the science of Genetics which includes the whole of the central nervous system as well as the brain.

"Bridges" between the levels are indicated by the bridge subjects listed in the bridge space between the levels. Peacocke reminds us that "no other part of the observed universe appears to include so many levels as do human beings, and level 4 refers only to human beings" (Peacocke 1996: 50), in terms of the culture of which they are a part, and what only they can know of all sentient beings, with their reflective consciousness of Levels 1–3.

The key feature about the hierarchy of levels is that at the levels of inanimate matter interactions between atoms and molecules take place according to the laws of physics, which provide boundaries which are not contradicted by chemistry but are incorporated and utilized by its laws, which are again not contradicted but transformed by the laws which constrain biological life. For example, if you knock an apple off the table it will fall to the floor – gravity! However, if you pick it up and eat it, you lift it. Gravity is not suspended but counteracted by the mind directing the muscles of the arm to lift the apple to your mouth.

Two principles are involved within the hierarchy of levels. First, there is "bottom-up" causation. The properties of materials constrain the next higher level in the hierarchy of knowing, in how they can be used. However there is also "top-down" causation. Your mind, because you were feeling hungry, directed the lifting of the apple to eat it. Other examples of "top-down" causation include the creation of all human achievement, the creation of cities, of music, art, literature, machinery from washing machines to flying machines. Top-down causation happens at all levels of sentient life, and this very fact modifies the "chance and necessity" model of evolution advanced by Darwin to a "choice/chance and necessity" model.

Focal Level 1: The Physical Basis of Human Being

Human beings are made of the stuff which originated in the supernovae explosions which occurred before planet earth was formed, and of which planet earth, like all other planets, is formed. A supernova is an expanding, exploding star which, at the climax of its explosion, is over one hundred million times brighter than our sun. Quarks, atoms and molecules are the subject matter of Physics and Chemistry. Larger complexes of inanimate matter are the subject areas of Geology, Cosmology, Astronomy and Astrophysics. Biophysics and Biochemistry study this matter, which is capable of forming molecular structures which can replicate themselves as cells.

The Bridge between Levels 1 and 2: Molecular Biology

DNA, the structure of which was discovered in 1953, is the chemical in a double helix in the nucleus of a cell that makes up its genetic matter. The cells replicate themselves as the organism grows, whether it is an amoeba or a human being. This is a key area of study in Molecular Biology, which is the bridge subject between Levels 1 and 2.

Focal level 2: Human Beings as Living Organisms

The evolutionary process is such that it develops along certain lines "towards increase in self-organization, complexity, information-processing and storage [memory], consciousness, sensitivity to pain, and even self-consciousness . . . necessary . . . [for] the cultural transmission of knowledge" (Peacocke 1996: 51).

The Bridge between Levels 2 and 3: Cognitive Science, Sociobiology and Behaviour Genetics

Cognitive Science is concerned to relate the process of information-processing of the organism to its actions. For example, if I want to repair the puncture on my wheelbarrow, I need to collect the relevant information. I require a new inner tube; I need to find out where there is a garage and how to get there. Then I process the information, in other words decide what to do: get the garage to do it or buy an inner tube and do it myself.

Information processing is concerned to try to understand how the human brain works. Scientists working in this area are well aware that to understand these links they need to take account of all the intermediate levels of analysis, such is the complexity of the task. It also underlines the reality of the hierarchical interdependence and inter-action of levels, sometimes "bottom-up" and sometimes "top-down". It also underlines the fact that the laws at one level of complexity may provide constraints, but also provide a platform for more complicated processing and the organization of *emergent realities*, e.g. a new species or a higher level of awareness of, say, the transi-tion from awareness of noise to awareness of music.

Sociobiology is concerned to study the biological basis, particularly the genetic, of social behaviour, and the relations between biological constraints and cultural change (Peacocke 1996: 55).

Behaviour Genetics has been a distinct discipline since 1960, and studies the genetic underpinning of personal behaviour. For example, what is the genetic contri-bution to moral awareness of different human beings? Almost no empirical research on this has been undertaken.

Focal level 3: Sciences concerned with Human Behaviour

Up to the 1960s, Behaviourism and Psychoanalysis dominated this field, though after the 1960s educationalists were particularly interested in the Gestalt school and Piaget.

By the late 1990s Peacocke reported:

there is a new openness in the behavioural sciences, not only in a "downward" direction via cognitive science to neuroscience, but also "upward" to all those studies and activities that regard human consciousness and its content as real and worthy of examination and inter-pretation. There thus appears to be occurring a re-habilitation from the scientific perspective of the reality of reference to humanistic studies – in which theology should be included, if only because of its concern with religious experience. (Peacock 1996: 59)

Psychology contains so many competing theories none of which dominate. Morea comments:

Thrown into the world I have become a puzzle to myself; scientific theory has failed to find a solution to [this puzzle of] St Augustine . . . Religion has traditionally taught that human beings are made in the image of God. This would explain why we have difficulty in under-standing personality. It would explain why, in our attempt to understand personality, we often experience wonder and awe. If human beings are made in God's image it would explain why – at the boundaries of our scientific knowing of human personality – we some-times sense beyond the mystery of human personality a much greater Mystery. (Morea 1990: 174; Peacock 1996: 61)

The Bridge between Levels 3 and 4: The Social Sciences

While the social sciences may help explore the effects of social factors on religious beliefs and behaviour, they cannot adjudicate on their truth claims. The fact that society is formed by persons, many of whom are driven by religious vision, suggests that there may well be a transcendental dimension to both personal and social life (Peacocke 1996: 62–3).

Focal Level 4: Human Culture and its Products

The fruits of human culture include all the results of human discovery and creativity in the sciences, in the arts, in individual, social and intercommunity human relations, including personal experiences of Transcendence. Peacocke suggests that an encounter with the transcendence in "the other" may take place through "the form of a work of the creative, imaginative arts, or of another person, or of God, the Beyond in our midst" (Peacocke 1996: 63). He finds support from Steiner, who has called all such encounters a "wager on transcendence":

> the wager on the meaning of meaning, on the potential of insight and response when one human voice addresses another, when one comes face to face with the text and work of art or music, which is to say when we encounter the other in its condition of freedom, is a wager on transcendence . . . The wager . . . predicts the presence of a realness . . . within language and form . . . The conjecture is that "God" is, not because our grammar is outworn; but that grammar lives and generates worlds because there is the wager on God. (Steiner 1989: 4; Peacocke 1996: 63–4)

TAKING STOCK: FORMS OF KNOWLEDGE

What Peacocke and Steiner have been saying above is extremely important, and we must cover the ground again more carefully in order to take stock of what each is affirming: first, Peacocke; second, Steiner.

First, what Peacocke is arguing by outlining the details of a "hierarchy of knowledge levels", can be explained graphically by using an illustration from a sermon by the Revd Harry A. Williams, then Dean of Trinity College, Cambridge, which involves the integrity of forms of discourse and the rejection of reductionism. They need to be discussed together.

In the passage below, Williams is illustrating the integrity of different forms of knowledge, and particularly the irreducible nature of "personal knowledge":

> Over the past few years I have got to know Fred very well. Like everybody else he has his good and bad points, not to mention his eccentricities. I am only too aware, especially when I am annoyed with him, that there is a great deal about him which can be explained in scientific terms. He is clever, but then his father is a professor, and he went to a school noted for the academic ability of its staff. He is moody, but then his mother is a manic-depressive. He can be lazy, but then a recent blood test showed him to have a calcium deficiency. He is always at ease in company, but then his grandfather was . . . [an actor]. In other words, genetics, physics, biochemistry, psychology and sociology can tell me a lot about Fred. They provide me with a valid description of him as far as they go, and they can go quite far when I want to explain him away. But, for all their validity and my irritations, the real Fred somehow eludes these scientific descriptions. I have to admit that he is much more than the sum total of these respective analyses. He belongs to the world of scientific enquiry all right

– at times he seems to belong to it with a vengeance – but he also belongs to a world where information about him becomes irrelevant because I have established communion with him. We had dinner together last night and talked until the early hours, and I never once thought of him genetically or biochemically or psychologically or sociologically. He was just him, and I enjoyed his company enormously. It made me realize that, although the Fred about whom I can be scientifically informed is real enough, he is far less real than the Fred I dined with, the Fred with whom I was in personal communion. Perhaps I could put it this way – I am very fond of Fred, but the Fred I am fond of is not just a collection of scientific information. He is more real to me than a statistical table. Indeed he is far too real for anything like adequate intellectual description at all. Yet I can know him, know him intimately, but here knowing him means being in communion with him. If I were a poet I might be able to throw out hints, think up one or two images, which might suggest to you what the quintessential Fred is like. If I were a painter, I might be able, in a portrait of him, to let you see a little of the mystery reality which lies behind his physical appearance.[7]

The passage needs analysing to see the different levels, from the hierarchy of knowledge described earlier, which can be used to understand "Fred", depending on the precise kind of knowledge which is required. When Williams talks about Fred, he says "I have got to *know* Fred". He did not say "know about Fred". Therefore he is using "personal discourse" to express "personal knowledge". Below is a list of the "modes of discourse" used, together with the words or concepts used from within the relevant "mode of discourse", or "language game".

Personal (significance)	"more real", *know Fred*, "more than the sum total",
Concepts	*personal communion* not information.
	annoyed, clever, moody, lazy, at ease,
	personal communion, enjoyed, fond, "poetic images",
	a portrait, 'the mysterious reality'.
Scientific (significance)	"less real", the Fred I can *"know about"*
Concepts	*Information* in the form of 'scientific knowledge', not
	personal communion.
Genetics	father, grandfather,
Biochemistry	calcium deficiency,
Psychology	manic depressive,
Sociology	noted school, professor, actor

This passage dramatically reverses the theory of knowledge that Russell, for example, embraced in the 1930s. For him, and the logical positivists, "science" was the only form of *real* knowledge apart from Mathematics, Logic and the principles of Reason. "Liking" or "knowing" a person were seen philosophically as emotional froth.

Now, at the beginning of the third millennium, we can confidently say that – personal knowledge is, to quote Williams, "more real" than "scientific knowledge", which is "knowing about" a person. However, having made that fundamental point, it is equally important to highlight Peacocke's emphasis. This is that forms of knowledge and the disciplines we use to express our understanding of different aspects of the physical and sentient world are each important in their own right, but are also complementary. Indeed, they often overlap or are woven together. But the higher form of knowledge cited integrates those others, which are discussed in a given situation.

REDUCTIONISM: A METHOD AND A PHILOSOPHY

A Method

Reductionism is used in one of two ways. It can describe *a method of studying* complicated systems by studying the composition and the behaviour of different parts of the whole. This is scientific analysis, exploring the basic elements of which a molecule, a mollusc, a monkey or a man are composed. This is sensible scientific methodology adopted by anyone engaged in exploration, whatever their philosophical outlook. This method may well be complemented by studying the object in question from a holistic perspective, as in the account of Fred above.

A Philosophical Outlook

Reductionism can also mean the making of *philosophical judgements*, or affirming *foundational beliefs* that are focused on the fundamental constituents of matter or basic principles of organization in the physical sciences and biology. As such, reductionism denies the integrity and independence of each of the forms of knowledge outlined in the hierarchy of knowledge earlier, in its search for the basic constituents of knowledge, which continues the nineteenth century search for the constituent elements in matter – an exploration that was largely completed a century ago.

Monod and Dawkins use the terms "chance" and "necessity", for opportunity and law. As mentioned in the last chapter, "chance" is a metascientific term implying directionlessness. By it Monod (1977: 80) introduced an atheist perspective which is irrelevant to science, particularly as he knew that a molecule can electively activate a reaction in response to information from several chemical sources, showing how an organism transcends physical laws. Dawkins adds his atheistic metaphysic to Monod by using the term "blind", though he recognizes good design (1991: 21; Polkinghorne 1998: 78).

Dawkins says that "the true utility function of life . . . is DNA survival . . . The universe we observe has precisely the properties we should expect if there is, at bottom, no design, no purpose, no evil and no good, nothing but blind, pitiless indifference" (1999: 123, 155). He is confusingly mixing scientific discourse, concerned with "utility function" and "design" with personal discourse concerned with "purpose, evil and good". He does not regard his relationships and work as having "at bottom, no design, no purpose, no evil and no good, nothing but blind pitiless indifference"; he does recognize human "capacity for conscious thought" and "genuine disinterested, true altruism" (1981: 215).

Dawkins's philosophical mistake is that of confusing modes of discourse: the mechanism of DNA can be analysed scientifically to see how it is constructed and works; human desire can lead to conscious purpose. A gate catch is made of iron which can be analysed scientifically to see how it is constructed and works, but its purpose is human, to keep animals out of the garden. The gate catch cannot have a purpose of its own. Again, Dawkins (1981: 206) only confuses his arguments by introducing a pseudo-scientific concept into religious discourse by calling God a "meme", a concept which leaps from brain to brain through mindless imitation. Ignoring the confusion, I strongly support Dawkins in drawing attention to unreasonable adherence to religious tradition, which is a central characteristic of what has previously been described as "an extrinsic or authoritarian religious orientation" (see Chapter 3, "Humanistic Psychology's Models of Being Human", p. 45; Dawkins 1981: 206–9).

The most complete refutation of Ultra-Darwinism, which is defended by Dawkins, is provided by Professor Keith Ward, the Emeritus Regius Professor of Divinity in the University of Oxford. Ward argues that "God is a hypothesis . . . [which] entails that any created universe . . . will be intelligible, morally ordered and goal directed. Consequently, a demonstration that this universe is not rationally ordered, or that it is non-purposive, or morally cruel or indifferent, will undermine belief in God."[8]

The reductionist thesis of Darwinians is universally rejected by all scientists who accept the hierarchical levels of knowledge between species and those experienced by humankind. This entails the upward emergence of new, novel and more complex organization and properties. Anyone can recognize the qualitative differences in life which distinguish different species, like an amoeba, a frog, a porpoise, a dog, and a human being. Amongst human beings there are qualitative differences in creativity and ways of knowing between Einstein, Constable, Mozart, Madame Curie, Martin Luther King, Mother Theresa and Nelson Mandela. Jesus of Nazareth is also in the list, too, as are all women and men. But He is singled out from the rest of us precisely because of the qualitative differences in His creativity and ways of knowing, talking, relating to others and behaving.

Language and Knowledge: Steiner's Quotation

We need only concentrate on the end of the quotation given earlier: "The wager . . . predicts the presence of a realness . . . within language and form . . . The conjecture is that 'God' is, not because our grammar is outworn; but that grammar lives and generates worlds because there is the wager on God" (Peacocke 1996: 63–4 citing Steiner 1989: 4).

We must not miss what Steiner is saying here. He is making the point that in all the modes of discourse, or forms of knowledge we have explored, like empirical, aesthetic, moral and personal, and all the disciplines within a form of knowledge, e.g. scientific discourse, we can distinguish physics, chemistry, biochemistry and the other sciences. We are quite clear that the language "represents a realness" in those aspects of the world and human experience which they outline. Indeed, he is saying that "grammar lives". I agree with that. I therefore agree with his inference that if we use grammar related to what is real in all the areas mentioned, it follows that if we have a universal human religious language, it is realistic to assume that it relates to "God who is".

I do not agree with Steiner "that grammar generates worlds", as I share the "critical realist" position of the scientist/religionists.[9] To say that the human mind "constructs" the world, does not accord with "bottom-up" thinkers in the sciences and arts, and those who reflect on the nature of personal communion. To say "that grammar generates worlds" is to ignore "bottom-up" causation and say that only "top-down causation" operates, rather than both, as asserted by critical realists.

Take an example of the folly of adopting a "top-down causation only" model. If I "construct" a picture of the girl I am talking to and then marry my "picture", I am likely to end up disillusioned when I find out what she is *really* like, and then what do I do? Of course she might be a critical realist and therefore read my thinking accurately, becoming "the model girl" I project onto her. If so, she is either a fool or a saint – you decide!

THE REVISED NATURE OF SCIENTIFIC KNOWLEDGE

Polkinghorne retired from being the Professor of Mathematical Physics at Cambridge in order to become an ordained priest in the Church of England. Since then, in his extensive writings on the nature of the relationship between science and theology and on how science affects the way he understands his Christian faith, he has offered the non-scientists amongst us a lucid understanding of the way the nature of scientific knowledge has changed in the twentieth century.

Classical physics, formulated by Newton, portrayed the world as a clockwork solar system within a framework of time and space which contained matter which moved. This was offered as a description of the world as it was understood. Such a picture is called *naïve* or classical *realism*, because it assumed an almost one-for-one correspondence between the different elements and forces which make up the world and the words or concepts used to describe them. This physics was governed by laws which *determined* the behaviour of matter and movement: there was no space for *freedom*. And lastly because it was concerned to analyse the nature of matter and movement, searching for the smallest particles and basic patterns of movement, it was *reductionist* (Barbour 1998: 166).

The Effect of Quantum Theory on the Nature of Knowledge

However this view of scientific knowledge changed radically in the first quarter of the twentieth century with the development of quantum field theory. Two things emerged: classical physics, at a deep level, offered a *clear* descriptive picture of the world, but quantum physics was *cloudy* (Polkinghorne 1998: 26). That is, it was not fully understood.

At the formal level, classical physics offered *one* theory of the behaviour of matter, but quantum theory required *two* complementary pictures. In studying light travelling it was discovered that it can behave sometimes as particles that are fired in packets or photons, and sometimes as waves of light. "A particle is a little bullet, and a wave is a spread out flappy thing."[10] If you know where an electron or photon is, you cannot know how fast it is travelling. This is Heisenberg's *Uncertainty Principle*. We can know either the position or the momentum (i.e. its velocity multiplied by its mass) of a particle of light, but not both at the same time. That is because when you make a measurement you interact with the quanta of light and disturb its performance uncontrollably.

Initially, Heisenberg's principle stated that what one could measure and therefore *know*, and what could not be simultaneously measured and therefore not *known*, is called "the uncertainly principle". This was an epistemological statement or a knowledge claim. But soon it was interpreted as a *principle of indeterminacy*, in other words, *what was the case is what was happening – it was what existed.* Polkinghorne calls this "epistemology models ontology". In other words, what we *know* tells us what *is*.

Our concern here is not with the physics, but with how this discovery changes our understanding of scientific knowledge. Newtonian physics still explains the world of our normal sense perception to us quite adequately, but quantum theory combines two theories, each of which offers a partial explanation of what is observed. They are remarkable because the "superposition principle" allows us to state that "a wave like state contains an indefinite number of particles" (Polkinghorne 1998: 31). The theories are thus not contradictory but *complementary.*

There is a third feature of quantum theory which affects our understanding of the nature of knowledge. When two quanta or packets of light have interacted, they become twinned, no matter how much they are separated. If one stays in Plympton and the other goes to the moon, and an astronaut measures the quanta light twin on the moon, it will interact with and change the Plympton light twin by "remote control". This is not common sense! This is called the *non-locality* of light quanta, which means their "togetherness in separation".

These insights from quantum theory are very important for our understanding of the nature of knowledge.

First, as Polkinghorne says, "the physical world is full of surprises" (1998: 32). No one with a background of Newtonian physics could have imagined the wave/particle duality of light before it was discovered. It well illustrates that "knowledge is not constructed" by the human mind and superimposed on the world around. That theory of knowledge, called *constructivism,* is overthrown by the fact that quantum theory is the result of postulating theories, testing them in measurement experiments, and reformulating theory until the observed events are mirrored by theory. This underpins the view of most scientists, including the scientist/religionists,[11] that *critical realism* is the only defensible view of scientific knowledge.

Secondly, quantum theory cannot be pictured. Dirac's "superposition theory" – that adds together states such as "his" and "her" – is neither *objective* nor *picturable,* but it is *intelligible* to scientists. In other words, this *reality* is not *objective.* Polkinghorne makes it very clear that this conclusion upset Einstein, who, despite being "the grandfather of the subject" with his theory of relativity, passionately defended the reality of the material world, based on "his mistaken belief that this reality required a classically objective character for all phenomena" (1998: 33). Einstein's failed attempt to defend the objectivity of knowledge of the world in the face of the united front of his scientist colleagues underpins Polkinghorne's point that *"reality is not based on objectivity but intelligibility"* (33).

Next *"there is no universal epistemology"*, in other words there are different kinds of knowledge for different aspects of the material world as well as for different aspects of human life, as we saw earlier in exploring the hierarchy of levels of knowledge. We may later feel this is relevant to other forms of knowledge e.g. if *intelligibility* is an adequate test for knowledge in some aspects of science, is it not an adequate test for some theological arguments for the existence of God?

Fourthly, Polkinghorne raises another point. Quantum theory is one theory which tells you when things will look like waves, and when like particles. But measurement affects the behaviour of quanta being measured.[12] The choice is made by scientists on the *kind of theory* they prefer. They prefer what they judge to be the *simpler* rather than the more *complicated* theory in this case. In other words, some scientific judgements are made on the grounds of *intelligibility* and *credibility* when the limits of the available data are exhausted. This further underlines the criterion of *intelligibility* when deciding on questions of *reality*, or what is really likely to be the case.

Fifthly, if it is the case that two complementary perspectives are needed to make quantum theory intelligible, might this not be true in other forms of knowledge? Perhaps the justification of "knowledge in the school curriculum" needs the complementary theories of Hirst and Phenix, as refined by their commentators (see Chapter 4). Polkinghorne (1998: 33) suggests that to give an adequate account of Jesus of Nazareth one needs the complementary models of "human" and "divine".

Physicists can say that they understand the history of the universe from "a thousand millionth of a second to an age of some million years". "Peak uncertainty is reached at the cosmic age corresponding to the Planck time 10^{-43} seconds. This is the epoch in which the universe would have been so small that quantum effects were important for cosmology." Perhaps very early, 10^{-35} seconds, the early universe "underwent a kind of 'boiling of space' [technically a phase change] which greatly blew up its size in a very short time" (Polkinghorne 1998: 35). This is called "inflation" and provides a natural account of certain properties of the universe, e.g. the close balance between the expansive effect of the "big bang", which began the creation of the universe, throwing galaxies apart, and the contrary effect of gravity, which pulled matter together. What quantum effects might have influenced the creation of the universe? From what agency?

And lastly, a major surprise is now called the *anthropic principle*. The big bang explosion caused the expansion of the universe, the speed of which was held in check by gravity. After the first three minutes, the universe was made up of two gases, three-quarters hydrogen and one-quarter helium. After about a billion years the gases condensed and caused a clumping of matter, which formed stars and galaxies. The chemical raw materials of life (oxygen, carbon, iron, etc.) were formed inside burning stars, some clusters of which exploded, scattering material which formed second-generation stars, like the earth. Hydrogen-burning stars like our Sun provided the energy for life on earth.

There are some amazing features of our universe:

1   If the big bang explosion had differed in power by one part in 10^{60}, human life would not have been possible. To give some meaning to that number, "suppose you wanted to fire a bullet at a one inch target on the other side of the observable universe, twenty billion light years away" you would have to be that accurate (Davis 1984: 179).

2   The big bang explosion was uniform in density of matter and rate of explosion, whereas explosions we initiate, e.g. bombs of all kinds, create in each case different and chaotic rates of explosion, so the big bang is unique (179).

3   Remote galaxies on each side of the sky, which are more than twenty billion light years away and therefore out of reach of any communication routes,

display similar structures and behaviour patterns (180). How can you explain co-operation without communication?

4   Within these processes we find evidence of *mathematical laws, complex order and fundamental constants* in nature (186).

Let us look at one of these *fundamental constants*. An atom of hydrogen is the same here as in a star in the western sky and a star in the eastern sky, now out of communication reach of each other. An atom is like a miniature solar system, with a small nucleus composed of protons and neutrons around which electrons orbit. The proton of a hydrogen atom is always 1,836 times as heavy as the electron (Davies 1984: 187). The proton, which is made up of three quarks, always has a positive electrical charge, which is always exactly the same. The neutron has a mass that is slightly less than the proton and is electrically neutral. The forces holding the protons and neutrons together are called nuclear forces. If the forces were weaker, the structure of the nucleus would become unstable and disintegrate. If they were stronger, they would have set up a chain reaction in which there would have been no hydrogen left soon after the big bang of creation. That would mean there could be no sun, no water, no life.

Penrose, a mathematical physicist, calculated that the odds against this universe appearing by chance were one to $(10^{10})^{30}$ (Davies 1984: 178). These hairline constraints which permit human life are together called *the anthropic principle*. This means that either there could have been a large number of worlds and the one we inhabit emerged by chance, or that the fineness of its construction indicates the influence of a Creator (Polkinghorne 1998: 38; Penrose 1999: 560–2, 583).

Besides Penrose, Barbour, Peacocke and Polkinghorne, many scientists like Davies and another thousand scholars who are actively working and publishing books in the field of science and religion (Stannard 1996: 198; Mott 1991; Graves 1996; Berry 1991), who are not also Christian theologians, find the anthropic principle compelling evidence that the world was created by a Designer (Davies 1984: 189; 1992: 214; Wilkinson 1993: 107–8; Polkinghorne 1998: 36–7).

Finally, Polkinghorne's discussion on complexity is very important. It takes account of relativity, quantum theory, chaos theory and recent evolutionary theory.

The two arenas are the inanimate world and the sentient world. In relativity, quantum theory and chaos theory, at the formative time of the split-second moment of the choice of a developmental path, there is an openness in causation, not fully understood, as indicated in earlier discussion. Also, in quantum theory, it is recognized that when one atom has associated with another, it is networked to its twin wherever it moves in the universe. An effect on one influences the other. This requires a serious consideration of a holistic or "whole view" approach to the system studied. In other words attention has to be given to the possibility of "top-down" causation.[13]

Chaos Theory and the Nature of Knowledge

In chaos theory, not only the butterfly effect, but also the fractal patterning processes in regression to infinity require that the student must at some time take a holistic or "whole view" approach to his or her study.

At the level of sentient creatures, "bottom-up" causation is self-evident. It is quite clear that physics and chemistry influenced the interaction which led to the formation of cells and therefore life and different kinds of species. Once formed, however, the *individual choice* of members of those species would be a factor in the development of their own species and the emergence of new species from them. In other words, a "top-down" influence is a factor. This has previously simply been labelled "chance", a "mechanical" factor which previously referred to environmental factors and/or genetic mutation. So former descriptions of evolution as the result of the interaction of *chance and necessity* are simplistic and often assumed to be mechanistic (Monod 1972; Dawkins 1981, 1991, 1999).

Chance, previously thought of as a two-dimensional, mechanistic factor consisting of genetic mutation and the environmental factors, is now a three-dimensional factor of which the third element is the "will" or "choice" of the creature itself, whether amoeba, mollusc or monkey, e.g. the monkey chooses where to eat, or the partner with whom to mate. So biological evolution is no longer understood as a closed mechanistic system governed by blind chance, but as an open system in which choice is a factor. Something similar is true for quantum theory. In the experiment where light is measured, the measurer affects the performance of the light. Likewise, in chaos theory the butterfly can cause a hurricane. Failure to forgive can cause a war.

In short, twentieth-century science showed by the end of the century that in each branch of science mentioned, systems which were thought to be closed and mechanistic at the beginning of the century were actually open systems. "Open" means open to the influence of creatures, worms, women or whatever, who help to create the shape and direction of future evolution.[14]

SUMMARY AND HOPES

What key insights can we gain from twentieth-century science about the nature of knowledge?

We live in a world in which Newtonian science satisfactorily describes much of our everyday lives. School textbooks still state that science involves making observations, making hypotheses and models, and testing them with experiments. This is our everyday world of sense experience of time, space, gravity and momentum.

But we now know that science is not the objective, but the subjective pursuit of knowledge. It is altruistic in its pursuit of truth, but it does not rest on a bedrock of objective facts, as Russell thought. Instead, as Popper ultimately admitted, it emerges as the result of a passionate intuitive personal search for truth through a swamp of changing intuitions, assumptions, observations and hypotheses.

Polanyi has shown us that "we know more than we can tell". Our "tacit knowledge", like how to ride a bicycle, we often cannot put into words. Our knowing is a complex procedure involving three sets of high-speed, parallel processing of focal and subsidiary awareness and their interaction.

Macmurray, Polanyi, Hirst, Phenix, Peacocke and others have shown us the hierarchy of different forms of knowledge, each containing a number of disciplines or subjects of study. Some subjects, like geography, are fields of study, as they include a number of disciplines, like geology, meteorology, cartography and many others. Tradition and discovery are elements in science, religion and other forms of knowledge.

Piaget has shown how the elements which make up schemes, or concepts, like a house which has four walls, windows, doors and a roof, are combined in practical or concrete ways of thinking. This way of thinking and knowing is included in the more powerful theoretical or formal ways of thinking which use and test hypotheses in everyday life.

Scientists use theoretical or formal thinking which develop and test hypotheses. Dirac, in his superposition principle, holds two hypotheses together to explain the "particle" and "wave" nature of light. This marks the end of mere objective science. The act of measuring light can change its behaviour. As Polkinghorne has said, "reality is not objectivity but intelligibility". So we may not live in an *objective* world – we cannot see light in the way Dirac describes it – but we do live in an *intelligible* world. What Dirac says makes sense of the evidence.

We have also found that the early mechanistic model of evolution has been found to be incomplete. Creatures choose their mates, where they eat and how they live within their biological, climatic and environmental constraints, though some argue that this is part of the evolutionary process. Some think that evolution involves chance, choice and necessity. Choice is a factor which changes the direction of evolution. This insight is supported by chaos theory, which has shown us how a butterfly's behaviour can cause a storm, or that for want of a nail a battle can be lost. This involvement of the choice of creatures introduces the top-down causation in the process of evolution, the causing of storms, the behaviour of light, and so on, to complement the bottom-up causation in the earlier stages of the evolution of the universe.

From quantum theory, we have also learned that if two quanta have interacted and one twin goes to the moon and the other to Plympton, they continue to interact with each other as well as with their new surrounding materials. This implies that the analytic scientific approach which has been so fruitful for so long now needs complementing by a holistic approach. That is, one needs to view a situation as a whole, and to interpret it as a whole, which is a difficult but necessary procedure.

In our global village – "the whole" – we can see that "objective scientific materialism" is a personally held exclusive faith, just as later we will see that the "extrinsic, authoritarian religious outlook" is a personally held exclusive faith. Both tend to operate in one form of knowledge. Both outlooks are being questioned by intelligible scientists and religionists, for within the hierarchy of forms of knowledge there can be complementarity. There is dignity in diversity if individuals can balance their focused specialisms with a concern for their place in multi-faith community life.

In these two chapters on science, we have explored the integrity of science as a balance between passionate personal curiosity, collecting evidence, intuition, model-building and the rigorous formulating and testing of hypotheses – a pragmatic procedure. This has also led to an increasing recognition that the insights of science are more adequately explored through a range of relevant forms of knowledge.

NOTES

1 Personal communication from Dr John Waldram.
2 Personal communication from Dr John Waldram.
3 Scott (1996: 56) uses the terms *schema* and *schemata*, which are more recently translated as *scheme* and *schemes*.
4 "Believe in order that you may understand" (Chadwick 1986: 42–3).
5 Scott (1996: 176–7) quoting "Lines Written above Tintern Abbey".
6 Peacocke (1993: 217; 1996: 49). In his table of four levels of knowledge, Peacocke has them in a vertical column arranged in ascending order, with Level 1 at the bottom and Level 4 at the top. In my diagram, I have arranged them in a cycle to illustrate the fact that exploration at all four levels is engaged in by people who have education and insight, and therefore start as persons who inhabit Level 4 before they engage in their own explorations and contributions to develop further understanding in any of the four levels they choose to work in. Peacocke kindly suggested that I should put the central circle of arrows pointing in both directions also to illustrate both "bottom-up" and "top-down" causation.
7 H. A. Williams, "Why I Am a Christian", a sermon preached in Great St Mary's Church, Cambridge, 19 January 1975.
8 Ward (1996: 98). For an excellent brief overview of "God and Evolution" see Ward (2002: ch. 9).
9 I refer to authors like Barbour, Davies, Peacocke and Polkinghorne. There are many others, e.g. Mott (1991), Graves (1996) and Berry (1991).
10 Polkinghorne (1998: 27). See this text for a fuller discussion.
11 I refer to authors like Barbour, Davies, Peacocke and Polkinghorne.
12 They are the Bohm or the Bohr solutions. Details do not concern the non-scientist. See Polkinghorne (1998: 28–34).
13 Polkinghorne (1998: 44). Polkinghorne refers to "neural net systems, linked arrays of centres that are capable of correlated influence upon each other".
14 Polkinghorne (1998: 44). Polkinghorne refers to "the spontaneous generation of order out of chaos, and the possibility of *active information* as a causal principle" (italics mine, to emphasize that the reference is to a sentient being, the only source of active information).

REFERENCES

Allen, R. T. (1992), *Transcendence and Immanence in the Philosophy of Michael Polanyi and Christian Theism*, Edinburgh: Rutherford House.
Allen, R. T. (2005), "Polanyi and the Rehabilitation of Emotion", in S. Jacobs and R. T. Allen (eds), (2005), *Emotion, Reason and Tradition*, Aldershot: Ashgate.
Barbour, I. G. (1998), *Religion and Science*, London: SCM Press.
Berry, R. J. (ed.), (1991), *Real Science, Real Faith*, Eastbourne: Monarch.
Chadwick, H. (1986), *Augustine*, Oxford: Oxford University Press.
Davies, P. (1984), *God and the New Physics*, Harmondsworth: Pelican.
Davies, P. (1992), *The Mind of God*, Harmondsworth: Penguin.
Dawkins, R. (1981), *The Selfish Gene* (1976), London: Granada.
Dawkins, R. (1991), *The Blind Watchmaker* (1986), London: Penguin.
Dawkins, R. (1999), *River out of Eden* (1995), London: Orion Books, Phoenix.
Gelwick, R. (1977), *The Way of Discovery: An Introduction to the Thought of Michael Polanyi*, New York: Oxford University Press.
Graves, D. (1996), *Scientists of Faith*, Grand Rapids, MI: Kregel Resources.
Monod, J. (1977), *Chance and Necessity*, Glasgow: Fount.
Morea, P. (1990), *Personality: An Introduction to the Theories of Psychology*, Harmondsworth: Penguin.

Mott, N. (1991), *Can Scientists Believe?*, London: James & James.

Peacocke, A. (1993), *Theology for a Scientific Age*, London: SCM Press.

Peacocke, A. (1996), *God and Science: A Quest for Christian Credibility*, London: SCM Press.

Penrose, R. (1999), *The Emperor's New Mind*, Oxford: Oxford University Press.

Polanyi, M. (1958), *Personal Knowledge: Towards a Post-Critical Philosophy*, London: Routledge and Kegan Paul.

Polanyi, M. (1960), *Knowing and Being*, London: Routledge and Kegan Paul.

Polkinghorne, J. (1986), *One World*, London: SPCK.

Polkinghorne, J. (1998), *Science and Theology*, London: SPCK.

Popper, L. (1959), *The Logic of Scientific Discovery*, London: Hutchinson.

Roberts, M., King, T. and Reiss, M. (1994), *Practical Biology for Advanced Level*, London: Nelson.

Rose, S. (1997), *Lifelines: Biology, Freedom, Determinism*, Harmondsworth: Penguin.

Russell, B. (1937), *The Scientific Outlook*, London: Allen and Unwin.

Scott, D. (1996), *Michael Polanyi* (1985), London: SPCK.

Stannard, R. (1996), *Science and Wonders: Conversations about Science and Belief*, London: Faber & Faber.

Steiner, G. (1989), *Real Presences*, London: Faber & Faber.

Ward, K. (1996), *God, Chance and Necessity*, Oxford and Rockport, MA: One World.

Ward, K. (2002), *God, Faith and the New Millennium* (1998), Oxford: One World.

Wilkinson, D. (1993), *God, the Big Bang and Stephen Hawking*, Tunbridge Wells: Monarch.

7 | Forgotten Knowledge

In the popular mind, apart from the work of people like Freud, it appeared as though one area of sensibility, emotion and feelings had been ignored since the time when the logical positivists seemed to gain a stranglehold on what counted as *knowledge*, i.e. from the1920s to the late1950s in the UK. While the positivist view of *knowledge* held sway, *emotion* was regarded as outside the realm of knowledge which could be verified or tested. It was held to be merely subjective and therefore not a source of knowledge at all, and was accordingly devalued. In this chapter we will explore how emotion relates to different forms of knowledge, the nature of personal knowledge, and the structures of personhood (human) and Personhood (Divine).

MACMURRAY: SCIENTIFIC, AESTHETIC, PERSONAL AND RELIGIOUS KNOWLEDGE

Just as Polanyi's discussion of personal knowledge has been largely neglected since the mid-1960s, so too has John Macmurray been largely neglected since the mid-1930s, except for a small trickle of interest in them both, which has recently been increasing with the broader recognition of the variety of forms of knowledge outlined earlier.[1]

Reason and Emotion

Macmurray's seminal work, *Reason and Emotion* was published in 1935. Macmurray is clear that he lived at a time when reason was contrasted with emotional life. *Reason* was just thinking and *emotion* was just feeling. Macmurray notes that, in current usage at the time he was writing, *Reason* referred to thinking, planning and calculating. "It does not make us think . . . of music, laughter, love . . . religion or loyalty or beauty . . . " (Macmurray 1972: 15). He says we must beware of thinking that "our emotions are unruly and fleshly, the source of evil and disaster, while reason belongs to the divine essence of the human mind which raises us above the level of the brutes into communion with the eternal" (16). Both these views are rejected by Macmurray. He says, "We must recognize, then, that if we wish to discover what reason is we must examine religion and art just as much as science" (19).

Scientific and Personal Ways of Knowing

Macmurray makes a very clear distinction between what he calls "information", "data" or scientific knowledge and what he calls "emotional reason" which is an element in what I shall call "personal knowledge", when discussing personal discourse or personal knowledge elsewhere in this book. The former, "information" on a person, he describes as *knowing about* a person; the latter, "emotional reason" is an element in *knowing* a person.[2] Aesthetic knowledge was an intermediate form of knowledge.

The Aesthetic Way of Knowing

Macmurray defines the aesthetic way of knowing as follows: "It consists of a critical appraisal of something through a continuous modification of feeling" (1961, cited by Mooney 1996). He feels that artistic expression is a one-sided pouring out of feeling without regard to whether or not anybody is listening. "The artist wants to give, not to receive; so the mutuality is lost, and his experience, though it remains intensely personal is one-sided, has lost part of the fullness of personal experience" (Macmurray 1972: 154). The artist is creative. But art and science are both abstracted from the full unity and wholeness of the personal (by his contemporaries) (see note 2).

Objective Values: Personal Knowing and Religious Knowing

Macmurray is concerned to defend objective values. For example, love can be subjective and irrational or objective and rational. "In feeling love for another person, I can either experience a pleasurable emotion which he stimulates in me, or I can love *him*. Do I love him, or does he keep me feeling pleased with myself?" (Macmurray 1972: 32). The capacity to love objectively is the capacity which makes us persons. It is the core of being human. Judgements of value are never intellectual alone, but always contain emotional elements. Intellectual knowledge is organized information about persons, not personal knowledge of them, and it does not reveal the human world as it is, because only personal knowledge can do that. Reason, the capacity in us which makes us human, can express itself in our thinking, but it must also express itself in our emotions, for that is what makes us human.

Art and religion are ways of living the personal life, for they are expressions of the demand within us to seek after beauty and after God. Art is individual and contemplative, and concerned to appreciate beauty in a person, objects or a scene. But contemplation must be replaced by personal communion in personal relationships, which is an act of self-transcendence.

Three modes of knowing are identified here: intellectual, aesthetic and personal. Intellectual knowledge is knowing *about* a person: it deals with information and facts leading to understanding about a person. Macmurray sets out the limits of this "scientific" extraction of information. He comments on the scientist who does the observing as a detached observer: "Because science is impersonal it is always worried about 'the observer'. It wants to find an 'absolute observer', an indifferent observer, that is to say an impersonal observer, a person who isn't a person." He then comments on what is observed and recorded: "This concentration on the object, this indifference to the person concerned, which is characteristic of the 'information' attitude, is often called objectivity. It is really only impersonality" (Macmurray 1972: 151).

This knowledge of co-ordinated information, which is at the heart of present scientific activity and writing, which is "knowledge about", is called scientific or sometimes, empirical knowledge by modern philosophers, and is so described here. A comment on Macmurray's use of the word "intellectual" is also necessary. Words change their meaning through time and usage, and when he wrote many may have assumed that "intellectual" related only to scientific, impersonal and objective reason. He said, "science is intellectual, while art and religion are peculiarly bound up with the emotional side of life. They are not primarily intellectual" (Macmurray 1972: 19). "Intellect" was defined in 1979 as "the faculty, or sum of faculties of the mind or soul by which one knows or reasons (excluding sensation and sometimes imagination: distinguished from feeling and will): power of thought; understanding" (*Oxford English Dictionary*). Today, I would have thought that "intellect" refers to the ability to apply the power of thought to obtain understanding to any subject matter; so, for example, I would not talk about "emotional reason" or "scientific reason", but rather reason applied to emotional or scientific matters. Macmurray sometimes uses "reason" in this sense (1972: 19). Back to Macmurray.

Aesthetic knowledge is both receptive and creative. Initially "the artistic attitude is that of the looker-on, admiring and loving what it sees, but not participating in the life that it contemplates, except in imagination, subjectively" (Macmurray 1972: 60). The artist, or anyone observing with the artistic eye, intuitively receives what one sees, relating it to prior aesthetic experience and knowledge. Aesthetic knowledge also leads to creativity. The artist who loves the subject of a painting, whether it is a landscape, a still life, a family group or one person, finds creative self-expression in making the effort to commit to canvas their reaction to the scene. It is a creative self-giving. If science or empirical knowledge is impersonal, aesthetic knowledge is "half-personal". The artist does not make a personal relationship with the viewer of his work. It is pure personal expression, without regard to its reception. It is that which the artist is impelled to do irrespective of consequences. Responses to works of art are variable, but also involve making the effort to empathize with the artist's vision, even if rejected. Millions of people love Monet's work, but what of the work of countless artists which has not been sold, kept or remembered by anyone after their death? And yet that creative self-expression, like the work every person does, is uniquely important as their personal, cognitive, responsive cry of "yes" to the gift of life.

You can look lovingly at another person, aesthetically appreciating their attractiveness and beauty. But that ceases to be simply aesthetic once your eyes meet, and a relationship begins. "You are now two people aware of one another: the emotional awareness is mutual, and with it comes self-consciousness . . . The artistic awareness must give way to another one, reason – emotional reason – if it is to persist . . . Contemplation must be replaced by communion."[3] This is the beginning of authentic personal knowledge, for the end of the process of "getting to know" a person is to know that person, which is true *personal knowledge*.

Religion: The impulse to communion with other Persons and with God

Macmurray holds that "the religious impulse is the fundamental impulse of reason – the impulse to communion . . . Religion therefore is reason in human nature creating the community of persons – recognizing and achieving the unity in all personal life. It is the force which creates friendship, society, community, co-operation in living" (Macmurray 1972: 61–2). Maturity in religion, in personal relationships, finds expression in St Paul at the end of his hymn to love in 1 Corinthians 13, "then shall *I know* even as also *I am known*" (63).

The great negation of religion is individualism. Religion is completely personal. "The personality of Jesus, it seems to me, in realizing the full sweep of religion in himself, created the possibility of its realization as the full life of mankind" (Macmurray 1972: 193). The philosopher faces the claim of religion that *reality* is personal. If reality is not personal, religion is an illusion.

In brief, "it is communion in conscious community which is the key word of religion" (237). "The capacity for communion . . . that makes us human . . . is evidenced at the highest level in the recognition, by the intuition of reason, that God is Love" (Macmurray 1972: 63; Mooney 1996).

From this analysis of different forms of knowledge, the base of the hierarchy of knowledge is empirical knowledge, from which emerge science, aesthetic, personal and then religious knowledge. Though there is clearly a complex interaction between emergence upwards, which gives the appearance of a bottom-up causation, there is also the exercise of the religious or humanistic imagination to effect, through the will, a top-down causation through the use of reason and emotion.

The Birth of Personal and Religious Knowing

Macmurray has already told us the moment of the birth of personal and religious knowing. In the section above, "Objective values: personal knowing and religious knowing", we were told: "You can look lovingly at another person aesthetically, appreciating their attractiveness. But that ceases to be simply aesthetic once your eyes meet, and a relationship begins." Aesthetic knowledge, then, can be the result of one's emotional response to whatever fills one with a sense of beauty, which is often tinged with awe and wonder – a view, watching children play, looking at a person who is unaware of you, watching squirrels searching for food, or ugliness in whatever form. We know what experiences give us aesthetic knowledge and aesthetic pleasure or distaste.

But what happens to two persons the moment their "eyes meet"? Well, they could look at each other as if the other was a detached object in a zoo, to be viewed. Perhaps you have been caught "looking" at the other person, and looked away in embarrassment because your motive for "looking" was not a suitable one, or it was not the right time for opening a relationship. It is something we do throughout our lives. But very often an emotional exchange takes place when eyes meet: a slight smile, a knowing smile "commenting" on something you have both just seen together, a word of greeting, a comment on the weather. This is true in personal relationships and also when "catching sight of ", and then "talking" to God. Each kind of relationship can be provoked or released by a great intuition. The writer of the Book of Exodus puts it this way, "the Lord would speak to Moses face to face, just like a friend" (Exodus 33:11), and millions of others, every year for more than two thousand years, have known just that experience, but we will talk about religious experience later.

MACMURRAY AND OTHER VIEWS ON EMOTION

Much has changed since the 1930s when Macmurray challenged the logical positivists' views that reason was about thinking about evidence which produced objective knowledge, and emotion was just about subjective feeling which had nothing to do with knowledge. Macmurray turned logical positivism upside-down, but few took any notice until about forty years later.

Macmurray regarded science as organized information. Such knowledge could only give us information about a person, and help us *know about* him. This information was not full, i.e. personal, knowledge. Macmurray defended love, when focused on another's welfare, as the objective capacity which makes us mature persons and is the heart of rationality. That will help us have real knowledge of a person: we will *know* them through being in communion with them. Judgements of value are never intellectual alone. They have an emotional dimension. Only personal knowledge, containing both elements, can reveal the human world as it is.

Midway between personal and scientific knowledge is aesthetic knowledge. Scientific knowledge, then, is acquiring organized information about the world, while aesthetic knowledge is the emotional outpouring of the artist in their "representation" of the world in, for example, a landscape painting, without any relationship or response from the viewers. That, however, is "half-real" knowledge, as aesthetic knowledge is "less complete" than personal knowledge. A full personal engagement with, and valued assessment of, the human world is personal knowledge – the fullest kind of knowledge available.

SOME HISTORICAL VIEWS OF EMOTION

It is helpful to place Macmurray's views into a historical context before exploring his relationship to the main modern view about emotion.

Dr Thomas Dixon, in his path-breaking study *From Passions to Emotions*, demonstrates that the concept or category of *emotion* did not exist until 1820, when it replaced such existing categories as *appetites, passions, sentiments* and *affections* (Dixon 2003: 109). The word *emotion* is unknown in Latin, Greek or in the New Testament. The New Testament talks of some categories as involuntary natural, animal-like responses to events or situations as "*sinful instincts, desires* or *passions*", part of the lower sensory appetite. These *sinful instincts, appetites, desires* or *passions* were signs of deficiency and imperfection, and were contrasted unfavourably with the cognitive powers of the soul (42). On the other hand, "the Christian idea of the *affections* (which can be both voluntary and virtuous, unlike the *passions*) could be summarized as being a recommendation of the cultivation of certain *thoughts of the heart*" (39).

In the thirteenth century Aquinas identified two kinds of passions, those related to things good or evil in themselves, and those related to *arduous* goods or evils, i.e. hard to attain or avoid (Dixon 2003: 44). Aquinas arranged his eleven *passions* into six stages involved in a person's response to *passion*. Concerning the lower sensory appetite, the individual can *move towards* a good or evil response to an object, situation or person, feeling (stage 1) *love* or *hate*, and then (in stage 2) *desire* or *aversion*. Sometimes the task is *arduous* (in stage 3), hard to attain *hope* and avoid sinking into *despair*. Stage 4 is also *arduous* – does one pursue the good with *courage* or *fear*? Stage 5 records the cumulative impact of a situation on an individual. Does it evoke *anger*? Aquinas thought there was no logical opposite to anger, for "no anger" is not an emotion. Stage 6 marks the individual's total response of either *sorrow* or *joy* (Dixon 2003: 44).

Aquinas held that "the natural opposite of *passion* was *action* [using] the intellective soul (reason and will)". Aquinas rejected Cicero's claim that all passions were diseases: for the passions "are not diseases or disturbances of the soul except precisely when they are not under rational control . . . once under the control of reason, these passions were tamed to become acts or movements of the will which were quite proper parts of virtuous living" (Dixon 2003: 52). "While Augustine had said that the root of all human wrong doing lay in the passions, especially sexual and violent ones, Aquinas looked at it from the other direction: 'the root of all human goodness lies in the reason'" (52). Both the New Testament and Aquinas held that responsible emotional response was guided by reason and the will and are cognitively related theories of the emotion.

The category of *the emotions* was invented by Thomas Brown in 1820 to describe *love, hate, jealousy, anger, joy, sorrow,* and so on (Dixon 2003: 109). The greatest exponent of *the emotions* in the nineteenth century, however, was William James, through his initial 1884 article, "What Is an Emotion?" In this article he outlined a reductionist viewpoint, offering a physiological explanation of emotion. "James' theory stated simply that the bodily changes that people had been inclined to call the 'expression' of an emotion were in fact the primary constituent cause of the emotion" (207).

James seems to have assumed that "emotions were feelings, rather than, for instance, judgements or voluntary acts" (Dixon 2003: 207). He thought emotions were caused by instinctive visceral or behavioural disturbances which were then felt as feelings or emotions. We have no space here to discuss the weakness of his very influential theory (204–30). James's reductionist position contrasts with the position he takes up in his book *The Varieties of Religious Experience* (1902), which we will study later, where he is sympathetic to the notion of "spirit", and that the religious life should be judged by its fruits rather than by its physical roots.

There is not space to note the long period since Brown in which "the feeling view of emotion" dominated the topic and which Macmurray criticized. We just need to know that it has had a long history.

The most persuasive modern view of the emotions is a cognitive account, an early account of which was proposed by Aristotle c. 360 BCE, i.e. over two thousand years ago. A cognitive account is one which holds that knowledge is involved in emotions, that there is emotional knowledge.

A 1990s THEORY THAT EMOTIONAL KNOWLEDGE
IS REAL KNOWLEDGE

I have just said that a cognitive account is one which holds that knowledge is involved in emotions. Feelings or emotions relate to a cognitive stimulus, i.e. to something one knows, like a view. "Emotion" is a multidimensional concept, which means there are lots of parts to it.

The two points made in the last two sentences can be illustrated by analysing one situation:

1   Susan sees a bear in a wood (an event – something she knows).
2   She thinks: "the bear will attack me" (she made an interpretation of its signif-icance in that situation – something she claims to know).
3   She may die: her goal is to survive (she made an appraisal or evaluation of the situation – something she claims to know).
4   She is afraid (the emotion of fear).
5   This is accompanied by physiological changes that lead to adrenaline secre-tions which make extra energy available if required, so she is ready to act, i.e. run away fast!
6   She is consciously aware of all that is happening: she knows what is happening (Power and Dalgleish 1997: 58).

All these six components are involved in the creation of the emotion of fear. The key point we are interested in is that emotion is not just a feeling, but is, as Macmurray suggests, a knowledge or cognitive-related concept. Indeed, the analysis above outlines a full "personal knowledge" concept because it includes all elements in a cognitive situation – cognition, emotion, physiological changes and conscious awareness.

This model surely vindicates Macmurray's claim, mentioned at the beginning of the section on "Macmurray and other views on emotion" above, that emotional knowl-edge is more inclusive than intellectual knowledge or aesthetic knowledge because it includes all relevant cognitive and affective information, thus becoming integrated in "personal knowledge".

Likewise, modern explorations of aesthetic knowledge defend the thesis that "the kinds of feeling which are central to involvement with the arts are necessarily rational and cognitive in kind . . . they are inseparable from understanding" (Best 1992: 202). Feelings may be spontaneous, "but without such understanding an individual would be incapable of such feelings" (202).

However, a fully integrated cognitive theory of emotion includes four ways in which information is processed, leading to the evocation of the emotion of fear, i.e. we know fear through four different ways of knowing.

1   Information of the bear is received visually – an analogue level.
2   It is also received at a cognitive level through thoughts and feelings which come into the mind as when Susan realizes the bear may attack her.
3   It is received through a schematic model level, a memory formed in Susan's early childhood of "bears attacking people".

4   It is received through an associative level as she associates "bear" with attack, death, fear, and running.

The arousal of the emotion of fear involves four representational systems, all giving knowledge in different ways (Power and Dalgleish 1997: 178). This simply adds to the view that personal knowledge is the most inclusive mode of knowing.

MULTILEVEL COGNITIVE SCIENCE THEORIES OF EMOTION

Revd Dr Fraser Watts, formerly a professional research psychologist, shows the relevance of multilevel theories of emotion related to, for example, the Levanthal's multilevel cognitive model of intellectual development with its sensori-motor level, conceptual level and level of forming schemes (Watts 2002: 86).[4] Each relates to different kinds of emotional response. More interesting to us are Interactive Cognitive Subsystems (ICS), which distinguish two systems called Propositional (PROP) and Implicational (IMP) Systems. Emotions emerge from meanings coded in the Implicational Systems, but not coded in words, suggesting that different kinds of consciousness, like religious consciousness, comes into being at this level. Watts emphasizes the importance of this source of meanings, many of which he feels lie at the heart of religion, indicating the unknowability of God (Watts 2002: 86–8). Some emotions may incline towards other forms of knowledge.

EMOTIONAL INTELLIGENCE

Many of us became aware that besides "intellectual intelligence", measured by IQ tests,[5] everyone has "emotional intelligence" through a book with this title by Daniel Goleman (1996). Reviewing research over the previous thirty years, he indicated that intellectual intelligence is often less important in a balanced life than emotional intelligence. "At best I.Q. contributes about 20% to the factors that determine life success" (Goleman 1996: 34; 2003, 2004). He explores "a key set of these 'other characteristics', *emotional intelligence*: abilities, such as being able to motivate oneself and persist in the face of frustrations; to control impulse and delay gratification; to regulate one's moods and keep distress from swamping the ability to think; to empathize and hope" (1996: 34).

Gardner, in 1983, outlined seven key varieties of intelligence: verbal, mathematical–logical ability, the spacial skills of an artist or architect, the kinaesthetic skills of athletes and dancers, the musical gifts of Mozart or the Beatles, and the two faces of "personal intelligences" – interpersonal intelligence and intrapersonal intelligence.

"*Inter*personal intelligence is the ability to understand other people: what motivates them, how they work, how to work co-operatively with them . . . *Intra*personal intelligence . . . is a correlative ability, turned inwards. It is a capacity to form an accurate, veridical model of oneself and to be able to use that model to operate effectively in life." Put another way, *inter*personal intelligence includes the "capacities to discern and respond appropriately to the moods, temperaments, motivations and desires of other people". The key to self-knowledge in *intra*personal intelligence includes "access to one's own feelings and the ability to discriminate among them and draw upon them to guide behavior" (Goleman 1996: 38–9).

Emotional intelligence is not only more important than intellectual intelligence but it is the essential equipment of, for example, a good leader or teacher, for "in any human group the leader has maximum power to sway everyone's emotions. If people's emotions are pushed towards the range of enthusiasm, performance can soar" (Goleman *et al.* 2003: 6). So laughter and empathy lift everyone's spirits and lessons become fun! (6–16)

EMOTION: THE MEDIUM FOR RELIGIOUS EXPERIENCE

Dr Mark Wynn (2003) explores the emotions as sources for religious understanding. His thesis adds new support to the work of Professor Keith Ward, formerly Regius Professor of Divinity in Oxford. I begin with two quotations: "[A] revelation which can call forth such a passionate commitment must be more than a set of theoretical truths proposed for our assent. It must enshrine a disclosure of value which can over-ride all selfish desires and all competing values" (Ward 1994: 30); "religious commitment . . . is a matter of fundamental vision and response" (Ward 1987: 161–2, abbreviated).

Wynn notes that most people probably think that feelings owe their origins to the thought to which they are attached, as we noted in the earlier statement of a 1990s account that emotional knowledge was real knowledge. Wynn questions this view, initially by quoting some studies which illuminate his general thesis, that feelings are themselves "about the world". Four of the studies he quotes will be briefly described.

One study, by Deigh, indicates that emotions like horror or fear can "run ahead" of thought to which they are attached, and sometimes cannot respond to reason, indicating that feelings can refer to the world directly without mediation of a verbally articulated thought. A second study, by Maddell, on music shows how feelings of tension can register "tension" in a piece of music. Likewise, a feeling of the desire for resolution can be evoked by hearing a dominant seventh (a particular combination of notes), even when the listener has no understanding of the concepts of music. This illustrates that feeling can register the character of the world independently of any verbalized understanding. A third study, by Goldie, illustrates that feelings have an intentional content not reducible to beliefs or desires. Seeing a gorilla in a zoo cage whose gate is open is an observation which can be expressed in words. However, it is the feeling of danger it arouses, which may reveal what one's circumstances really are and what action is required. The intentional content, "to run away", comes from feelings or emotion alone ahead of the formulation of reasons for that behaviour.

Wynn then turns to his main subject: feeling and belief in religious contexts. He reminds us that in 1902 William James wrote that feeling is the source of religious belief, as we mentioned before: "I do believe that feeling is the deeper source of religion, and that philosophic and theological formulas are secondary products" (James 1928: 431).

The first study that Wynn quoted above, showing how feeling can "run ahead" of conceptualization, offers support to James. Wynn hints that emotions are like scientific paradigms: "paying attention to certain things [as we do when our emotions are engaged] is a source of reasons, but comes before them" (2003: 50). A fourth study, by Gaita is of some psychiatrists whose belief that "their patients were fully their equals" was shown to be superficial when one day a nun visited the ward. The way she talked to the patients, her facial expressions and demeanour towards them, showed the psychiatrists that unlike them, she believed they were equal from her heart. Perhaps "the revelatory force of affective responses of this kind" found conceptual expression in the nun's religious belief that the patients "were created in the image of God", though this is not Gaita's view. Gaita holds that "the revelatory force of a nun's example cannot be checked against any independent standard (e.g. the idea that human beings are created in the image of God)". Wynn comments that it can be argued that "it is the language of God as parent in particular that enables the nun's affective response and associated behaviour. That language may not be an independent check on the affective response, but the response is in some way informed by it."[6]

In his conclusion, Wynn notes that feelings may offer "a pre-conceptual appreciation of the world's significance", or they may help to deepen a conceptual understanding of the world, as in the gorilla example. This may be expressed in beliefs, intentional action or a felt musical correction or completion respectively: sometimes emotion alone leads to actions, belief or desire. In the nun's case, her "behaviour is grounded in something she believes in her heart". Sometimes belief inspires a high quality of action. Wynn (2003: 54) comments: "These approaches offer different, though compatible accounts of the relationship between feeling and belief in religious contexts . . . just as affects should not be considered as add-ons in a theory of the emotions, so they ought not to be considered as mere add-ons in any developed account of the religious life."

BECOMING A PERSON THROUGH DIALOGUE

Talk of personal knowledge lacks the focus of what a person is. One becomes a person through contact with other persons. Initially, as a newborn baby, contact may be through sucking, through touch, hearing, taste and smell. Real personal contact, and "becoming a person", begins when the mother's eyes meet those of the infant, through smiles, facial expressions, the tone of voice and the quality of touch. A relationship begins through the mother communicating through all five senses.

A person becomes a person through conversation, *through dialogue* (McFadyen 1990: 113–50). While the child begins to become a person within a parent-child relationship, the aim of a parent is, in time, to shed authority and become a friend. Dialogue, by then, should have become a conversation between two individual, independent, free and separate human beings searching for mutual understanding.

Other World Faiths emphasize dialogue not just in personal and social development, but in relationships between different religious communities. For example, the editor of the *Sikh Messenger*, when talking about living in peace, reminded his listeners that, five centuries ago, the founder of Sikhism, Guru Nanak, knew that the

formula for preventing . . . [a] downward spiral of human depravity lay in a single word, *Dialogue*. It starts with his very first sermon. In it he taught that God isn't interested in our different ethnic or religious labels, but in our piety, tolerance and concern for others. The Guru followed this up by entering into dialogue with the different faiths of his day, respecting them as different paths to the same one God.[7]

Dialogue is the basic concept of what it is to become a person, because we are what we are as a result of our relations with others, parents, friends, teachers and all the other persons one meets, including those of different World Faiths. As relationships are central to being human, an individual seeks to become a centred being through commitment to open and honest discussion with others. Becoming human is a moral achievement, just as trust in a conversation involves a moral dimension to the exchange. A person is constituted by the "sedimented" sequential deposits of memories of discussions, from which they select what is necessary to their maturing centred being (McFadyen 1990: 17–18).

This model of what it is to be a person occupies a midpoint between individualism and collectivism – between "only I matter" and "I am only a member of the community". This model is not centred on independence or dependence, but on interdependence and emphasizes the social aspects of becoming a person in the mode of personal discourse. But a person can only be fully understood through using the hierarchical model which involves all modes of discourse, physical, chemical, biological, sociological, psychological, aesthetic, moral, personal and religious. One must not confuse the unitary notion of "an embodied soul" or the dualist notions of "body–spirit" or "mind–spirit" as anything other than models used in religious discourse as different attempts to relate the spiritual or religious domain of human activity to the biological or psychological aspects of personhood.

The Basic Nature of Personal Being – the Image of God

In the mode of religious discourse, the Jewish–Christian insight is that a person is created "in God's image". This defines the religious understanding of a person's being for most people in the Old and New World (Greeley 1992: 57–70; Hay 1990: 79–84). A "person" can be defined at many levels, as we have noted in the earlier exploration of the "hierarchy of ways of knowing" – by physics, chemistry, biology, psychology, sociology, or as persons in relation with other persons, communities and God.

To say that a person is created in the image of God is to state the nature of a person's being. Philosophers call this an ontological basis of human nature. They mean that this "picture" of a human being is adequate to encompass the development of moral and religious maturity. A person is not merely a product of animal evolution, though that is the case at the biological level. The structure of being a human allows reciprocal human relationships and dialogue with God. This is based on the fundamental religious intuition, backed up by experience, that not only human life is personal, but that the whole of creation is the result of God, who is both Mother and Father, sharing the Divine personhood with human beings.[8]

Human parents know what this means. After they have given their children life and nurture as a foretaste of personhood, they "stand back" and give them freedom to become themselves, with the opportunity to grow in wisdom and maturity as interdependent persons.

A Reductionist View of the Basic Nature of Being a Person

Richard Dawkins talks about becoming, or being, a person in quite different terms. He would say that "we are survival machines – robot vehicles blindly programmed to serve the selfish molecules known as genes" (Dawkins 1981: x). In other words, we are programmed "daleks"!

However, Dawkins goes on to say that ideas evolve from the "soup of human culture" (1981: 206) in which we are nurtured. These ideas pass from brain to brain in the process of cultural transmission. He coins the term *meme,* which is the unit of imitation. "Memes propagate themselves . . . by leaping from brain to brain . . . [through] imitation" (206). An idea-meme is "an entity which is capable of being transmitted from one brain to another".

Dawkins likens the brain to a "computer" in which memes compete for memory space, so only the most valuable memes win the struggle for survival. The God-meme survives because of the threat of hell-fire (207–13). Apart from drawing on religious ideas that were culturally acceptable up to the early years of the twentieth century (Richardson and Bowden 1991: 143, 248, 333–4), Dawkins is a reductionist who thinks of a person as nothing more than a mechanical meme replicator, which is programmed socially for survival but is capable of meme mutations effected through mindless Freudian slips of the tongue.

No sensible religious person denies that the idea of the meme is an interesting picture or model for the thoughtless transmission of culture. But a model taken at the biological level of understanding humans, which can properly speak of the biological nature of persons, using biological concepts and discourse, takes no account of human thought, which operates at a more complex level of organization, recognized by the hierarchical levels of knowledge. Talk of meme mutation is a reductionist, mechanistic account of the soaring achievements of a person's creative imagination seeking to comprehend the different kinds of "secrets" discovered and creatively "modelled" by people like Curie, Einstein, Turner, Mozart, Christ and His disciples, and leaders of other World Faiths including humanism. It cannot account for people of the twentieth century like Bonhoeffer, Luther King, Gandhi and Nelson Mandela, who are concerned to promote the human understanding of the natural and social world around us and the coherence of and reciprocal relationships within the global human family. Their central interest, like that of most people of goodwill, was and is not "cultural survival" but the search for truth, beauty, justice and harmony within the person and between members of the global human family.

The strength of the religious view of the nature of a person, which is misunderstood and dismissed by Dawkins, is that it alone takes account of all aspects of personhood, including immortal longings, which for Christians find fullest expression in interpreting Christ "as the highest unsurpassable instance of a self-disclosure to man on God's part . . . " (Kerr 1997: 163). That means that Christians see Christ as the human person most filled by what they think they know about the nature of God.

But Dawkins really accepts this in principle. He makes clear that he does not accept the gene base or the meme base as anything other than the basic drive-mechanism to survive biologically or culturally. He says, "I am not advocating a morality based on evolution . . . Let us try to *teach* generosity and altruism" (Dawkins 1981: 3). Clearly he will not be inconsistent and prevent *religious education* from moving the focus from his "memes" of "hell-fire" and "blind faith" to "rational faith" focused on love. Love is expressed through the altruism found in Jesus' teaching: "love God and your neighbour", expanded in St Paul's exposition of the nature of love in 1 Corinthians 13. While this form of religion is life-enhancing, there is no denying that some forms of religion, perhaps held by blind faith, can be survival mechanisms, which are evil.[9] On the other hand, no human endeavour ever began, such as Dawkins writing *The Selfish Gene*, doing one's homework, getting married, starting a new job, without *prevenient faith*, i.e. "the faith which goes before". This is one aspect of "rational faith". There are other aspects, but this is not the place to explore fully the nature of "rational faith", which is the foundation of thoughtful religious people. Indeed, in one form or another, "rational faith" is the foundation of all persons of goodwill who contribute to the creation of wealth and welfare in the community. Dawkins deserves the other half of this answer to his dismissal of religion. Here are some illustrations of a different mode of knowing from normal "personal discourse". Personal discourse can be illustrated by typical comments like, "I trust John", "Mary is a good friend", "Fred is loyal".

The mature religious mode of knowing is based on the recognition that life, our abilities, our social and environmental circumstances are gifts and opportunities which come from God. This recognition, whether it comes soon or late, evokes a response to the Love of God which finds expression in worship and extended and active concern for others. Sometimes mature religion, focused on worship, is described as total commitment, an act of will, based on sufficient evidence and reasons.

"Secular" examples of worship include hero worship, whether of a sport star, pop star or other celebrity (Eyre 1997). The love between bride and groom evokes worship, as stated in one marriage service. Examples of secular worship are important as they underpin the belief that worship is a natural characteristic of a healthy person. The question is, what is the ultimate satisfactory focus of worship? Dawkins does not think it is the meme.[10] Professor Keith Ward (1996) clearly shows how Dawkins's own work points towards the existence of God.

The Restoration of Broken Relationships

The claim that humans are "created in God's image" is balanced by the recognition that humans do not live up to their shared and individual ideals. Most World Religions accept this. Relationships between humans and between a person and God can become diseased, fractured or broken. A Christian insight is that these states of affairs can be ended and good relationships restored through recognizing the structure of good relationships, recognizing one's mistakes in impairing a relationship. This can involve identifying elements in the teaching and lifestyle of Jesus of Nazareth that focus on areas of necessary change in one's own life.

In particular, changing one's ways of behaving from self-centredness to concern for the other, forgiving, accepting back, or apologizing and humbly going back. This is all so easy to say and difficult to do until one sees the astonishing example of one who forgives those who betray and desert him just before He faces death by crucifixion. And his remorseful disciples came to *know* he accepted them back. They recognized that Jesus *still* loved them, and so they believed that God did as well. Such good news spreads like wildfire. It led to the formation of the communities of forgiven and restored individuals that became the Church.

The Gifts of Life and Relationships Restored

God's graciousness towards human beings is expressed in various ways, as in the free gifts of undeserved life, undeserved forgiveness, undeserved sustenance, undeserved guidance. A proper human response is gratitude, a recognition of our status as offspring of His creative love and our status of "learning children" in His human family. Our relationships are nurtured through open dialogue with God, through worship, prayer and dialogue, with the written and oral word in the gathered worshipping community, all of which help to empower the individual for personal creative living and promote inclusive community life.

NOTES

1　Interest in Macmurray: e.g. (note the date of reprinting) Macmurray (1972 [1935, reprinted in 1962]); Conford (1996); Costello (2002). Interest in Polanyi: e.g. Gelwick (1976); Scott (1996 [1985, reprinted in 1996]); and journals like *Appraisal* and *Polanyiana*.

2　Cf. Macmurray (1972: 18); Costello (2002: 169). Macmurray uses the term "personal knowledge" in the same inclusive way as Polanyi uses it, i.e. all knowledge, e.g. science, art, religion, is personal knowledge. Cf. Macmurray (1972: 191): "science is an activity of persons, it falls within the personal life and must be dominated by it."

3　Macmurray (1972: 61). Macmurray assumes that "religion . . . is the force which creates friendship, society, community, co-operation in living" (62). I am calling this "personal knowledge", which I judge to be a prior stage to "religious communion" and "religious knowledge".

4　Cf. Piaget's multilevel model of cognitive development in Chapter 4.

5　IQ or Intelligent Quotient tests were designed in the 1940s. One of their uses in the UK was to divide children into intellectually able and less able groups in order to decide who should go to grammar schools or secondary modern schools. These types of school have largely been replaced by comprehensive schools.

6　M. Wynn, personal communication, 22 June 2004.

7　Indarjit Singh, *Thought for the Day*, BBC Radio 4, 20 August 1999.

8　Of course I am sensitive to the fact that ethical humanists provide a noble model of "a person" and that some major World Faiths like Hinduism, Islam and Sikhism also share aspects of the model explored here.

9　Bowker (1987). For example, Mr Al-Mihdar, whose gang kidnapped and killed western hostages in Yemen, "shouted defiantly that his group had done everything in the name of God and that he had no regrets" (*Independent*, 14 January 1999, p. 4). Allport (1969: ch. 3). Cf. Northern Ireland, Rwanda and Kosovo. Fromm (1971: 34–5).

10　Dawkins (1981). Dawkins lives in the scientific world of materialism and the philosophical world of the logical positivists and scientism. For a brief, and more detailed criticisms, see: Bowker (1995: 1–118); Ward (1996); Polkinghorne (1998: 76–9).

REFERENCES

Allport, G. W. (1969), *The Individual and His Religion* (1950), London: Macmillan,

Best, D. (1992), *The Rationality of Feeling*, London and Bristol, PA: Falmer Press.

The Bible: Contemporary English Version (1997), Swindon: British and Foreign Bible Society.

Bowker, J. (1987), *Licensed Insanities*, London: Darton, Longman & Todd.

Bowker, J. (1995), *Is God a Virus?*, London: S.P.C.K.

Conford, P. (ed.), (1996), *The Personal World: John Macmurray on Self and Society*, Edinburgh: Floris Books.

Costello, J. E. (2002), *John Macmurray: A Biography*, Edinburgh: Floris Books.

Dawkins, R. (1981), *The Selfish Gene* (1976), Oxford: Oxford University Press.

Dixon, T. (2003), *From Passions to Emotions*, Cambridge: Cambridge University Press.

Eyre, A. (1997), *Football and Religious Experience: Sociological Reflections*, Religious Experience Research Centre, Westminster College, Oxford OX2 9AT.

Fromm, E. (1971), *Psychoanalysis and Religion* (1950), New Haven and London: Yale University Press.

Gelwick, R. (1977), *The Way of Discovery: An Introduction to the Thought of Michael Polanyi*, Oxford: Oxford University Press.

Goleman, D. (1996), *Emotional Intelligence*, London: Bloomsbury.

Goleman, D. and the Dalai Lama (2003), *Healing Emotions*, Boston, MA: Shambhala Publications.

Goleman, D. and the Dalai Lama (2004), *Destructive Emotions*, London: Bloomsbury.

Goleman, D., Boyatzis, R. and McKee, A. (2003), *The New Leader*, London: Time Warner.

Greeley, A. M. (1992), "Religion in Britain, Ireland and the U.S.A.", in R. Jowell, L. Brook, G. Prior and B. Taylor (1992), *British Social Attitudes: The 9th Report*, Aldershot: Dartmouth.

Hay, D. (1990), *Religious Experience Today*, London: Mowbrays.

James, W. (1928), *Varieties of Religious Experiences* (1902), New York, London: Longman, Green & Co.

Kerr, F. (1997), *Immortal Longings*, London: SPCK.

Macmurray, J. (1961), *Religion, Art and Science*, Liverpool: Liverpool University Press.

Macmurray, J. (1972), *Reason and Emotion* (1935, reprinted in 1962), London: Faber & Faber.

McFadyen, A. I. (1990), *The Call to Personhood*, Cambridge: Cambridge University Press.

Mooney, P. (1996), "Macmurray's Notion of Love for Personal Knowing", *Appraisal* 1 (2), p. 57–67.

Oxford English Dictionary (1979), Oxford: Oxford University Press.

Polkinghorne, J. (1998), *Science and Theology*, London: SPCK.

Power, M. and Dalgleish, T. (1997), *Cognition and Emotion: From Order to Disorder*, Hove: Erlbaum.

Richardson, A. and Bowden, J. (eds), (1991), *A New Dictionary of Christian Theology*, London: SCM Press.

Scott, D. (1996), *Michael Polanyi* (1985), London: SPCK.

Ward, K. (1987), *Images of Eternity*, London: Darton, Longman & Todd.

Ward, K. (1994), *Religion and Revelation*, Oxford: Clarendon Press.

Ward, K. (1996), *God, Chance and Necessity*, Oxford: One World.

Watts, F. (2002), *Theology and Psychology*, Aldershot: Ashgate.

Wynn, M. R. (2003), "Religion and the Revelation of Value: the Emotions as Sources for Religious Understanding", in T. W. Bartel (ed.), (2003), *Comparative Theology: Essays for Keith Ward*, London: SPCK.

8 | Is All Knowledge Relative?

WHO SAID ALL KNOWLEDGE WAS RELATIVE?

In the 1970s a movement called postmodernism emerged that suggested that all knowledge was relative. It is true that there is very little agreement on what are the presuppositions or basic beliefs of modern philosophy. Like all other human beings, philosophers have their own assumptions about what matters. Groups of philosophers differ from each other.

POSTMODERNISM REJECTS . . .

Postmodernism is so called because it rejects modernity and modernism. It also thinks that "difference" is what matters: all knowledge is relative. However, even that statement needs to be qualified, for Lyotard claims that the postmodern is really modernity taking account of the "unpresentable", which has previously not been expressed (Ward, G. 1997: 587). Let us explore modernity and modernism to get a clear picture of what postmodernists dislike in order to understand them better.

Modernity

Modernity is a term used to describe the quest for objective truth in the sciences and in the arts. It began with the rise of capitalism, technology and individualism in the late sixteenth century and developed through the eighteenth century Age of Reason, which is sometimes called the Enlightenment. This quest found expression through the work of the philosophers of the Enlightenment, like Kant, who placed emphasis on reason and individualism at the expense of tradition. They sought to develop objective science, universal law, autonomous morality and independent art. In theological studies it led to the enquiry, through historical and literary criticism, into the origin and development of Scripture and the objective search for the "real meaning" of the text. Postmodernism tries to deny the possibility of achieving such an aim (Barton 1997: 35–6).

If the thinking of "modernity" can be pictured as being concerned with precise forms, such as the circle, the cube, the spiral and the double helix, then postmodernity can be pictured as a rhizome, an underground root without shape or form, sending tentacles and shoots in all directions (Ward, G. 1997: 585).

Modernism

The modernist movement, with its concern for individualism and personal self-expression, heavily influenced philosophy, literature, architecture and the many schools of art which developed at the beginning of the twentieth century. It reached its greatest influence in the 1920s, but its views were important up to the end of the Second World War.

Modernism picked up the concerns of modernity in the 1890s, which found initial expression in the modernist movement in the Roman Catholic Church. This criticized traditional views about revelation, tradition and doctrinal teaching in the Church. It attempted to remove dogma from traditionalism in order to express religious belief in modern thought forms. Sixty-five proposals for change were condemned by the Pope in 1907 in opposition to the movement. Modernism in all the churches was concerned with the evolving nature of Christian faith and a search for God within human experience.

The First World War increased uncertainty concerning the historic creeds and concerning values. T. S. Eliot echoes this mood in *The Wasteland* (1922). He pictures the individual soul and society in pieces, in despair, seeking reintegration and a new centre. Modernism was a search for a new, reformed Christian world-view with purpose, design, a centre and a Presence, asserting "people matter most", which is an emphasis and an assumption shared by humanists.

Postmodernism also opposes foundationalism, essentialism, realism and transcendental standpoints characterized by any world-view like Christianity, Judaism or humanism. These three "isms" need to be described briefly so that one can more easily understand the nature of postmodernism.

Foundationalism

Foundationalism is the view that knowledge exists in two tiers.

The first tier is *basic beliefs*, which are those not justified by other beliefs. For example, "I see that tree" – a perceptual belief. This foundational belief has a reliable origin: I saw it! Foundational beliefs can sometimes be reformed in the light of new evidence. Other foundational beliefs include those based on memory, e.g. 'I had a cup of tea just now,' and beliefs based on other people's thoughts and feelings, e.g. 'Barbara was ever so nice after I broke her vase.' Plantinga takes "God exists" to be a basic belief (Peterson *et al.*: 1991: 124). Religious experience can similarly be a source of basic belief as it is similar to sense perception (127).

The second tier is *derived beliefs* inferred from evidence. For example, the grass on the lawn is short. Inferred belief: it has been cut recently. June is not angry today. Inferred belief: her boyfriend apologized to her yesterday.

Essentialism

Essentialism is the view that objects have essences. For example, what is it to be a human? If you answer that and gain the support of others concerned with answers to that question, you know the essence of being a human! You need to define the characteristics which all humans can have, otherwise your definition of "the essence of being human" fails. While you are doing that, do notice that *Homo sapiens* is the species to which we belong.

However, to claim that a member of the species *Homo sapiens* is a *human* being, or a person, is to make a *value judgement*. It is not a biological fact. One usually becomes a human being through the processes of nurture, education, socialization, revelation, inspiration and idealistic aspiration. Through such processes we *become human,* and become *persons*. Take one example: if someone steals from you, they are immoral. They may apologize. If so, they *understand* what it is to be moral; they *are* human! But if the person who steals *does not understand* that taking something without permission is stealing, they are amoral, i.e. they have no understanding of morality. It has been said that Lee Harvey Oswald, the man who is alleged to have assassinated President Kennedy, was amoral. He is alleged to have shot at people as a weekend sport. If that is true, Oswald never became "truly human".

Naïve realism

Naïve realism is the view that our senses give us real and direct knowledge of a real world. Thus there is often a one-to-one correspondence between a word and an object or idea, e.g. "I see Doris and Ruby"; "I feel the atmosphere is wonderful." As a view of reality naïve realism collapsed when it became obvious that knowledge of the world involves interpretation prior to description.

Critical realism

Critical realism acknowledges that the mind is an intermediary between what is perceived and how it is first interpreted and then described. With the advent of scientific theories of relativity, quantum theory and chaos theory, most scientists accept this view of the world. For example, as we have seen earlier, we can observe a lighthouse as a source of a beam of light, but the light itself is understood, paradoxically, as being both a wave and a particle at the same time.

Grand narratives or metanarratives

Postmodernism rejects all grand narratives or metanarratives, which are the different names used for world-views. Societies or groups of people in the past have tended to accept one or other world-view or grand narrative, like Christianity, Marxism or humanism. Of course, in democracies it has been possible for individuals and groups to hold one or other of different world-views, or a mix and match amalgam.

143

At the beginning of the twentieth century, Christianity was the world-view which our western society used to describe its beliefs and its communal identity. However, it was still in rivalry with the eighteenth century Age of Reason (often called the Age of Enlightenment), another world-view whose values were science, reason, logic and progress. Free-thinking rationalists holding these values aimed to free humanity from misery, religion, superstition, irrational behaviour and unfounded belief. Scientism, another grand narrative, was embraced by some at this time, as scientific knowledge was seen to be reliable knowledge. It is basically a mechanistic world-view asserting that methods so successful in the natural sciences are equally applicable to all areas of human experience: "matter and mechanisms matter most"!

Structuralism

Structuralism was a term coined in the 1950s, but it legitimately includes the work conducted earlier in the century by Freud in psychoanalysis, studying the unconscious mind and other aspects of the structure and activity of the mind and instincts, and the work of Skinner, studying the mechanisms underlying human behaviour, called behaviourism.

Structuralism gained its name from the work of Lévi-Strauss, an anthropologist concerned with the study of the underlying structures in different cultures and their social groups. He argued that society was organized according to a particular method of communication. For example, the use of myth and story in religion, when studied, would indicate the underlying structures in a society "controlled" by religion. The focus of the movement was on the nature of underlying structures, or mechanisms in society.

Structuralists were less interested in subjectivism and history, i.e. the feelings of people and the story of their society. Their methods have been found to be useful in modern biblical studies (Barton and Morgan 1998), though weaknesses have been revealed in that area of study (Young 1990: 8–9).

POSTMODERNISM EMBRACES . . .

Postmodernism is disenchanted with world-views and has tended to become an umbrella term for other movements disenchanted with large-scale organization, like post-structuralism, or movements that prescribe and control the meaning and use of language, like deconstruction.

Post-structuralism

Post-structuralism as a movement grew out of the 1968 riots of French students and is largely a French phenomenon. It was pioneered in the 1970s by Foucault, who was less concerned with social organization than with the use of language. He embraced some Freudian ideas. He was concerned with power, for that was the basis for the construction of knowledge, which is a key feature in social relationships. He felt that Marxism was authoritarian and outmoded (Sarap 1993: 87). Foucault rejects the traditional conception of power as vested in centralized state control, from which it is filtered down to regional and local centres of organization.

Post-structuralists are very much against totalizing systems and reject the idea that there is any progress in history. They are aware of the increasing pressures towards conformity and are critical of them. They are obsessed with the subjective and the "small story", as opposed to the grand narrative. They find reason oppressive. They have moved away from scientific value-free perspectives towards a concern with the individual and human rights. They stress plurality and difference, the individual and the subjective. They have been more concerned with their ethical problems than with political action, and have often sought solace in aesthetics and religion (Sarap 1993: ch. 4).

Deconstruction

Deconstruction is another movement of the 1970s, with Derrida as the central figure. It sees itself as both the successor of existentialism and the major critic of structuralism. It challenges structuralism by questioning the reality of what is present. For example, a narrative, a letter to you, may talk about Plymouth, but because "Plymouth" means different things to different people, you cannot be sure of the reality indicated by the word. For instance, for one person it could be a place of happy memories where they had their honeymoon, while for another it was where they were arrested and sent to prison for twenty years. Plymouth means different things to different people.

Derrida argues that all western thought starts with a centre to every idea. The central idea of "Plymouth" is that it is a city in a specific location in Devon, nothing more.

Take another example, God. Many early church paintings of God, or Christ for that matter, picture the Deity as despot and judge. The central ideas are order and justice. These central meanings Derrida calls *logoi*. But central meanings can be swamped with associated meanings, as with the honeymooner's and the prisoner's idea of Plymouth above. Moreover, the central idea can change. For example, many regard the picture of God as judge as primitive and have replaced it with the central ideas of Father and Mother (Luke 15:11–31; Matthew 23:37; Psalm 36:7).

Let us examine a more technical example: *the signifier*, i.e. the narrative, is *a sign*. For example, the Gospel of Mark indicates what is *signified*, the narrative world of Mark, which is *the determinate referent*. It seems like jargon, does it not, but do note the use and value of the technical terms needed to make these distinctions. Now the deconstructionist emphasized *the signifier* – the narrative

> which is such only in relation to other signifiers, rather than in relation to a determinate signified, for the signified is itself a signifier in another sign. In other words, the text is a bottomless series of references to other references which never comes to rest in a real or determinate referent. Thus the text itself subverts the very meaning it creates. (Brown *et al.* 1995: 1159)

In practice this means that a text can be interpreted freely by a reader. The absurd logical consequence of this position is that it celebrates individualism, relativism and the breakdown of communally shared meanings. The end point of this process has been expressed by Lewis Carroll through the character Humpty Dumpty, who said, "When I use a word . . . it means just what I choose it to mean – neither more nor less" (Carroll 1998: 124).

However, more needs to be said. The longing for a central idea can lead to the creation of binary opposites. Derrida points out that for two thousand years, most of western culture has centred on Christianity and Christ. Derrida was born into an assimilated Jewish family in Algiers and grew up as a member of a marginalized dispossessed culture. So if Christians were central, this has meant that Jews and Muslims were marginalized. Whites marginalize blacks. If God is claimed to be male, women are marginalized, as feminist writers correctly emphasize. Statues can do the same thing, i.e. produce binary opposites and marginalize some people. A "white" Christ can marginalize a black Christian.

Now where these original central ideas, called by Derrida *logoi*, are the focal points of an extended narrative, Derrida regards them as *magical thought-terms*. This pejorative phrase needs to be replaced by the factual phrase that these central ideas were *the foundational thought-terms* at the centre of western thought, and, as such, have total legitimation. By that I mean that every system of thought has foundation beliefs or assumptions without which it cannot exist. It therefore has a legitimate right to them, and unhelpfully calling them "magical thought-terms" is quite misleading. Any system of thought needs rigorous rational criticism. Misleading terms indicate weak criticism. Had Derrida called the *logoi* "mythical thought-terms", that would have been acceptable, for it simply indicates the literary forms through which the *logoi* are expressed.

However, Derrida's criticism of the possible binary consequences of a central idea has two implications. First, such usage needs careful historical investigation, and where it has caused spiritual and political disempowerment, it needs political attention. Secondly, the exploration of the extrinsic religious orientation indicates that such a way of holding a religious belief system may be responsible for much of Derrida's legitimate criticism. This would give a boost to the need for the intrinsic religious orientation to figure more prominently, with greater lucidity and understanding, in religious and educational teaching in faith communities and educational establishments. The "intrinsic religious orientation" was discussed in Chapter 3, p. 45.

Derrida is clear that *logoi* are important, as they are the basis of the discussion of the philosophical foundations of different philosophical theories. They are also the important basis of our understanding of literal and metaphorical use of language.

The strength of Derrida's work is that it has highlighted the stipulative and normative nature of language use, with its social dangers. It highlights the fact that central ideas can change and/or collect subsidiary ideas, which may add to, alter or replace the original meaning of a word. In other words it indicates the central importance of words which express central ideas and their power.

However, to say that words cannot have a fixed referent, and that their meanings change according to what a reader brings to a text, is to retreat into the position of the nominalists, who assert that only words or names are real, and not that to which they refer (Audi 1995: 181–2). In everyday contexts, for postmodernists and the rest of us, words relate to specific objects, ideas and relationships. Examples include "instruction booklets, that come with household equipment, legal documents, personal letters conveying information, shopping lists or cookery books" (Barton 1997: 41).

Another important quality of postmodernism is concern for what "is excluded from or excess to the discourses of knowledge" found in universities and schools, and also discussion of religious themes, which will be discussed later in this chapter (Ward, G. 1997: 587).

POSTMODERNISM AND ITS WEAKNESSES

Postmodernism is a term which has existed for the last fifty years. It initially referred to a movement in architecture. In the last twenty years it has referred to experiments in literature, music and dance in different ways. The third use of the term began with Lyotard, who became Professor of Philosophy in Paris in 1968. For practical purposes here, the terms postmodernism and post-structuralism are taken to mean the same thing.

As we have already noted, Lyotard criticized and rejected all grand narratives or metanarratives like Christianity, Judaism, Marxism, Fascism, nationalism and humanism. All are criticized for exercising control, although postmodernists do not explain how religious or secular democracies are guilty of this charge. Instead of two cultures based on two opposing classes, there are many cultural, social, sport clubs and societies, reflecting many different kinds of groupings and allegiances. Lyotard thinks that the individualized and fragmented society is here to stay. He regards the different kinds of knowledge and their attendant language games as further evidence of the fragmentation in society. However, this illustrates a failure to see how language games are not rival interpretations of life but interrelate as integrating aspects of one "human language game", as we have seen earlier.

Lyotard favours "the little narrative". He is not even interested in the metanarrative of "the emancipation of the working class". He regards the "social bond" as constructed from the overlapping networks of different groups with the individuals as the nodes which mark the points of overlap (Lapsley and Westlake 1988: 206–7; Sarap 1993: ch. 6).

Postmodernism can be criticized for a number of reasons. For a start, many of its advocates assume that the grand narratives will break up, but no reasons are given why they will or should. They fail to indicate why we should reject all grand narratives, though they have no difficulty in disapprovingly pointing out the manipulation of power in Marxism and Fascism.

Advocates of postmodernism reject totalizing social theories, master narratives, as being too simplistic and ignoring differences in society.

They fail to recognize that the very concept of postmodernism presupposes a new grand narrative which is "little narratives rule, okay", which replaces existing ones like Christianity and humanism. Lyotard attacks philosophy as an imposition of truth, and says that histories, like grand narratives, are dogmatic and to be rejected.

Postmodernists take away the dynamic of the grand narrative, which has been the basis of the development of social thought. In its place they have few constructive suggestions and are unable to identify with the culture to which they belong and which is subject to their destructive criticism.

Their antagonism to totality and their advocacy of fragmentation and pluralism means a tendency to accept relativism: one small group is as good as another one. This is not moral sense in a community. Postmodernists sometimes fail to recognize the refined insights gained through religious communities and their developing understanding of the breadth and direction of religious revelation and experience, and the humanists' beliefs in the moral value of education and social progress.

Postmodernism offers no recognition of the strengths of critical realism in both the sciences and in Biblical Studies. For example, its dismissal of historical criticism as a search for what the Bible "really means" fails to take account of a number of factors operating when a believer reads the Bible.

First, the reader takes account of what the text says. Secondly, to consult a commentary is to search for an insight into what the writers originally meant. Thirdly, it is inevitable and necessary that a commentator brings not only their scholarship to a text, but their own religious experience (Barton 1997: 43). Fourthly, the commentator needs to bring their authentic contemporary religious experience to their work of discerning the contemporary meaning of the text (Richardson 1997: 75–6). Fifthly, it is recognized in psychotherapy that emotional or "felt meanings" become a means of knowledge when they are effectively symbolized, i.e. put into words (Watts and Williams (1988: 72–3). This is what is happening to each individual involved in the sequential process of understanding just described. Sixthly, every reader of the Bible needs to bring their religious experience to the text, which will add a particular nuance to that person's understanding (Howcroft 1998: 28–9). Readers will differ, and to that extent postmodernism does highlight that interpretations have a personal dimension. But that is different from the postmodernist radical claim that the interpretations form separate understandings rather than variations on shared understandings.

Persons involved at each level of this process accept that interpretation is involved in each stage of these activities. But that also means accepting the discipline of testing the meaning of the text and the nature of one's own experience against the shared insights of the religious tradition of a faith community.[1] This recognition of the centrality of emotional knowledge at the heart of religious understanding indicates that the task of biblical criticism and interpretation is very similar to the task of understanding music and its performance (Young 1990: 45–6).

The practice in scientific work is similar, where a research worker examines data in the light of existing scientific theory or interpretation and offers a modified interpretation which is tested by the scientific reader against his or her experience and interpretation, and is tested against the experience of the scientific community. This is the practice of critical realism, which is shared by both serious scientific thinking and practice, and serious religious thinking and practice (Sarap 1993: 187).

My final criticism is that "we urgently need to provide individual and social groups with public 'spaces', in which they can deal with subliminally felt experiences and learn to understand these experiences on a more conscious, critical level". Is this not a request for space in time, and the social opportunity to celebrate experiences of awe and wonder – to explore the shared religious and spiritual needs of what it is to be human? Dawkins certainly celebrates experiences of awe and wonder (Dawkins 1998:

17). Perhaps these continually upwelling human spiritual yearnings will assure the future of religious grand narratives, hopefully in dialogue. Indeed, we will later find that these upwelling spiritual aspirations are well in evidence.

For the last criticism I quote the philosopher of education, Dr Charles Bailey: "The essential criticism of at least some postmodernism, it seems to me, is that even post-modernists must necessarily presuppose the force and validity of reasoned justification if they want us to take any notice of their arguments."[2]

Postmodernism offers useful cautions to those who properly subscribe to a partic-ular grand narrative, so that such adherence must not countenance the exclusion of communities of goodwill who do not accept mainstream foundational views.

Part of the value of postmodernism is that it allows for the expression of what is unconsciously "repressed" in the subject disciplines which emerged through the modernist movement. For that reason, it focuses on what has been excluded from those major discourses of knowledge found in the universities and school curricula.

Postmodernism includes serious discussion of religious themes. Derrida, when exploring his notion of "difference" can write eloquently on Eckhart, a fourteenth-century German mystic. Levinas, a Jewish philosopher, "cannot separate God from his account of our . . . responsibility for the wholly other". Lyotard, in criticizing "totality" and highlighting what has previously been "unpresentable" has been drawn into analysing "the sublime" (Ward, G. 1997: 587). Ward notes that "the postmodern God is emphatically the God of love". Ward concludes his review of postmodern theology by quoting a comment of Hélène Cixous, a Jewish pupil of Derrida: "When I have finished writing . . . all I will have done will have been to attempt a portrait of God" (598).

But this is not the end of this discussion of the movements reviewed in this chapter which seem to be advocating relativism. The suggestion that what anyone understands of an object, a situation, a narrative, a person is true for them, and therefore knowl-edge for them, has been rejected here. This is because solipsism, individualism and autonomous subjectivism leads ultimately to a breakdown of communication and therefore of community. However, it does raise a very important related topic, namely, cognitive freedom, which must be discussed.

COGNITIVE FREEDOM

Cognitive freedom means the individual's freedom to know. The central argument advanced so far is that there are certain forms of knowledge, each with their own central concepts and distinctive logical structure, with criteria for relating cognitive freedom to experience according to the satisfaction of the relevant related community, such as a science, art, music or religion. Cognitive freedom includes emotional freedom (see Chapter 7).

However, the value of Derrida's work is to remind us of the extra complicating fact that, in the study of knowledge, the relevant community is not always the arbiter, in detail, of what counts as knowledge. Secondly, different subject areas offer differing degrees of freedom to an individual. Roughly speaking there is less room for freedom in the sciences and considerably more room in the arts, in personal knowledge and in religious knowledge. But first let us remind ourselves of the three-phase model of knowing we are using uniformly across all ways of knowing, or forms of knowledge. It starts with perception, which is then interpreted, and then evaluated or valued as to its significance, and then described.

Let us take some examples. Turn back to the discussion around table 2.1 (p. 28) of "what counts as a stool or a coffee table". In Chapter 2 the discussion centred around the critical moment when most people decided that at a certain point in the change of size and height of the object being viewed, it was no longer a stool but a coffee table. The point of that discussion was to illustrate that the perception of the object was being interpreted by allocating either the concept of stool or coffee table to it. But it is perfectly possible for two individuals to disagree. A particular model at the centre of the gradually changing models can be labelled and used by one person as a coffee table, and by another person as a stool. They each *know* what the object is: they also disagree!

Another example. In the early 1900s, bills through Parliament were either for England or for Wales and Monmouthshire. My father and his brothers and sisters were born and brought up in Ebbw Vale in Monmouthshire. The language spoken, the syllabuses taught in school, the flavour of community life then, were all totally English. My father always said he was an Englishman. My uncle, a champion gymnast of Wales, always said he was a Welshman. They each *knew* they were right!

Usually there is little or no freedom in matters physical or scientific. There is no cognitive freedom in knowing what time the train will leave the station according to the timetable. There is no cognitive freedom in mathematics or science in school.

There is greater cognitive freedom in music. When does sound become music or noise? There are a huge number of factors which lead one person to say this piece or that style of music is great music, and that they *know* that is the case (for them and the community which agrees with them). The same is true in art and in religion.

In personal relationships there is enormous cognitive freedom. You choose friends, like or dislike colleagues, and then someone gets you to see them in another light and you change your mind. Or you end up falling in love with someone you initially found noisy and overbearing! But you do *know* who your friends are, the others whom you have to treat very tactfully, and the one you want to marry.

Freedom and Determinism

So far we have explored postmodernism as the latest philosophical fashion which advocates relativism. Relativism holds that any world-view is just one among many, no better or worse than any other. Relativism must ultimately be rejected because it denies the legitimacy of a hierarchy of shared values. These are discerned and established by communities after a difficult and long journey, like faith, national or international communities and associations.

The search for a global hierarchy of values has begun through dialogue (Kung 2003: 8–19). For example, in international affairs the United Nations Association attempts to handle issues like global warming, third world health, debt relief and fair trade, while Inter-Faith dialogue takes place in such organizations as the World

Congress of Faiths and the World Government of Religions. We have explored the different degrees of freedom that we have in different areas of study. On the whole, there is less freedom to interpret in the physical sciences than there is in the humanities, which are concerned with personal and religious knowledge.

Now we need to draw together some strands of thought that have been frequently mentioned in different contexts. Often we have discussed materialists, like the biologist Dawkins. A central argument of many materialists is that ultimately everything – natural objects, creatures and humans – is basically constituted as different kinds of clusters of matter, only differing in the degree of complexity of the organization of each object or species.

An implication of this for many materialists is that all aggregates of matter, i.e. of different materials, behave in a way which is *determined* by built-in processes, e.g. of survival or reproduction. Extreme determinism holds that, for example, human behaviour is totally determined by our genes. Determinism as a theory is found in all fields of study. In religion, for example, Calvin, a contemporary of Luther and a leader of the Protestant Reformation of the sixteenth century, held that some people were destined to be saved and others damned. Hard luck on us if we were in the wrong group, we had no choice! This determinist doctrine is called predestination, i.e. our destination is predetermined.

Either form of extreme determinism or predestination – scientific or religious – regards human beings as programmed robots. Before showing the weakness of determinism as a description of what a *human being* is, we must recognize one fact. Namely, it is theoretically possible for a person to be totally indoctrinated and programmed so that they *do* behave as robots, like those charged with Nazi war crimes who tried to defend themselves by saying they were just obeying orders. We have already seen that existentialism recognizes this possible state of existence, of drifting along with the crowd, as inauthentic existence. Many existentialists regard inauthentic existence as being "less than human". We only become human when we exercise our sanctioned freedom, share in authentic existence, and aspire to become our "true selves".

Extreme determinism makes no sense to *Homo sapiens* because all forms of extreme determinism fail to recognize the ascending hierarchy of forms of knowledge, with the increasing degrees of freedom conferred by the ascent up through the higher forms of life.

Such a generalization, however, must be qualified. A human being's behaviour, as a baby or as a very old person, the former's powers being embryonic and the latter's fading rapidly, is very largely determined, or better to say limited, by food and appreciation. The greatest human freedom lies at the height of the development of a person's powers, which may appear at different times of life with different peaks for different skills or qualities developed. For example, one may peak as a tennis player at eighteen years, as a research scientist at forty years, and as author at eighty years.

We have already mentioned, when discussing cognitive freedom above, some examples to illustrate how the hierarchy of forms of knowledge affects the growth of freedom, which is only experienced by forms of life – the higher the form the greater the potential freedom.

Now we must look at the relationship between how far a person is determined, or limited, and how far a person has freedom. Clearly I cannot fly like a bird; I cannot change my age, my age changes me; my genes are inherited, and through them my intelligence potential, my temperament and other aspects of my character have certain "limits". Perhaps in my brain "there may be dedicated neural machinery in the temporal lobes concerned with religion", sometimes called the "God spot", though this is shown to be too simplistic.[3]

I am also influenced by social factors acting on me, both from my upbringing and since. These influences come from the ideas and beliefs passed on to me, from the way I am treated, loved or abused in my family, school, workplace, and from the social groups I inhabit. If I internalize all the views passed on to me, then they form my assumptions for living; they also provide the framework and therefore the limits of my outlook. I can change these assumptions, e.g. by throwing off the religious assumptions handed down by my parents and adopting materialist assumptions (Luscombe 2000: 116).

In other words there is no doubting the fact that every person is born within biological, psychological and sociological limits or constraints. Freedom is sanctioned choice (Walsh 1960), i.e. choice within limits which continually vary through time and space. And in different areas of knowledge we have seen different degrees of freedom. Now we have to nail down the nonsense of extreme determinism in relation to *Homo sapiens*, whether in science, theology, history, geography or elsewhere in the humanities. We will do this by showing that a driver in a racing car is structurally the same as a human being. Three situations will illustrate the range of responses for the human being and the racing driver.

A formula one racing driver in his car is similar to a human being when alive. A dead human being is like an empty car: it cannot "start", because it has no mind to tell it what to do and where to go. Both the "living car" and the "living person" are in a real sense psychosomatic unities. Both the car and the driver have a physical body; both have mind: the car has the driver and he has a working brain.

If a person's physical options are controlled by mechanisms of the physical body, behaviour is determined. If your arm is paralysed, it cannot have free movement: a case of physical determinism. If you brainwash your daughter to marry someone, her behaviour has been causally determined by your immoral behaviour: psychological determinism. But if your daughter chooses her husband, career, place of employment, between options she may be influenced by social factors from the past or she may react to them: she has freedom within limits, or sanctioned choice.

That is true for most human situations, for it is not causal mechanisms which determine action in these situations, but *human purposes and reasons* which exercise choice and implement action. This disinction becomes clearer in the case of the racing driver. Consider three races.

In the first race, the driver's car got a puncture which threw him off the track into the safety fence, out of the race: the mechanism determined the result in this case as there was no degree of freedom in which a decision could take effect, so it was a "robotic response": a case of mechanical determinism.

In the next race our driver saw a tyre burst in the car ahead, one of a series of driving situations which his training had prepared him for, and reacted to it automatically to ensure his safety. He had purposely conditioned his reflexes for such a situation, and one can say his response was predetermined. One could say it was a case of social and psychological determinism.

In the third race, the rules were clear. Drivers had only to drive, *nothing* else. Accidents would be attended *only* by the trained and equipped accident squad equipped with life-saving and fire-extinguishing equipment. Our driver was in second position, with the lead car far ahead of the field, when the first car crashed. The second driver knew from experience that the crashed car could very soon catch fire, much sooner than the accident squad could arrive. He stopped his car off the side of the track, ran over to the crashed car and quickly dragged the dazed driver out and away from it before it exploded. That was a free choice within the limits of time, space and options available – sanctioned choice.

If you listen to skilled commentators covering a motor race, they are usually quite clear about when there is a mechanical failure, or when there is human error, or skilled success. Mechanical failure is determined. Human performance is purposive: the result of free choice within the limits of the race.

Finally, let us repeat the point made earlier, that to call a member of the species *Homo sapiens* a human being is to award that person a diploma on the grounds that that person has begun the process of perceiving, interpreting and valuing. The end of the search to become mature has been expressed eloquently by Rudyard Kipling in his poem *If*, which was voted the nation's favourite poem in 1995, with twice as many votes as the runner up (Rhys Jones 1996).

> If you can keep your head when all about you
>   Are losing theirs and blaming it on you,
> If you can trust yourself when all men doubt you,
>   But make allowance for their doubting too;
> If you can wait and not be tired by waiting,
>   Or being lied about, don't deal in lies,
> Or being hated, don't give way to hating,
>   And yet don't look too good, nor talk too wise:
> If you can dream – and not make dreams your master;
>   If you can think – and not make thought your aim;
> If you can meet with Triumph and Disaster
>   And treat those two impostors just the same;
> If you can bear to hear the truth you've spoken
>   Twisted by knaves to make a trap for fools,
> Or watch the things you gave your life to, broken,
>   And stoop and build 'em up with worn-out tools:
> If you can make one heap of all your winnings
>   And risk it on one turn of pitch-and-toss,
> And lose, and start again at your beginnings
>   And never breathe a word about your loss:
> If you can force your heart and nerve and sinew
>   To serve your turn long after they are gone,
> And so hold on where there is nothing in you
>   Except the Will which says to them: "Hold on!"
> If you can talk with crowds and keep your virtue,
>   Or walk with Kings – nor lose the common touch,

> If neither foes nor loving friends can hurt you,
>> If all men count with you, but none too much;
> If you can fill the unforgiving minute
>> With sixty seconds' worth of distance run,
> Yours is the Earth and everything that's in it,
>> And – which is more – you'll be a Man, my son!

Even though I share the national judgement on *If*, I wish to make four comments. First, this is a national vote in favour of the argument that to call a member of the species *Homo Sapiens* a human being, or in his period language a Man, is to make a value judgement – to award a diploma requiring very stringent criteria. Secondly, a religious leader must face this test. Thirdly, I have met more women than men who would make me think of this poem. Lastly, it is clearly a goal of what it is to be a mature human being, freely chosen and pursued, a goal shared by the religious and the humanist.

Freedom and Licence

The "forms of knowledge" to which "freedom", used literally, belongs are personal discourse and moral discourse. The use of "freedom" in any other form of knowledge is analogical. When critics of zoos argue that lions and other wild animals should be returned to their natural environment, they quite rightly base their argument on "freedom". But "freedom" is used analogically when talking about animals, and means "without physical restraint".

Most human beings who are not behaviourists or materialists[4] distinguish between human "freedom", which is a moral and personal concept, and "licence". Existentialists, for example, regard freedom as the context in which a person can become his or her "true self " – a personal goal which is moral. To engage in murder, stealing and lying as a misuse of power for personal gain is not exercising freedom but taking licence, which all communities ban in law. Humans cannot, like animals, have "freedom without physical restraint", for that then becomes "licence". In exercising freedom as a human being one is engaged in the task of distinguishing between optional judgements or behaviours which favour human flourishing, of both the individual and the community.

In the *Concise Oxford Dictionary* these distinctions are clearly made. Thus the first use of "free" is defined as "not in bondage to another, having personal rights and social and political liberty". It is also used to define an action which is not illegal. In the dictionary the first use of 'freedom' is defined as 'personal liberty, non-slavery, civil liberty'. Mention is made of the four freedoms – freedom of speech and religion, and freedom from fear and want. These are all "freedoms", whose purpose is to provide space for human flourishing in its many life enhancing forms. All these 'freedoms' are concerned with human development in a moral, interpersonal and interdependent set of relationships in community, which are at the heart of spirituality and at the heart of "being human". The second and later dictionary definitions of "free" are analogical uses, like loose, unrestricted, not confined, unimpeded. Similarly other uses of "freedom" are analogical, like "ease in action", or "unrestricted action".

Freedom is the fruit of human knowledge and responsibility, a gift held out for us to grasp and use as we seek to develop global inclusive theories or "grand narratives". This will involve the different but interrelated modes of discourse as we seek to improve the quality of life in our global village in the twenty-first century.

RELATIVISM: A CONCLUDING COMMENT

Professor Hilary Putnam, the very distinguished American philosopher, says that "no matter how sceptical or relativistic philosophers may be in their conversation, they leave their scepticism and their relativism behind the minute they engage in serious discussion about any subject other than philosophy". Putnam advocates the maxim "that we should give weight in philosophy to the ways we think and talk and go on thinking and talking" (Putnam 1998: 135, 139). It seems that in everyday life, like the rest of us, they are critical realists.

Putnam makes a further point, "As a practicing Jew I am someone for whom the religious dimension of life has become increasingly important . . ." (Putnam 1998: 1). He has also reminded us that Wittgenstein had a very respectful attitude to religious belief, and described himself in conversation as having a "religious temperament". Both tried to defuse relativism and scepticism, which we explored in Chapter 5 and in this chapter, for they are what "keep us from trust, and perhaps even more important keep us from compassion" (179).

NOTES

1   Levels of interpretation are more fully discussed in Chapter 13.
2   A personal communication.
3   S. Connor, "'God Spot' Is Found in the Brain", *Sunday Times*, 2 November 1997. See Dr Fraser Watts's judgement in Watts, 2002: 78–82.
4   Behaviourists and materialists live in, and are self-locked in, the reductionist world of scientific discourse, denying the validity of all higher forms of discourse. Their world-view is scientism, an ideology derived from, but nevertheless separable from, scientific methodology. Scientific methodology is accepted by all human beings who hold a humanist or religious world-view which is holistic and concerned with the integrity and interrelationship of all forms of knowledge.

REFERENCES

Audi, R. (1995), *The Cambridge Dictionary of Philosophy*, Cambridge: Cambridge University Press.
Barton, J. (1998), "Biblical Commentaries", *Epworth Review* 24 (3), 35–6.
Barton, J. (ed.), (1998), *The Cambridge Companion to Biblical Interpretation*, Cambridge: Cambridge University Press.
Brown, R. E., Fitzmyer, J. A. and Murphy, R. E. (eds), (1995), *The New Jerome Biblical Commentary*, London and New York: Geoffrey Chapman.
Carroll, L. (1998), *Alice's Adventures in Wonderland*, London: Macmillan.
Dawkins, R. (1998), *Unweaving the Rainbow*, London and New York: Allen Lane, Penguin Press.
Howcroft, K. G. (1998), "Reason , Interpretation and Postmodernism – Is There a Methodist Way?", *Epworth Review* 25 (3), 28–9.

Kung, H. (2003), "A Global Ethic: Development and Goals" in A. Race (ed.), (2003), *Interreligious Insight: A Journal of Dialogue and Engagement*, Vol. 1, No. 1, pp. 8–19, c/o World Congress of Faiths, 2 Market Street, Oxford OX1 3EF.

Lapsley, R. and Westlake, M. (1988), *Film Theory: An Introduction*, Manchester: Manchester University Press.

Luscombe, P. (2000), *Groundwork of Science of Religion*, Peterborough: Epworth Press.

Peterson, M., Hasker, W., Reichenbach, B. and Basinger, D. (1991), *Reason and Religious Belief*, Oxford: Oxford University Press.

Putnam, H. (1998), *Renewing Philosophy* (1992), Cambridge, MA: Harvard University Press.

Race, A. (ed.), (2003), *Interreligious Insight: A Journal of Dialogue and Engagement*, c/o World Congress of Faiths, 2 Market Street, Oxford, OX1 3EF.

Rhys Jones, G. (1996), *The Nation's Favourite Poems*, London: BBC Worldwide.

Richardson, N. (1997), "Biblical Interpretation and Christian Experience", *Epworth Review* 24 (1), 75–6.

Sarap, M. (1993), *Post-Structuralism and Postmodernism*, New York and London: Harvester Wheatsheaf.

Walsh, W. (1960), *The Use of Imagination*, London: Chatto & Windus.

Ward, G. (1997), "Postmodern Theology", in D. F. Ford (ed.), (1997), *The Modern Theologians*, 2nd edn, Oxford: Blackwell.

Watts. F. and Williams, M. (1988), *The Psychology of Religious Knowing*, Cambridge: Cambridge University Press.

Watts, F. (2002), *Theology and Psychology*, Aldershot: Ashgate.

Young, F. (1990), *The Art of Performance: Towards a Theology of Holy Scripture*, London: Darton, Longman & Todd.

9 Religion and Transcendence

This chapter explores some of the many definitions of religion from different disciplines which are interested in religion, and provides an explanation of the different uses of the word 'transcendent'. Subsequently we shall be able to investigate the links between religion, transcendence, spirituality and humanism.

Our exploration will illustrate that human beings have almost complete freedom in interpreting religion and transcendence responsibly, from atheism through agnosticism to religious belief when considering the "top end", i.e. the inclusive end, of the hierarchy of knowledge. So all "stances for living" or "lifeways" include all forms of knowledge, with the obvious exception that secular lifeways do not recognize religious knowledge. It is worth bearing in mind that *lifeways* at the top end of the hierarchy of knowledge have "many degrees of freedom" when interpreting. Compare them with those forms of knowledge with "very few or no degrees of freedom" available in interpreting some areas at the "bottom end" of the hierarchy of knowledge, like empirical knowledge – the things we experience through our five senses. These separate areas of empirical knowledge include distinct skills like weaving, plumbing, sailing, carpentry, each with their own vocabularies. It all counts as knowledge, justified by the criteria related to the appropriate mode of discourse, or form of knowledge.

"TO BE" AND "TO BECOME"

In Chapter 5 we found, through Professor Rose, that organisms, including humans, can only be adequately understood by taking account of five levels of explanation and interactive processes. The development of a human lifeline, like that of all organisms, involves the capacity to "build, maintain and preserve itself . . . [that is, it has] simultaneously *to be* and *to become*" (Rose 1977: 18, my italics). We must be careful about the use of "to be" and "to become" here. Rose is only speaking as a biologist about the development and maintenance of an organism – *Homo sapiens*. That only involves survival of the individual and the species.

But just as the biologist requires at least five levels of explanation to give an account of the species, a student of modern society needs to understand all the levels of understanding outlined by Polanyi, Hirst, Phoenix and Peacocke. These are summarized in the table included in Chapter 6 to describe what a person is. To be and to become a *person* or a *human* being – to become authentically yourself – is a cultural, moral and spiritual achievement.

Then you are awarded the diploma, which has moral, spiritual and many other dimensions, of being a *person* or *human* being. That too, like looking at a coffee table, involves perception, interpretation and evaluation as to its significance – it is a value judgement, that you have risen above the biological level of existence to the human level of being a person.

This involves Macmurray's half knowledge – aesthetics, and 'complete' knowledge – personal knowledge, which includes all the 'lower forms' of knowledge. Polanyi, Hirst and Phoenix, besides accepting aesthetic and personal knowledge, include moral and religious levels of feeling, knowing, understanding and relating to others, though in different ways. Thus there is an enormous difference in the meaning of 'to be' and 'to become' or 'being' and 'becoming', held by a materialist biologist like Rose (Rose 1977: 306), a humanist like Fromm (Fromm 1978), and a Christian theologian like Paul Tillich. Tillich talks of God as *Being*, and not as *a* Being: that is in the mode of religious discourse. However, when Tillich challenges an individual to explore the religious life of commitment with the *Courage to Be* (Tillich 1962), he is saying that a person's courage is the power of mind to overcome fear and anxiety, caused by an awareness of one's possible non-being. A religious or mystical experience can be a source of the courage to be – that is to share as a follower in the power of Being, empowering one to become one's true self (Tillich 1962: 43–4, 153–6, 173).

We must then take careful note of how the *uses* of these same phrases, 'to be' and 'to become' have quite different *meanings* when found in three different forms of knowledge- biological, personal and religious. For example, Tillich, focusing on God-talk, says, "Theology, when dealing with our ultimate concern, presupposes in every sentence the structure of being . . . " (Tillich 1957: 24), and as we have just noted earlier, he talks of God as *Being* itself. So personal discourse is a primary mode of language *used* in religion figuratively or analogically. When talking of the individual seeking the courage to be, the language *used* is sometimes just personal or both personal and analogically religious.

We will return to this point of the relationships of the modes of discourse or forms of knowledge to each other, with some practical examples, analysed, when we examine some individual religious experiences later.

OBJECTIVE AND SUBJECTIVE WRITERS

It is necessary at this point to encourage the reader to be alert to the viewpoint or assumptions of a writer on religion (or any other topic for that matter). There is a fairly wide range of approaches.

At one end of the spectrum are social scientists, theologians and philosophers who study the experiences of individuals and societies, who approach the material they are studying as reporters, accepting it at face value, reporting the claims of the individuals and societies without influencing it with their own personal opinions. These writers 'look in from outside' to explore how the beliefs of the individual, group or society influence the life of that individual, group or society, and the way they express their beliefs in their own ways and with their own feelings. The writer tries empathetically to communicate their feelings and ideas. S/he seeks to represent them 'photographi-cally'. This method of study is called phenomenology (Smart 1997: 1–2).[1] We will explore this later in this chapter. The study of history can take the same approach to, for example, the history of the Second World War. Phenomenological studies try to be descriptive and non-judgemental.

There has been a development since the 1939–45 war in some subjects, such as history and theology, towards developing comparative critical studies from a global perspective. History may be more advanced than theology in attempting to view their subject from a global perspective, though there has been an increasing exploration of a global theology.[2]

At the other end of the spectrum there are writers who seek to interpret what they study on the basis of the assumptions which underlie their own outlook on life. Such writers take a subjective position. There would be considerable differences between nationalistic histories of the 1939–45 war written by an English, or an ex-Nazi author. The subjective view may be expressed by an individual personally, and/or as the repre-sentative of a community.

Likewise, there will be considerable differences between a conservative Muslim, an orthodox Jew and a conservative Christian writing about the nature of God. These would be theologies produced for use inside a particular group of Mosques, Synagogues or Churches. Histories and Theologies like these tend to be inward looking and exclusivist: these writers think that they express 'true belief' and the others writing contrary histories or theologies are 'outside their communities'.

There is however an inter-mediate position between 'objective' and 'subjective' perspectives which can be called the 'inter-subjective' perspective. "Recent scholar-ship has found useful the notion of 'a community of readers'. The notion implies that Christians together are responsible for making meaning with the Scriptures."[3] The 'inter-subjective' method can be used in any community.

A particular group of scholars who need mentioning again, as they have been referred to in earlier chapters, are those who believe that science is the only source of knowledge. They hold that matter is not just the basis of all physical and living forms, which most people accept, but that all forms of individual behaviour or social organi-zation are *nothing but* chemical or biological interactions. Such scholars are *reductionists*, that is they seek to reduce the level of understanding to a level lower down the hierarchy of knowledge.

We have mentioned reductionists before in Chapters 2, 5, 6, and 7. They are not concerned with the complex inter-relationships of all the different kinds of knowing, which is a view they do not hold. Rather, their view is based on a materialist theory of knowledge, such as "all statements of what is the case must be tested scientifically with evidence, to see if they can be proved to be true or false."[4] So all matter must be analysed to the smallest components of which it is constituted, as that is what they regard as most 'real'.

USES OF RELIGION

The word 'religion' comes from the Latin word *religio* which means obligation or bond.[5] Religion, then, is about ultimate commitment. So Ferre concludes that, "Religion is one's way of valuing most comprehensively and most intensively" (Ferre 1968: 69). But this definition must now be rejected because atheists and agnostic humanists rightly claim that they have a way of valuing that is most comprehensive and intensive. They do not like to be called 'religious' for that is what they are rejecting.

We will here *use* 'religion' in accordance with the primary contemporary *use,* as defined by the Concise Oxford Dictionary: "Religion – particular system of faith and worship (*the Christian, Muslim, Buddhist religion).*" One could now modify Ferre's definition and say that ' a lifeway, or stance for living, is one's way of valuing most comprehensively and most intensively.'

The primary, literal, *use* of 'religion' then needs to be defined as related to world views which are identifiable religions, historical religions like those of Greece, Rome and Egypt, and modern religions like Buddhism, Christianity, Hinduism, Islam, Judaism, Sikhism, and Chinese and Japanese religions. This is the core meaning of the term.

However, before we turn to definitions of religions, we need convincing, if we are from non-religious backgrounds or persuasions, that religions are earthed in daily life, concerned with questions like, 'what is it to be a *human* being?' 'what is my place in the universe?' Religions must consider and treat what can go wrong in life, as well as consider all kinds of situations and experiences human life encompasses. An imaginative study thirty years ago did earth the notion of religion in daily life, to be described in the next section: implicit religion. Otherwise, why shouldn't non-religious people laugh off religion as 'just like astrology' – pure superstition? After all, don't materialists and humanists (a person can be both) have legitimate complaints about 'religion' when it often looks as though all religious people inhabit the world of Alice in Wonderland, like the White Queen who said, " . . . sometimes I've believed as many as six impossible things before breakfast"! (Carroll 1998: 100).

IMPLICIT RELIGION

The Revd Canon Professor Edward Bailey, when a curate in the 1960s, used interviews to explore what really mattered to people. He used four definitions to embrace implicit religion: 'commitments', 'integrating foci' 'intensive concerns with extensive effects' and 'human depths' (Bailey 1998: 22–4).

Bailey found that for most people their sense of self is a sacred focus, held in awe. One commitment was the daily round of tasks – the ritual dimension. People were very pleased with any day filled with good relationships and jobs well done – the ethical and social dimensions. Another commitment might be a regular drink at the local public house, with its ritual dimension – having a drink. The social dimension found expression in communal solidarity. The mythic and ethical dimension found expression in the idea of "being a man" (Bailey 1998: 48–58).

The key point to note about this analysis is that it highlighted the fact that everyday life has its rituals, e.g. holidays, birthday parties, wedding anniversaries, a regular Sunday lunch. Everyday life has its myths, e.g. for a farming family, what it means to be a farmer expressed through past and present stories of farming which contain and express the beliefs or doctrines which are the teachings about what good farming is about. The social dimension finds expression in activities that farmers share, like farming organizations, meetings, clubs, shows and markets. One expression of the ethical dimension is the handshake clinching a deal, prior to written confirmation. The experiential dimension of farming is ultimately the key source of the other dimensions already mentioned.

WHAT IS A RELIGION?

We must first look at some of the different definitions of religion and the reasons for them. A definition needs to outline the borders of the area of study concerned. This indicates what needs to be explored, interpreted and explained. The unity of the subject under consideration can then be explored. We next need to notice that a definition will serve the interests of the person offering it, so, for example, different kinds of social scientists will each offer very different definitions: a psychologist is concerned with the effect of religion on the individual; the social psychologist is concerned with the place of the individual in society; the sociologist is concerned with the viewpoint of society or the social group; and the philosopher stands back from all these groups and examines the underlying assumptions of all subject areas.

Modern scholarship focuses attention on five styles of the definition of religion: *experiential, substantive, functionalist, family resemblance* and the *essentialist approach* (Clarke and Byrne 1993: 6–7).

An *experiential* definition focuses on some common form of experience which is central to religion.

A *substantive* definition is usually focused on the main substance or area of religion, often thought to be beliefs and practice, e.g. "a particular system of faith and worship".[6]

A *functionalist* definition is concerned less with *what* is believed, and more with *how* one believes. In other words, what role does religion have in the lives of people? Pragmatism uses this sort of definition: religion is what works for the benefit of all people, so it must be true (see Chapter 3). Many of the social sciences approach religion in this way, focusing on how religion meets the needs of the individual, the individual in society, or the needs of society.

The *family resemblance* method avoids the preoccupation with necessary and sufficient conditions of what constitutes religion found in the preliminary and final definitions of religion of the three styles of defining religion just described. It is called the phenomenological study of religion and will be outlined below.[1]

The *essentialist approach* asks questions like "Is there an inner core of religion shared by all religions?", i.e. "Is there an essential structure shared by all religions?"

Before we look at these in turn, it is worth noting that these five perspectives can sometimes overlap when considering the approach of a particular writer.

Subject-based views of religion

Religious experience is fundamentally the experience of an individual and as such is the province of the biographer and the psychologist. William James, the father of the modern study of religious experience, offers one psychological view which centres on the individual's experience. He suggests that religion refers to the feelings, experiences and acts of individuals when they are alone and feel themselves to be in the presence of God.[7]

A social-psychological view is one from the perspective of the individual in society. From this perspective, here is a definition concerned with the function of religion; it focuses on the effects of religion. Religion is defined as "whatever we as individuals do to come to grips personally with the questions that confront us because we are aware that we and others like us are alive and that we will die." (Batson *et al.* 1993: 8). This focuses on questions concerning meaning like "What is the meaning and purpose of life? How should I relate to others? How do I deal with the fact that I am going to die? What should I do about my shortcomings?" (9). While religious belief is greatly influenced by our social environment, we need also to take account of the "depth of the individual's psyche to uncover the source of religion" (51).

Durkheim, the father of the sociology of religion, has defined religion as a system of beliefs and practices related to sacred things which unites those who accept them into a moral community called a Church.[8] In this social definition, religion is regarded as society deified. This definition implies that society is the object of worship, not God; Durkheim thought that religion is a purely human construction, which makes him a *reductionist*, as he reduces all religious ideas to functional ideas to promote harmony in society. In other words religious claims are really social facts, nothing else.

A modern sociologist, Peter Berger, takes the view that "society defines us, but is in turn defined by us." (1963: 128–9). The second emphasis here is expressed when he explores religion from an anthropological starting point and identifies five signals of transcendence (Berger 1971: ch. 3), experiences which he interpreted as manifestations of the divine. These will be discussed later in the chapter when we consider "transcendence". Berger recognized the existence of pluralism, a range of competing outlooks on life, including individualistic outlooks. He saw that in the modern situation pluralism undermined religious traditions (Berger 1980: xi). For that reason he suggests that there are three possible methods of upholding a religious tradition: *deduction, reduction* and *induction*.

I   The method of *deduction* starts by assuming the authority of a tradition, and then one *deduces* what is true from the tradition. If Jews accept the authority of the Torah, or Christians accept the authority of the Bible, they turn to those texts to answer questions concerned with the nature and purpose of what it is to be a human being.

2   The method of *reduction* is the process of secularizing the tradition, which is what Durkheim did when he affirmed that religious claims are really the claims of society and *nothing but* the claims of society.

3   The method of *induction* begins with individuals having special experiences which they intuitively interpret as religious experiences. In other words, one

infers the existence of a divine presence as a result of reflecting on one or more important experiences of a particular kind (Berger 1980: 136). This inductive model of sensing religious experience is central to all the great religious traditions of the world. In Judaism the experiences of Abraham, Moses and prophets like Amos, Hosea, Micah and Jeremiah were central in the forming of the tradition and the Scriptures. In Christianity the experience of Jesus is central, and in Islam that of Mohammad.

Induction, the method of reflecting on experience, was an important model for those engaged in Christian theology in the nineteenth century, and has been used in Protestant theology ever since (Schleiermacher 1988). Its thesis is that traditions, scripture and rituals emerged from, and are based on, exemplary religious experiences.

It is now possible to return to Berger's first point, "society defines us, but is in turn defined by us". Berger shows how this is true of religion. The tradition defines us, which in turn is defined by the religious experiences we have which break into everyday life (Schleiermacher 1988: 47). Here, Berger is referring only to those experiences which in some way change or refine the direction of religious social aspiration. This illustrates how the structure of a belief system, which includes *experience, tradition, scripture* and *reason*, can develop and change. Berger calls his belief system a Nomos.

As the result of social processes, the individual becomes a person, attains an identity and carries out the projects which constitute their life. A society is formed through turning the physical and mental acts of people into objects or events, e.g. writing a newspaper or giving a concert. The products of a person's work become a reality outside the individual. The demand for these products arises from the needs of the individual and the society in which they live, and they contribute to the society.

The society of which they are a part transcends the individual. A society is *sui generis*, i.e. it cannot be reduced or separated into its constituent parts without destroying the concept of "a society", which is a unity greater than the sum of its parts. Children become products of, and therefore members of, society by internalizing the patterns of activity and attitudes around them. A society is held together by a Nomos, a fabric of meaning which may be religious or secular e.g. Christianity or humanism, expressed in stories which describe the way the world is seen or understood, e.g. Genesis 1–3. This fabric of meaning, frequently expressed in cultural and religious stories, transcends the individual's life and places it within society (Berger 1973: 62).

Berger defines religion as "the establishment through human activity of an all-embracing sacred order which will be capable of maintaining itself in the ever present face of chaos" (1973: 59). This provides a religious Nomos or fabric of meaning for a society which embraces it or accepts it as the "community's account of the way the world is".

It will be a useful exercise to compare the views of two philosophers of religion: Feuerbach, a late nineteenth-century German materialist philosopher who was a major influence on Marx, and Keith Ward, the Emeritus Regius Professor of Divinity in Oxford.

Feuerbach defined religion in a reductionist way by saying that theology is disguised anthropology. In other words, talk about what God is like is really disguised talk about what humans are like. He regards religion as a form of self-consciousness. "Religion being identical with the distinctive characteristics of man, is then identical with self-consciousness – with the consciousness man has of his own nature." (Clarke and Byrne 1993: 21).

Ward provides a philosophical definition of God which expresses well what the 69 – 95% of the English-speaking populations believe on each side of the Atlantic. "The educated theist sees God as a self-existent being of supreme perfection, the source of all other beings which are generated for the sake of their goodness" (1996: 96).

Different Kinds of Definitions of Religion

We have already noted that there are different styles of definition of religion. Those mentioned included substantive, experiential, functionalist, family resemblance and essentialist definitions (Clarke and Byrne 1993: 6, 16).

(1) A *substantive* definition outlines the limits of existence of what is defined and focuses on its content. One example is that of Spiro, who regards religion as an institution which is formed by groups who interact together with whatever they consider to be the Divine (see Spiro's definition of religion in Clarke and Byrne 1993: 6). The aim here is to focus on the area of corporate worship and service of God.

(2) The famous *experiential* definition of religion stated by William James has already been quoted. Another example is that of W. L. King, who attempts "to identify the religious via the 'depth dimension in cultural experiences at all levels'" (quoted by Clarke and Byrne 1993: 6).

(3) *Functionalists* focus on how people believe and its effects, rather than on what they believe. One such definition defines religion as "any system of thought and action shared by a group which gives the individual a frame of orientation and an object of devotion" (Clarke and Byrne 1993: 7). Pragmatism, outlined in Chapter 3, is a good example.

Some scholars note that "the three approaches listed so far all share one common assumption: that we ought to seek to lay down necessary and sufficient conditions for the use of the word 'religion'" (Clarke and Byrne 1993: 7). A necessary condition for being a member of a theistic faith would be "one must believe in God". An example of "sufficient conditions" could be a Christian minister who said that one did not need to believe in all the clauses of the Nicene Creed to become a church member. One could be a member by believing the Apostles' Creed, or simply that "Jesus is Lord", i.e. he is "the human face of God", guide, inspiration, friend and a means of reaching God's forgiveness.[9]

(4) The *family resemblance*, or *phenomenological* approach to religion, denies the need for necessary and sufficient conditions. This is a valuable method for studying the structures of different World Faiths and secular lifeways, and exploring resemblances between them. Professor Ninian Smart's analytic model of religion, developed over thirty years, contains seven dimensions:

❖ The *doctrinal* or *philosophical* dimension outlines the beliefs central to a faith.
❖ The *mythic* or *narrative* dimension outlines the stories or narratives of a faith.

✧ The *ritual* or *practical* dimension includes communal activities like worship and meditation.

✧ The *experiential* or *emotional* dimension focuses on religious or transcendent experiences.

✧ The *ethical* or *legal* dimension lays down laws or principles of desirable behaviour.

✧ The *material* or *artistic* dimension is seen in the design of forms of worship and buildings.

✧ The *organizational* or *social* dimension: trained leaders are needed for all these activities (Smart 1997: 10–11).

This approach is essential to every member of a multi-faith society, like Britain or the United States, for each person needs to be not only a member of the lifeway in which they are brought up or adopt, but also to inhabit the phenomenological approach to lifeways. This will promote respect, friendship, dialogue and understanding of those who inhabit other lifeways. This has been the policy of religious education in state schools in Britain since the 1970s. So Britons under forty years of age have some knowledge of this and for most over forty this is all new.

While the "family resemblances" approach to defining or describing religion is concerned to offer an inclusive and comprehensive picture, another approach asks whether there is something which lies at the heart or core of one and all religions.

(5) This fifth approach is sometimes called *the essentialist approach* to the study of religion. In this approach scholars seek to answer the question, What is the core or essence of a religion? One good example indicates that religion rests on a central emotion, *awe*, which is the focus of a central activity, *worship* (Wilson 1971: 33). Rudolph Otto provides a more complex, but better-known example. The essence of religion is its inner core: "There is no religion in which it does not live as the real innermost core . . . " (Otto 1959: 20). That core was the "numinous", the awe-filled sense of the Sacred, which Otto described as the "tremendous and fascinating mystery" which can be stirred into life in the person. This approach has been criticized because it is detached from any specific theistic belief; alternatively it can be attached to any theistic belief.[10]

Critics of the essentialist approach often prefer the "family resemblance" approach, but Smart is an example of one who combines both. He assumes that experience finds expression through the other dimensions mentioned.

An important portrait of religion is given by Bishop John Robinson in his famous book *Honest to God*; he states, not that "God is a person", but that "God is personal". This means that "reality at its very deepest level is personal, that personality is of *ultimate* significance in the constitution of the universe, that in personal relationships we touch the final meaning of existence as nowhere else". Robinson continues:

> To believe in God as love means that to believe in pure personal relationship we encounter, not merely what we ought to be, but what is, the deepest, veriest truth about the structure of reality . . . Belief in God is the trust, the well nigh incredible trust, that to give ourselves to the uttermost in love is not to be confounded but to be 'accepted', that Love is the ground of our being, to which ultimately we "come home". (Robinson 1963: 48–9)

TRANSCENDENCE

Uses of Transcendence

In Chapter 1 we explored Michael Paffard's account of the effects of his teaching of Wordworth's poetry, which expressed the poet's experiences of nature as a result of roaming in the Lake District in England. Paffard's students described similar experiences, which he summarized as "transcendental experiences".

He explained that these transcendental experiences were "outside of " or "greater than" normal experience, being of an intimate spiritual relationship with nature or with a person. Paffard speaks as an agnostic, so for him transcendence indicates an inclusive whole in which a person experiences nature in a profoundly spiritual way. It can be bipolar: experience = person + nature; it is sometimes a mystical union with nature. Transcendence means "rising above" both the person and nature, and is thus an experience of self-transcendence.

Self-transcendence is experienced when a person seems to be lifted out of herself/himself into a relationship which transcends individuality, e.g. feeling part of a beautiful view; feeling love for all people as a result of love for one's spouse (for all the world loves a lover). That person has the feeling that within the present situation God is experienced as an all-encompassing Presence.

Transcendence is mainly used by religious people to indicate a key quality of the nature of God. People who believe in God affirm that God exists as *Being* who transcends time and space yet is available in the depths of a person's being.[11] Professor Keith Ward pictures God as perfect, eternal, transcendent, and so above and outside of time, and yet is involved in time, able to respond to the free acts of human beings. God, as Creator, is the all-powerful and all-knowing ground of all creatures and human beings.

Aspects of God's transcendence are disclosed through, for example, our understanding of the order in creation, illustrated in the pattern of cause and effect. Such understanding rightly evokes "appropriate attitudes of dependence, reverence, awe and thankfulness in the believer" (Ward 1974: 158). One does not try to deduce from existing facts and relationships that God exists; rather one says that the experience of God as the Being who transcends the world, life and relationships is a dimension of personal consciousness which cannot be reduced to anything else (80).

There is another aspect of the concept of God which relates to transcendence. One often talks of the 'transcendent depth' of life, indicating its supreme *value*. For example, one speaks about two people having a deep and rich relationship. "Deep" here is used to mean of very great value, so to talk of the "transcendent depth" of life is to talk of its sacred value. God is the greatest value humans can hold, and humans are of the greatest value to God (88).

Summarizing the use of transcendence as an attribute of God, we have seen that it relates to God in two ways: to the *Being* of God; and to the *value* of the world and life which exists within God's shadow (Ward 1974: 83).

A particular use of transcendence was suggested in the 1970s, and was called the way of transcendence, which involved following Jesus without necessarily believing in

God. It was suggested to those who were attracted by the life of Jesus but who found it difficult to believe in God. "Jesus is the very incarnation of the way of transcendence" (Kee 1971: 211).

Transcendence is often used to indicate moral commitment. Kant (1724–1804) accepted that the two pieces of evidence for the Transcendent were the "starry heavens above" and "the moral law within". Kant's formulation of the moral law is based on the Categorical Imperative, which is what you *must* do, not because of the consequences, but because *you believe it is right* and it is your *duty* to do it. He said, "Act as if the maxim for your actions was to become through your will a universal law of nature" (Vardy and Grosch 1994: 69). For Kant the moral imperative was autonomous and unconditional (Kant 1960: bk II, ch. 2, ss. 4 and 5). This is because Kant believed that "a good will" alone creates virtue through thought and action. To be moral one must have freedom of will to choose to make or not to make a moral judgement, or to do or not to do a moral act. However, everyone wants to be happy, but this is often not possible for the person whose intent is "duty". As the greatest good (*summum bonum*) is virtue combined with happiness, Kant assumed that God existed and that there must be life after death, because these two assumptions provided the basis for the only situation in which a just God could ensure that a virtuous person was ultimately happy (Acton 1970: ch. 10).

However, Paul Tillich disagrees with Kant's view that morality is autonomous, i.e. freely chosen of one's own will. He argues that one aspect of the moral imperative "is the command to become what one potentially is, a *person* within a community of persons" (Tillich 1969: 11). He then argues that the unconditional character of this duty is religious (15). The will of God then becomes for us our essential being with all its potentialities, which are created by God, who declares them to be "very good". Kant thought of the "will of God" as something external that an individual would have to obey, thus making a person's commitment to virtue heteronomous, which means under the law of another, like a child under parents' rules.

Tillich holds that the will of God is not an external will imposed or embraced, but the individual's inner drive to become their true self. It is an awareness of belonging to a dimension that transcends our limited freedom. To fail to will this is to leave an unsatisfied spiritual hunger, or even to risk the possibility of personal disintegration. It follows that because one is human one frequently falls short in this and many other matters. "The moral imperative" makes demands of us and cannot accept our falling short; it offers no forgiveness and acceptance. Only grace offers forgiveness and acceptance (Tillich 1969, ch. 3).[12]

This argument of Paul Tillich shows how a successful case can be mounted to claim that the ultimate moral demand on us is a religious requirement, a demand from the *transcendental voice within*. In other words, religion is the foundation of moral duty and action.[13] When I first read Tillich's very small book, I thought he was cheating: suddenly moral language becomes religious language – cheat! But I read it again and kept thinking! When I first saw a dragonfly emerge from a grub on a reed in a pond, I did not believe it! I thought biology was about gradual change! A grub changing into a butterfly is actually a metamorphosis, a total change in the level of being.

For some religious thinkers the moral categorical imperative, when uncovered, is a voice from the Deep Within, the voice of God. Yes, that is an interpretation: we now know that all judgements are based on interpretation. Only you can recognize it, just as only you can recognize toothache! Only you can decide when a sound is a noise or music! This shows the emergence of one level of knowing from another, the emergence of the religious form of knowledge from the moral form of knowledge. But this is only *one* source of human knowledge of God.

For Kant, the categorical imperative, the moral demand, is a signal of the creativity of Transcendence: it is a "well pleasing to God" signal from without.[14] For Tillich the moral voice within the individual is the Transcendent voice within. I prefer to talk of the Transcendent voice within as a Divine resource – "the voice of a Person waiting to be called upon" by an individual in their upward moral struggles towards authentic being.

This debate cannot be left without adding comments from John Hick. He recognized that Kant's great contribution to our understanding of morality is that it is "based in the structure of our human nature" (Hick 1989: 98). This means of course that it is an aspect of our human nature that "generates the invisible dimension of moral value" (98). It follows then that if moral awareness is a dimension of what it is to be fully human, it is open to both a religious, but also a non-religious interpretation.

So, one key point in this discussion is that morality, whether conceived as autonomous (Kant), as "an element of" or "one voice within" religion (Tillich), is a dimension of life which is self-transcendent. Those who hold a religious or a humanist world-view can agree to that.

However, I find that the human drive for the greater good, which transcends the goal of personal advantage or personal happiness, is a community goal that, for the humanist, is a desired *consequence* as the necessary context for peaceful life in a community. But the moral imperative discerned through reasoned analysis of the general good is intrinsically good and needs no further justification. So the moral dimension of human nature can be seen as one of the gifts of God in his work of creating the world, or it can be a dimension of human nature which is able to seek a Divine relationship leading to a quality of life which is a foretaste of 'life in the kingdom of God' in this life or the life hereafter.

Thus God's gift of moral awareness needs space to work in, and that space is the complementary gift of human freewill, for the moral imperative can only operate in an environment where freedom of action is possible, otherwise it is responding to a desired, imposed or expected consequence.

Another use of transcendence refers to naturalistic phenomena, i.e. natural things and events, which are expressions of self-transcendence that point beyond self towards God. Two writers who use the concept in this sense are Peter Berger and Keith Ward.

We have noted earlier in the chapter, when discussing subject-based views of religion, that Berger, as a sociologist, argued that society is a human product. Berger, as hinted earlier, discerned five "signals of transcendence" in society: "By signals of transcendence I mean phenomena that are to be found within the domain of our 'natural' reality that appear to point beyond that reality" (Berger 1971: 70).

They are, first, based on the recognition of, for example, the consistent pattern of order found in the presence of love, loyalty, hope and faith in family and community life, which leads to "the argument from ordering", e.g. a mother reassuring a child awake in the night that everything is all right. This is the belief that the created order of society paralleled a divine order which transcended it and supported it.

The second signal is children's play, which, because of the recognition that the intention of play is "joy", an eternal quality, leads to the development of "the argument from play" that the game in time shares this eternal quality of life – joy – which is a characteristic of the Divine.

The third signal is "hope", for human life is always focused on the future: the desire to finish a piece of work or attain a goal in spite of adversity or suffering, defying, almost denying, the power of death. Such is "the argument from hope" which is religious.

The fourth signal leads to "the argument from damnation" – the recognition that total moral depravity of itself places a man outside the moral order which transcends the community. For example, the murderer of an innocent child receives immediate and total condemnation. Such a deed calls for damnation, for religion supports judgements of damnation as well as ultimate judgements of redemption. I used to ask my class to imagine they were watching a film in the cinema. In it the villain has murdered an innocent child – he seems totally wicked, unredeemable. Is he no longer human? (He seems to have lost his humanity through his amoral behaviour. Is he nothing but a biological member of *homo sapiens*?) He runs away from the police and is half a mile ahead of them. An old lady opens her front door onto the pavement as he is about to run past, and begins to stoop down to pick up a bottle of milk. He picks it up, puts it into her hand and runs on. Five minutes later he is caught, and is tried and put into prison. You go home happy. Why?

The fifth signal, humour, leads to the "argument from humour". Humour is a signal of self-transcendence in which an individual recognizes that an incident can be viewed from two frames of reference. For example, in company one might slip on a banana skin and fall into an embarrassing position on the ground. You can't help laughing in the "human" context, for in your mind you can see how funny you look on the ground. But in the "biological" context you are in great pain, and find nothing funny about the fact that you have broken an ankle. Laughter relativizes both the embarrassment and the pain, implying both will be overcome – a signal of redemption.

Laughter is also a signal of joy. Each of these signals of transcendence also involves a stepping outside of the natural world and being open to the religious and metaphysical hinterlands which surround us. Berger was convinced that the rediscovery of the supernatural would help people regain the ability to handle tragedy, triviality and human affairs seriously so that they could laugh, play and live with a new fullness in the light of the "comic relief of redemption" (Berger 1971, ch. 5). By this Berger means that awareness of the supernatural allows one to laugh and play while still treating the moment with appropriate moral seriousness.

In conclusion, Berger is drawing attention to signals within everyday life which focus on the temporal but evoke an awareness of the eternal. As such they are signals of transcendence.

Keith Ward finds signals of transcendence in important events in life through the silence of personal contemplation, and in situations of maternal care through the expression of a mother's love and attention towards her children (Ward 1974: 48–51).

Transcendence can be discovered in the beauty, elegance and discerned purpose of the universe (Ward 1994: 101). Transcendence is mediated through personal encounter, events which are seen as providential, and guidance recognized as inspirational (230). But all these events disclose the Divine through the conviction that a religious interpretation makes most sense of the situations to the subject of these experiences.

In this chapter, beside glancing at visual images of transcendence, we have explored five meanings of transcendence which often overlap. These meanings show a greater focus on religious use than on secular or purely humanistic use.

We first met "transcendence" in Chapter 1, where we noted that Paffard described experiences of nature mysticism, whether religious or humanistic, as transcendental. As both religious and humanistic interpretations are associated with mystical experiences, "transcendence" is used as the central point of reference in this exploration. For Paffard "transcendental" refers to an experience "in another dimension" which may or may not be religious, or to an outlook on life which transcends the individual and the particular. It is the case that for some people the former, "an experience", leads to the latter, "the adoption of a related outlook on life".

The second use of "transcendence" is related to a quality of the nature of God. The third use is related to Kee's description of the way of Jesus as the way of transcendence, which even an atheist could follow. The fourth use is related to "the moral demand", or duty, as transcendent. The fifth use drew attention to natural phenomena which have been regarded as signals of transcendence, that is, of God's presence in the world. A sixth use of transcendence – a humanist use – follows in Carl Rogers's account of humanism, outlined in the next section.

TRANSCENDENCE, SPIRITUALITY, RELIGION AND HUMANISM

To clarify the focus on "transcendence", it will be helpful to explore its relationship with the main closely related concepts, "spirituality", "religion", and "humanism".

We have already noted that any attempt to explore "transcendental experience" must recognize that to describe any experience in this way indicates that it is an experience "in another dimension". Further, it usually encompasses more than the individual who has the experience. Frequently, it will be set within a personal framework of ultimate meaning and value.

Spirituality will be a major quality of the experience and its related outlook on life. Spirituality has been defined as "the way in which one mobilizes oneself religiously in the total and actual living out of one's daily activities . . . there are two complementary forms in which persons live out their lives . . . lifeway and lifework" (Lee 1985:7). "Lifeway" refers to a person's overall lifestyle pattern; "lifework" refers to a person's career. This position has massive support.

One book regards spirituality as the sole province of Christianity and the other world religions. Crossroads Publishing's impressive encyclopaedic series of books on world spirituality limits their discussion to the world religions, but there is at present no acknowledgement that humanism or other non-theistic creeds may have a spiritual dimension which feeds and develops human spirituality.[15] This massive support shows that the use of "religiously" in the definition above is central to the religious person.

It is commonplace that the ideal religious life, as expressed in all World Faiths, is centrally concerned with spirituality, as the evidence above indicates. But the word "religion" can be used in two senses: *literally* (its core meaning when describing something to do with religions and the religious life) or *metaphorically* or *pictorially* (i.e. in a peripheral or parasitic way). The following passage illustrates this second use:

> On the 8th August 1996 Alan Shearer returned to his home town to be welcomed by 20,000 fans even though it was raining. Newcastle United had paid £15 million for him, the highest transfer fee yet paid in English football. The television newsreader told us: "They gathered in their thousands outside their place of worship." Twenty years ago Bill Shankly, the manager of Liverpool United, talked of football as the religion of the fans. (Ayre 1997)

What meaning can be attached to the use of the words "worship" and "religion" used here? The use of these words indicates that the value the fans attach to their football transcends most of their other values: football is of very great importance to them, indicated by the choice of these words used metaphorically. The word "religion" comes from the Latin word *religio*, which means "obligation", "oath", "absolute commitment". Football is of very great importance not least because it establishes the significance of that community if the team is successful, and also the significance of each individual who identifies with the team as "our team".

However, the use of the word "religiously" in the definition of spirituality at the beginning of this section is meant literally. The religious definition above is unhelpful to those who embrace a humanist outlook on lifeway. Humanism is, for some people, a spiritual lifeway. Let us explore one account of ethical humanism which illustrates this.

Eric Fromm focuses on the "being mode" contrasted with the "having mode". He lists the qualities that the New Person will have, indicating a humanist outlook, but nevertheless articulating one form of spiritual life which is self-transcending. The qualities are:

- Willingness to give up all forms of *having*, in order to fully *be*.
- Security, sense of identity, and confidence based on faith in what one *is*, on one's need for relatedness, interest, love, solidarity with the world around one, instead of one's desire to have, to possess, to control the world, and thus become the slave of one's possessions.
- Acceptance of the fact that nobody and nothing outside oneself can give meaning to life . . . [so that one may become] devoted to caring and sharing.
- Being fully present where one is.
- Joy that comes from giving and sharing, not from hoarding and exploiting.
- Love and respect for life in all its manifestations in the knowledge that . . . life and everything that pertains to its growth are sacred.
- Trying to reduce greed, hate and illusions . . .
- Living without worshipping idols and without illusions . . .
- Developing one's capacity for love as well as one's capacity for critical, unsentimental thought.

✧ Shedding one's narcissism and accepting the tragic limitations inherent in human existence.

✧ Making the full growth of oneself and of one's fellow beings the supreme goal of living.

✧ Knowing that to reach this goal, discipline and respect for reality are necessary.

✧ Knowing that no growth is healthy that does not occur in a structure . . .

✧ Developing one's imagination . . .

✧ Not deceiving others, but also not being deceived by others . . .

✧ Knowing oneself, not only the self one knows, but also the self one does not know . . .

✧ Sensing one's oneness with all life, hence giving up the aim of conquering nature . . . but trying rather to understand and co-operate with nature.

✧ Freedom that is not arbitrariness but the possibility to be oneself . . .

✧ Knowing that evil and destructiveness are necessary consequences of failure to grow.

✧ Knowing that only a few have reached perfection, but being without ambition to "reach the goal" in the knowledge that such ambition is only another form of greed, of having.

✧ Happiness is the process of ever growing aliveness . . . (Fromm 1978: 170–2).

Within this account, faith, love, joy, giving, sharing, growing are included in the listed spiritual qualities that are of importance to the writer.

Here is another account of humanism. Carl Rogers, after forty years as a central figure writing on humanistic psychology, espouses evolutionary humanism:

> I hypothesise that there is a formative directional tendency in the world . . . This is an evolutionary tendency towards greater order, greater complexity, greater inter-relatedness. In humankind this tendency exhibits itself as the individual moves . . . to knowing and sensing below the level of consciousness, to a conscious awareness of the organism and the external world, to a transcendent awareness of the harmony and unity of the cosmic system, including humankind. (Rogers 1980: 133)

Later in his book Rogers explores the topic of building person-centred communities. He says:

> Another important characteristic of the community forming process, as I have observed it, is its transcendence or spirituality. These are words that in earlier years I would never have used. But the overarching wisdom of the group, the presence of an almost telepathic communication, the sense of the existence of 'something greater,' seem to call for such terms.

A participant in one of his community-building workshops put it like this:

> I found it to be a profoundly spiritual experience. I felt the oneness of spirit in the community. We breathed together, felt together, even spoke for one another. I felt the power of the "life force" which infuses each of us – whatever that is. I felt its presence without the usual barricades of "me-ness" or "you-ness" – it was like a meditative experience when I feel myself as a centre of consciousness, very much part of a broader universal consciousness.

And yet with that extraordinary sense of oneness, the separateness of each person present has never been more clearly preserved. (Rogers 1980: 196–7)

This account is full of qualities cited to promote spirituality and transcendence, both specified goals.

The Spirit of the Child

The final contribution to this discussion of religion, spirituality, transcendence and humanism comes from David Hay and Rebecca Nye. They are aware of the relationship between spirituality and religion, which they have explored through using a word association test with many people.

"Religion" is associated with churches, mosques, Bibles, prayer books, religious officials, weddings, funerals, and so on. It is regularly linked with boredom, narrow-mindedness and being out of date. More seriously, it is linked with fanaticism, bigotry, cruelty and persecution. These last serious allegations are characteristic of an extrinsic religious orientation.[16]

"Spirituality" has warmer associations: love, inspiration, wholeness, depth, mystery and personal devotion like prayer and meditation. These qualities are characteristic of an intrinsic religious orientation. In saying this, however, I am not claiming that spirituality is synonymous with one kind of religious orientation, for the latter emerges from the former, as Hay and Nye make clear (Hay and Nye 1998: 162).

Sir Alister Hardy, formerly Professor of Zoology at Oxford and committed Darwinist, argued that "religious experience has evolved through the process of natural selection because it has survival value to the individual" (9). Hay, who finds considerable overlap between religious experience and spiritual experience, supports this view with considerable evidence from the 1986 Gallup Omnibus survey in Britain, which indicated that about half of those surveyed had had such an experience (16).

In their exploration of spirituality, Hay and Nye find evidence to suggest that "ordinary experience of wonder, which when it is profound enough, shifts imperceptibly into spiritual or religious awareness" (31). Likewise they affirm "that morality has its source at a deeper level than specific religious adherence, since it arises in the first place out of spiritual insight".

Hay and Nye outline an analytic structure of categories of spiritual sensitivity (59). These categories were found in the research programme based on eighteen boys and girls who were 6–7 years old, and twenty who were 10–11 years old (100). Rebecca Nye, who did the research, analysed 1,000 pages of transcribed interviews and discovered one key category that enveloped all seemingly relevant data, which she called *relational consciousness* (113).

Relational consciousness was characterized by an unusual level of consciousness or perceptiveness in relational matters, focused on "I–Others", "I–Self", "I–World" and "I–God". Ruth provides an example of relational consciousness:

Six-year-old Ruth's conversation included a sensual description of heaven. She referred to the key elements in her spiritual response as "waking up" and "noticing", both of which suggest that a different quality of consciousness was crucial to her experience. The relational component in this was a strong feeling of connection to the natural world as something which was full of gifts for her and deserved her respect and love in return. The sense of intimacy also had reverberations in her relationship with herself, as seen in her self-conscious perception of a symmetry between her own joy and the joyful leaping of the lambs.[17]

Children's spirituality involves a sensitive awareness of four levels of parallel and interactive functioning: cognition, emotion, action and sensation. Nye lists the dimensions of relational consciousness. The main contexts were four kinds of consciousness: child–God, child–people, child–world and child–self-consciousness (Hay and Nye 1998: 120). Within the dimension of child–God consciousness children might relate a religious experience, relate what they thought it would be like to have a religious experience, or indicate the meanings and feelings evoked by the concept of God. An example of child–God consciousness is provided by Beth, aged ten. Her account of her prayer experiences following a period of religious doubt:

> So I just half believed in God and half didn't, and then I had to pray extra hard to get his love back because I had been really mean to him. I had not prayed to him for ages, and so I was really mean to him, so I had to give him extra love, and I felt really good after that, but when I wasn't praying I felt really, really bad. (Hay and Nye 1998: 121)

In the light of this research the authors recognize not only that spirituality is about love, inspiration, wholeness and mystery, but that it is focused on an awareness of a holistic relationship with the rest of reality (6, 142). Relational consciousness is a form of emotional consciousness which is prior to intellectual activity.[18] It points towards a communal direction in direct opposition to the individualistic culture we live in. The social pressures of secularism, materialism and individualism are hostile towards relational consciousness and so can "sabotage children's sense of community" (Hay and Nye 1998: 155).

Hay and Nye go on to say that children, as a result of these social pressures, cut their links with religion and "create a 'macho' defensiveness against their own spiritual awareness" which becomes orphaned (156). Nye's research shows that "by the time they are 10 years old, a substantial number of children living in a secularized community harbour a shyness or embarrassment about anything closely linked to religion" (162).

Hay and Nye conclude: " . . . the purpose of spiritual education is the reverse of indoctrination. The task of nourishing spirituality is one of releasing, not constricting children's understanding and imagination."[19] Social integration arises from "a widespread awareness that relational consciousness is the bedrock of a free and humane society. In such a society the primary task of education is the nurture of the spirit of the child" (Hay and Nye 1998:175).

CONCLUSIONS

It has been possible to show that there is a relationship between "transcendence" and "spirituality" for both religious and humanistic lifeways or outlooks on life. It is quite clear from the account of "being" given by Fromm and the account of evolutionary humanism from Rogers that both include all four realms identified by Nye, a view that is shared by Fisher (Fisher 1999: 32). This can be summed up by Fisher's diagram of Spiritual Well-Being, shown in figure 9.1.

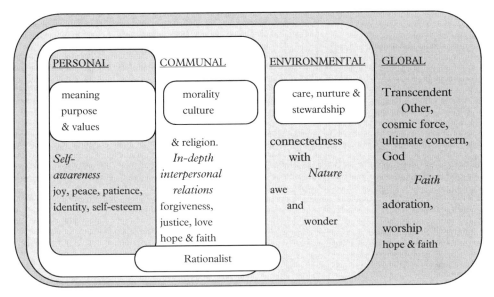

Figure 9.1   Spiritual Well-Being – expressed by the quality of relationships in each DOMAIN

Fisher describes the four areas as domains. In the personal domain one intrarelates with oneself in relation to meaning, purpose and values in life. The human spirit creates self-awareness, which is related to self-esteem and identity. In the communal domain, the quality and depth of interpersonal relationships between self and others relates to morality, culture and religion. This includes love, justice, forgiveness, hope and faith in humanity. Stewardship of the environmental domain can surpass care and nurture for the physical and biological, evoking a sense of awe and wonder. For some this reaches a sense of unity with the environment, even nature mysticism as described by Wordsworth. The global domain involves the relationship of self with some-thing or some-One Other (i.e. ultimate concern, cosmic force, transcendental reality or God). I think it is possible for the humanist, avoiding theistic claims and language, to embrace a global perspective, though Fisher regards it as a religious category: "[In the figure 9.1], the knowledge aspects of each domain are written in bold type at the top of each cell. The inspirational aspects are written in italics, in the centre of each cell. The expressions of well-being in each domain are written in roman type at the bottom of each cell." (32)

Fisher (1999) notes that different individuals may inhabit different domains. Spiritual well-being can be experienced as personalist, communalist, environmentalist or globalist. The rationalists embrace the knowledge, not the inspirational aspect of the first three domains. That confirms that rationalists are intellects on stilts. Lacking emotional literacy, they are hardly fully-rounded human beings. Just as an amoral person lacks a dimension of being fully human, so a rationalist who ignores the emotional side of his personality is, to that extent, not fully human (18–28). We must remember that to call someone a human being is to award them a diploma, attesting to some level of interpersonal, moral, emotional, aesthetic as well as intellectual competence and spiritual insight, leading towards holistic personal maturity.

Our central concern is still the nature of knowledge. In this chapter, we began, from previous chapters, with Tillich's insight that moral knowledge at its apex can be transformed into religious discourse, a higher level of knowing, just as a caterpillar is transformed first into a chrysalis and then into a butterfly. In his book Hay (1998) has claimed two things. First, that the roots of morality are deeper than religion, emerging from the spiritual depths of a person's being. Secondly, that there are times when wonder can shift imperceptibly into spiritual or religious awareness.

Finally we have noticed that Fisher, in the mode of personal discourse, has detected that there are five levels of knowing: rational, personal, communal, environmental and global. The least valued mode of knowing, overlapping three domains in part, is that of the rationalist, who seems only concerned to process information into intellectual knowledge. Then we have the four ascending levels of knowing, three in the extended realm of personal knowledge: personal, communal and environmental. Personal knowledge becomes religious knowledge through the introduction of God-talk in the narrative, and through religious experience, which is the hypothesis we can test by examining the data collected to support it.

At this point we need to recap the content of Chapter 1 in terms of the kinds of experience which it contains, so that we know where to begin our investigation into religious experience.

Let us take an imaginary situation. Four people are in a car approaching a car park at the top of a mountain pass. They stop their car. A magnificent scene opens up before them, taking their breath away. They silently "take it all in". After a very long silence, the first to speak says, "I just feel part of this wonderful view"; the second says, "I am lost on cloud nine sharing this view"; the third says, "This beauty speaks to me of the care of the Creator"; and the fourth says, "This is nature at its very best, but there is nothing more than that." What are they each saying? All say the view is beautiful: they all interpret the view aesthetically. The third person also interprets the view from a religious perspective, and the fourth also from a naturalist, possibly humanist, perspective. The first two only offer an aesthetic view but they each also have a metaphysical view. We do not know what it is; one might be religious and the other humanist, or both might be the one or the other. From the work of Paffard it is clear that given the "neutral" stimulus he used, i.e. not religious, it has been possible to discover four groups which parallel these four people.

Published experiences, (which will now be explored), fall into the same four groups. We shall only study religious experiences and spiritual experiences which are expressed philosophically in naturalistic and humanist terms, affirming that the world is all there is, in other words, they believe there is no God. We shall not study examples of those who only give an account in aesthetic terms, like the first two sightseers quoted above, but accounts of those who also affirm their underlying metaphysical position, either religious or naturalist/humanist.

NOTES

1   N. Smart,. 1997, London: Fontana, p. 1f. "The word 'phenomenology' derives from the philosophical traditions of Husserl. . . . Among religionists it means the use of *epoche* or suspension of belief, together with the use of empathy in entering into the experiences and intentions of religious participants. .. (It) is the attitude of informed empathy. It tries to bring out what religious acts mean to the actors."

2   A great pioneer is Wilfred Cantwell Smith. See E. J. Hughes (1986), *Wilfred Cantwell Smith: A Theology for the World*, London: SCM.

3   Personal communication from Professor Adrian Thatcher.

4   'Materialism', 'empiricism', 'logical positivism' are different ways of describing this outlook.

5   *Oxford Shorter English Dictionary* (1975), p. 1788.

6   *Concise Oxford Dictionary* (1983), p. 377.

7   W. James (1968), (1902), p. 50: "the feelings, acts and experiences of individual men in their solitude, so far as they apprehend themselves to stand in relation to what ever they may consider the divine."

8   E. Durkheim (1976), p. 47, "A religion is a unified system of beliefs and practices relative to sacred things, that is to say, things set apart and forbidden – beliefs and practices which unite into one single moral community called a Church, all those who adhere to them."

9   *The Methodist Worship Book* (1999), pp. 61, 135, 151; Rom. 10: 9, 1 Cor. 12; 3, Phil. 2: 11.

10 Ninian Smart has long had sympathy with Otto's theory but has observed that it does not seem able to accommodate atheistic forms of religion, for example, Theravada Buddhism. Smart takes the view that his "hypothesis of two (or more) strands of basic religious experience has much greater explanatory power", justifying it on the grounds of its greater fruitfulness in this respect. The two strands he suggests are first, Otto's emphasis on the numinous experience of the Sacred, already mentioned. Secondly, the contemplative experience at the heart of Theravada Buddhism, which focuses on *nirvana*, a state of purified consciousness in which there is no distinction between subject and object, leading to an awareness that 'there is nothing'.

11 J. Robinson (1963), pp. 48–9.
    H. R. Price (1969), p. 465.
    F. C. Coplestone (1974), p. 160.
    I. T. Ramsey (1973), p. 58.
    J. Macquarrie (1975), p. 159.
    J. Hick (1989), pp. 5–6.

12 This gap, between the person we should have become and the person that we are, is met by the healing power of which the Christian Gospel speaks (Romans 7). It is the grace, or acceptance of us as we are by God in Christ which removes the failure with its paralysing guilt, releasing power to restore broken relationships, and attempt once more to realize more closely our true nature (Tillich 1969, op. cit., ch. 3).

13 The argument here has convincingly demonstrated how it can be successfully shown that religion is the foundation of morality. There is also a natural law argument which depends on a religious foundation (see Vardy and Grosch,*op. cit.*, ch. 4). There are other arguments which show morality is the basis of religion, morality is independent of religion, religion can be reduced to morality, or morality reduced to religion, but these arguments are not our present concern; see, for example, C. W. Kegley (1971), 'Relations of Religion and Morality', and T. E. Jessop (1971), 'Hume, David' in J. Macquarrie (ed.), (1971).

14 I. Kant. (1960), p. 158."I take the following proposition to be a principle requiring no

proof: *Whatever, over and above good life-conduct, man fancies he can do to become well pleasing to God is mere illusion and pseudo- service of God."*

15 C. Jones, G. Wainwright and E.Yarnold (eds) *Study of Spirituality*, e.g. four volumes on *Christian Spirituality* and one volume devoted to *Classical Mediterranean Spirituality* : *Egyptian Greek and Roman.*

16 D. Hay, with R. Nye. (1998), p. 6. See ch. 3 above on "extrinsic religious orientation".

17 D. Hay, with R. Nye (1998), *op. cit.*, p. 114; cf. J. W. Fisher, 'Helps to Fostering Students' Spiritual Health' in *International Journal of Children's Spirituality*, Vol. 4, No. 1, 1999, p. 31.

18 Cf. J. Park, "Emotional Literacy: Education for Meaning" in *International Journal of Children's Spirituality*, Vol. 4, No. 1, 1999, pp. 19–28.

19 D. Hay, with R. Nye (1998), *op. cit.*, p. 163; cf. P. A. Burke. 'The Healing Power of the Imagination' in *International Journal of Children's Spirituality*, Vol. 4, No. 1, 1999, pp. 9–17.

REFERENCES

Acton, H. B. (1970), *Kant's Moral Philosophy*, London: Macmillan.

Bailey, E. (1997), Implicit Religion in Contemporary Society, Netherlands: Kok Pharos Publishing House.

Bailey, E. (1998), *Implicit Religion: An Introduction*, London: Middlesex Press.

Batson, C. D., Schoenrade, P. and Ventis, W. L. (1993), *Religion and the Individual*, New York and Oxford: Oxford University Press.

Berger, P. L. (1973), *The Social Reality of Religion* (1969), London: Penguin.

Berger, P. L. (1963), *Invitation to Sociology: A Humanistic Perspective*, New York: Doubleday.

Berger, P. L. (1971), A Rumour of Angels (1969), London: Pelican.

Berger, P. L. (1980), *The Heretical Imperative* (1979), London: Collins.

Burke, P. A. (1999) "The Healing Power of the Imagination". *International Journal of Children's Spirituality* 4 (1), 9–17.

Carroll, L. (1998), "Through the Looking Glass" in *Alice* (1896). London: Macmillan.

Clarke, P. B. and Byrne, P. (1993), *Religion Defined and Explained*, London: St Martin's Press.

Concise Oxford Dictionary (1983), Oxford: Oxford University Press.

Coplestone, F. C. (1974), *Religion and Philosophy*, New York: Gill and Macmillan.

Durkheim, E. (1976), *The Elementary Forms of the Religious Life* (1915), London: Allen and Unwin.

Eyre, A (1997), *Football and Religious Experience: Sociological Reflections*, Religious Experience Research Centre, University of Wales, Lampeter SA48 7ED..

Ferre, F. (1968), *Basic Modern Philosophy of Religion*, London: George Allen & Unwin.

Fisher, J. W. (1999), "Helps to Fostering Students' Spiritual Health", *International Journal of Children's Spirituality* 4 (1), 29–49.

Fromm, E. (1978), *To Have or to Be*, London: Cape.

Hay, D. with Nye, R. (1998), *The Spirit and the Child*, London: Fount/HarperCollins.

Hick, J. (1989), *An Interpretation of Religion*, London: Macmillan.

Hughes, E. J. (1986), *Wilfred Cantwell Smith: A Theology for the World*, London: SCM Press.

James, W. (1968), *The Varieties of Religious Experience* (1902), London: Fontana.

Jones, C., Wainwright, G. and Yarnold, E., (eds), (1987) *Study of Spirituality*, 4 vols, Oxford: Oxford University Press.

Kant, E. (1960), *Religion Within the Limits of Reason Alone* (1793), New York: Harper & Row.

Kee, A. (1971), *The Way of Transcendence*, London: Pelican.

Lee, J. M. (ed.), (1985), *The Spirituality of the Religious Educator*, Birmingham, AL: Religious Education Press

Macquarrie, J. (1975), *Thinking about God*, London: SCM Press.

Macquarrie, J. (ed.), (1971), *A Dictionary of Christian Ethics*, London: SCM Press.

The Methodist Worship Book (1999), Peterborough, Methodist Publishing House.

Otto, R. (1959), *The Idea of The Holy*, (1917), Harmondsworth: Penguin.

Park, J. (1999), "Emotional Literacy: Education for Meaning", *International Journal of Children's Spirituality* 4 (1), 19–28.

Price, H. R. (1969), *Belief*, London: Allen and Unwin.

Ramsey, I. T. (1973), *Religious Language*, London: SCM Press.

Robinson, D. (1983), "Spirit", "Spiritual", *Concordance to the Good News Bible*, Swindon: British and Foreign Bible Society, pp. 1113–16.

Robinson, J. A. T. (1963), *Honest to God*. London: SCM Press.

Rogers, C. R. (1980), *A Way of Being*, Boston: Houghton Mifflin.

Rose, S. (1997), *Lifelines: Biology, Freedom, Determinism*, Harmondsworth: Penguin.

Schleiermacher. F. D. E. (1988), *On Religion: Speeches to Its Cultured Despisers*, (1821), trans. R. Crouter, Cambridge: Cambridge University Press.

Smart, N. (1997), *Dimensions of the Sacred* (1996), London: Fontana.

Sykes, J. B. (ed.), (1983), *The Concise Oxford Dictionary*, Oxford: Oxford University Press.

Tillich, P. (1957), *Systematic Theology*, Vol. 1, Chicago: Chicago University Press.

Tillich, P. (1962), *Courage to Be* (1952), London: Fontana.

Tillich, P. (1969), *Morality and Beyond* (1964), London: Fontana.

Vardy, P. & Grosch, P. (1994), *The Puzzle of Ethics*, London: HarperCollins.

Ward, K. (1974), *The Concept of God*, Oxford: Blackwell.

Ward, K. (1974), *Religion and Revelation*, Oxford: Oxford University Press.

Ward, K. (1996), *God, Chance and Necessity*, Oxford: One World.

Wilson, J. (1971), *Education in Religion and the Emotions*, London: Heinemann.

PART III

The Integrity of Religious and Mystical Experience

Empirical Studies

SEVEN USES OF THE TERM "RELIGIOUS EXPERIENCE"

Religious experiences are those which are interpreted as experiences that mediate the Divine or Sacred. Before we can outline accounts of religious experiences, what was collected, how it was analysed and interpreted, we need to recognize that different people use the term "religious experience" in different ways. We will first describe seven of the many different uses. The first use refers to a personal religious experience, which often occurs when people are alone. It is often a relationship between two "persons", you and God. This is a primary religious experience (Clark 1967: 23), two examples of which are Moses and the burning bush (Exodus 3) and Paul on the Damascus Road (Acts 9), a numinous experience with focal awareness.

Secondary religious experience relates to the formation of habits of regular attendance at a place of worship, where a person routinely carries out obligations which were made under solemn circumstances, often within, or related to, that place of worship (Clark 1967: 23–4). The religious experience is of a pattern of worship and its related way of life.

The third use refers to tertiary religious experience, so called because its content is intellectual learning, as when a child or an adult learns aspects of the religious life in, for example, a church discussion group, Sunday school or church service. Such an experience may not involve personal commitment if attendance is a parental demand, or it may involve commitment of a kind, as in the case of an adult concerned with a personal search for meaning in life (23–4).

The fourth use relates to a personal religious experience when it occurs within, or in relation to, a religious community, such as a conversion experience, for example that of John Wesley:

> Wednesday, 24th May 1738: In the evening I went very unwillingly to a society in Aldersgate Street, where someone was reading Luther's Preface to the *Epistle to the Romans*. About a quarter before nine, while he was describing the change which God works through the heart in Christ, I felt my heart strangely warmed. I felt I did trust in Christ, Christ alone for salvation; and an assurance was given me that he had taken away *my* sins, even mine, *and saved me from the law of sin and death*. (Parker 1906: 43)

We will also refer to many accounts of conversion experiences of ordinary persons.[1] This is a primary religious experience related to a faith community (Clark 1967: 24).

The fifth use of religious experience relates to religious mysticism, which develops from a subsidiary awareness of a Sense of Presence, a bipolar experience, which sometimes develops into a mystical union with Being.

The sixth use of religious experience relates to religious mysticism, which results from the discipline of contemplative meditation within a religious tradition in order to attain mystical union with the Divine.

The seventh use of religious experience is the most general use and refers to all experience interpreted from a religious perspective. Advocates of this position argue that a religious person cannot have an experience which is not religious because "religious experience" is *all* experience when *interpreted* from a religious perspective. Professor Nicholas Lash takes this view, saying that "the symbolic, linguistic, affective resources . . . the *public* world of culture and institutions" are *given* by prior experience and are the primary influences in the formation of our religious outlook, not *individual* religious experiences (Lash 1988: 58).

Within this classification of "religious experience", we have outlined seven ways of understanding the concept of religious experience which provide the foci of this study, though treatment will involve different levels of analysis.

However, before proceeding further, it is useful to look at other classifications so as both to clarify the foci of this study and to indicate other uses of the term. These will indicate areas outside the scope of this study.

"Religious experience" is a term which is sometimes used to cover such activities and areas of study as paranormal experiences; charismatic experiences (Donovan 1979), which include glossalalia (speaking in tongues) and Pentecostal experiences (Meissner 1984: 9); prayer; faith healing (Meadow and Kahoe 1984: ch. 8); possession; mediumship (ch. 9); stages involved in the mystical life; meditation techniques and programmes (ch. 10); and techniques for facilitating religious experience, like fasting, sleep deprivation, ecstatic dancing, rhythmic drum beating and drug enhancements (Wulff 1991: ch. 2).

This chapter will focus on the seven uses of "religious experience" outlined earlier, namely:

I   personal religious experiences involving focal awareness between a person and God, often when the person is alone – a bipolar numinous experience;

2   personal religious experience in the form of regular worship and commitment to the attendant religious way of life;

3   religious experience which is really cognitive experience, i.e. knowledge about a religion, which is often gained in a class within a religious community;

4   a conversion experience, often related to membership of a religious community;

5   mysticism as that form of religious experience which frequently grows out of a binary experience, one of a person experiencing the tacit or subsidiary awareness of the Presence of God, sometimes leading to a mystical union with the Divine;

6   mysticism as that form of religious experience which develops through the

7   discipline of contemplative meditation towards a closer union with God;
all experience interpreted from a religious perspective, better called one's religious outlook.

The treatment adopted here proceeds chronologically, telling the story of the exploration of religious experience, initially by chronicling the findings of major figures in the field. This means that the seven uses of religious experience and accounts of religious mysticism – the sixth use, will be discussed out of sequence. For example, the work first explored, that by Starbuck, focuses on *conversion*, the fourth use. Other works often explore more than one of these uses. James, for example, uses the term in the first, fourth and fifth ways. Sometimes the uses overlap or even merge, so we have to read carefully what a writer is saying. This method has been followed for two reasons: it indicates how a writer is influenced by their predecessors and it outlines the main foundation material from which the synoptic account outlined later is constructed.

Strictly speaking, Underhill and Buber should not be included in this chapter, with its focus on empirical studies. They are included because they outline part of the conceptual structure assumed by many of the other writers quoted.

STARBUCK: CONVERSION EXPERIENCES

The first major psychological study of religious experience focused on the fourth use, religious conversion, and was conducted in 1899 by E. D. Starbuck. His results were published in *The Psychology of Religion*. He was concerned to explore the religious conversion experiences of about 200 male and female subjects, all Protestants.[2] He found that a third of the males and half of the females had revival conversions. The peak age of conversion for both sexes was sixteen years; he observed that it followed the age of puberty.

Starbuck found that adolescence was a period of "storm and stress", doubt and alienation, and was the period in which the birth of a "larger self" may well take place. Conversion was frequently its midwife. Starbuck found that amongst those converted, the period of "storm and stress" was a fifth of the length of those not converted (1899: 224). Starbuck detected two periods of transition, from childhood to adolescence and from adolescence to maturity (392). The first movement was "that which transfers the centre of activity from self-interest to interest in the whole of which the self is but a part". Conversion achieves this quickly and less painfully. The second movement leads to "mature religion . . . [and] shows a strenuous advance forwards losing the self in service" (393).

Starbuck reaches his key conclusion: "We have then three precepts, representing three stages of growth: in childhood, conform, in youth be thyself, in maturity lose thyself" (Starbuck 1899: 415). William James, who was massively indebted to Starbuck, summarized Starbuck's essential insight: "Religion thus makes easy and felicitous what in any case is necessary" (1968: 68).

We need to note carefully that "religious experience" here refers to an experience which occurred within a situation in which there was a social stimulus within a ritual context, without which the experience would not have occurred.[3] Secondly, the definition of religion assumed is pragmatic, which means he believed that the truth of any idea was shown in its effects (Flew 1979: 265). In other words he was a *functionalist* or *pragmatist*. Religion is true because it achieves its fruits or goals.

JAMES: BIPOLAR, CONVERSION
AND MYSTICAL EXPERIENCES

"Religious experience" is a term which was not really used until the beginning of the twentieth century. It was popularized by William James as a term which referred to a person's discrete experience, often of brief duration, which was understood as an experience of the Presence of God in some form. James was, however, greatly influenced by Starbuck, whose work we have just described.

William James was the father of the modern study of religious experience. His book, *The Varieties of Religious Experience* was published in 1902, and popularized the use of the concept of religious experience. James was a pragmatist, i.e. truth was what worked. He offered a phenomenological account of religious experience, as he described and analysed religious experiences to demonstrate what they were like. Religious experiences were the phenomena being studied. His work was based on quoting a large number of personal statements. He purposefully selected "extremer examples as yielding the profounder information" (James 1968: 465). He is concerned to describe personal religious experiences, conversion experiences and mystical experiences.

Because his focus of attention is on the individual's experience, his definition of religion focuses on its experiential dimension of religion, "Religion . . . shall mean for us the feelings, acts and experiences of individual men in their solitude, so far as they apprehend themselves to stand in relation to whatever they may consider the divine" (James 1968: 50). James is concerned with the meaning and validity of religion and not with its nature and origin. He states two criteria of spiritual value, rejecting "origin" as a legitimate criterion. Superior states of mind either give us immediate delight or produce good fruit. Ideally they do both.[4] You can identify such people by their fruits.[5]

This important mistake of trying to defend human excellence by an appeal to human origins is often used. It is a *genetic fallacy*. It is used in Scripture, e.g. concerning the birth and infancy of Moses, Jeremiah and Jesus. There is no doubt amongst the faithful of the supreme value of any or all of these spiritual giants, but their greatness lies in what they did and said. To say that they had no choice but to be what they were because of their human origins is both a *non sequitur* and trivializes their spiritual achievements. Their achievements are greater because they made free, courageous and sacrificial choices at each stage of their lives. To say they were destined for greatness by their birth is to turn them into computer-driven robots. The argument from human origins is always fallacious. That is different from legitimately saying that a person's origins, e.g. good home, able and loving parents, excellent education or religious experience may be important. Genetic, experiential and social factors can contribute to the development of a person by providing them with great opportunities. However, such opportunities can be wasted as well as used well.

James developed two models for the understanding of religious experience. The first model is "the religion of healthy mindedness" (James 1968: 92–136). James assumes here that life's chief aim is "happiness". So some people are optimistic and adopt "a way of feeling happy about things immediately". In this way they assume everything is good.[6] They tend to be extrovert and unreflective. They are usually associated with a fairly liberal form of theology (Clark 1967: 155–6). There is a tendency to ignore or fight evil, believing that things are getting better and better, and if only we concentrate on this we will find peace. The outlook is focused on religious growth.

The second model of religious development occupies a larger part of the book, the religion of the "sick-soul", the "divided self". If the healthy are the once-born, the sick souls are the twice-born, for their internal struggles are overcome through conversion and the development of spiritual stature – saintliness. The sick soul is joyless, dreary, discouraged and dejected (James 1968: 153), so renunciation and new birth in conversion provide deliverance from meaninglessness and all that divides an individual, rendering him unhappy and listless (189). "To be converted . . . to receive grace . . . are so many phrases which denote the process gradual or sudden, by which a self hitherto divided, and consciously wrong, inferior and unhappy, becomes unified and consciously right, superior and happy" (194).

Religious maturity develops the marks of saintliness (269–70, 333–4). For James "personal religious experience has its root and center in mystical states of consciousness" (366). He sets out four criteria of "the mystical state of consciousness":

- ❖ Ineffability, or what cannot be put into words;
- ❖ Noetic quality, or its intellectual content;
- ❖ Transiency, that is, it is a brief experience; and
- ❖ Passivity, that is, it happens to you. (367–8).

James describes the stages by which an individual enters more deeply into a mystical union with God (368–413; cf. Happold 1975). He quotes one example of a mystical state from the memoirs of Meysenbug:

> I was alone upon the seashore as all these thoughts flowed over me, liberating and reconciling; and now again as once before in distant days in the Alps of the Dauphine, I was impelled to kneel down, this time before the illimitable ocean, symbol of the Infinite. I felt that I prayed as I had never prayed before, and knew now what prayer really is: to return from the solitude of individuation into the consciousness of unity with all that is, to kneel down as one that passed away, and to rise up as one imperishable. Earth, heaven, and sea resounded as in one vast world-encircling harmony. It was as if all the chorus of all the great who had ever lived were about me. I felt myself one with them, and it appeared as if I heard their greeting: 'You too belong to the company of those who overcome'. (James 1968: 381)

This example classically illustrates how an individual can lose all sense of self, which is then replaced by an almost out-of-body sense of mystical union with God, nature and everyone.

James is concerned with the truth of these mystical experiences and offers three conclusions. First, he says that the individual's mystical experience is totally authoritative for that person, who is then invulnerable to criticism. Secondly, this authority does not transfer to another, who must personally listen to and evaluate the mystical experience(s). Thirdly, this kind of experience breaks down the singular authority of rational knowledge, adding a supersensual form of knowing which shows there are differing ways of knowing, each important.[7]

James sums up the characteristics of the religious life as including three main beliefs: the visible world is part of a spiritual universe which gives it its value; our true goal is union with this higher spiritual universe; and spiritual energy flows through the inner communion with the source of spiritual energy, sometimes called God.[8]

Based on an analysis of his model of the "sick soul", James discerns a common nucleus in all religions.

> It consists of two parts: (1) An uneasiness, and (2) Its solution. (1) The uneasiness . . . is a sense that there is something wrong about us as we naturally stand. (2) The solution is a sense that we are saved from the wrongness by making proper connection with the higher powers. (James 1968: 484)

James feels that the individual recognizes that "with the wrong part there is a better part of him, even though it is but a helpless germ". He recognizes "his real being with the germinal higher part of himself", and "further he becomes conscious that this higher part is coterminous and continuous with more of the same quality which is operative in the universe outside of him" (1968: 484).

The question arises, does this "more" exist? "In answering this question . . . the various theologies . . . all agree that the 'more' really exists" (485). As a psychologist how can he account for "the more"? "Let me then propose as a hypothesis, that whatever it may be on its farther side, the 'more' with which in religious experience we feel ourselves connected is on its hither side the subconscious continuation of our conscious life" (James 1968: 487).

The "more" on its farther side is expressed in "overbeliefs"[9] that are distinctive of different religions. James regards the 'more' as an aspect of the better self which is continuous, perhaps through the collective unconscious, and is a channel through which saving experiences come.[10]

In his book, James accomplishes four tasks. He offers an extensive description of examples of religious experience which he feels fit one of two models of religion with their particular understandings and outlook; he outlines a developmental path towards spiritual maturity, or sainthood, which finds expression in mysticism; he provides an account which embraces the theory of an "inner core" common to all religion, so providing an *essentialist* or *perennialist* definition of religion; he postulates a psychological theory of the nature of the means of communion with the "more" through the unconscious, by means of which saving experiences reach the individual. Watts calls James a *perennialist* as he "assumes that religious experiences provide direct access into the true nature of reality" (Watts 2002: 89).

Thus far he spoke as a psychologist, but when he continues by saying that "God is the natural appellation, for us Christians at least, for the supreme reality, so I will call this higher part of the universe by the name of God", for "God is real since he produces real effects" (James 1968: 491) he exchanges his psychologist's hat for his theological hat to indicate the nature of his personal overbelief. This is a perfectly legitimate double role which everyone plays, for everyone can make judgements in both ways of knowing, psychological and theological, for they are different modes of discourse representing different kinds of knowledge, and need to be identified as such, not fused or confused. It is, however, quite reasonable to suggest that the psychological analysis gives support to a particular theological judgement, for complementary modes of understanding at best harmonize in synoptic tiered levels of understanding.

UNDERHILL: MYSTICAL EXPERIENCES

Evelyn Underhill's book *Mysticism,* written in 1911, focuses on disciplined contemplation within a religious tradition, drawing attention to devotional manuals which promote the mystic way towards the goal of union with God. Even though we will not focus on this understanding of religious experience, it is of interest to us because it offers a developmental model of a conversion in five stages, and overlaps with our fourth meaning of religious experience – conversion.

The individual first experiences an awakening of self. This is followed by a time of self-condemnation that is experienced as purgatory and is paralleled by a process of moral and spiritual improvement or purification. The third stage of the process of spiritual maturation is illumination, introversion, ecstasy and rapture. This is the time when at last the individual "sees", discovers or has the secret of the spiritual life revealed to them. However, the spiritual high point reached provides spiritual resources for facing evil, persecution, temptation, despair or ennui, which is mental weariness from lack of interest or belief that life has any point at all. This is called "the dark night of the soul". The final stage is reached when the spiritually mature individual is able to use the spiritual resources gained to meet and transform the brief or extended periods of "dark night" testing experiences into a balanced outlook and a unitive or integrated life.

These insights of Underhill's are quoted because they anticipate and in some sense parallel insights, which we will mention later (Chapter 18) into the process of creative thinking, which also goes through a number of stages. First, one has a problem, which one struggles with but cannot solve or resolve. Then the problem is put to the "back of one's mind", and one attends to other immediate interests, thus allowing the problem a period of "incubation" in the unconscious mind. Thirdly, often out of the blue, a solution appears – illumination! Then one explores the new insights to test their truth, validity and usefulness.

BUBER: BIPOLAR EXPERIENCES

Martin Buber was concerned with primary religious experience. He wrote his book *I and Thou* in 1937. He contrasted the world of I–It, in which the ego treats the world as an object to be manipulated, used and assimilated, and the world of I–Thou, in which the ego confronts the world as subject. The "I" requires the constructive presence of Otherness in order to form that genuine community of internal relationships which foster love, compassion and tenderness. The Thou of the world is truly God. Buber regards the I–Thou principle, which he calls the dialogic principle, as an expression of the existential meeting of man and God. That means they can "talk" to each other. This is sometimes called prayer.

GLOCK AND STARK:
BIPOLAR, CONVERSION AND MYSTICAL EXPERIENCES

In 1965, Glock and Stark offered their own sociological definition of religion in their book *Religion and Society in Tension*. They view religion as "the social scene" with its "system of symbols, beliefs, values, and practices focused on questions of ultimate meaning" (Glock and Stark 1965: 14). They suggest that religion has five dimensions, the experiential, the ritualistic, the ideological (beliefs), the intellectual (knowledge) and the consequential. They analysed 3,000 questionnaires which they had sent out to Roman Catholic and Protestant church members in San Francisco Bay.

They thought that they detected four kinds of experience: (1) "A Confirming Experience" was characterized by a feeling, knowing or intuition that one's beliefs were true, based on the sense of sacredness or awareness of a presence. (2) "A Responsive Experience" might be salvational, miraculous or sanctioning: people felt saved, they interpreted some experience as miraculous guidance, or as punishment, or deflection from inappropriate goals. (3) "The Ecstatic Experience" involved a sense of awareness of the Divine and a feeling that one enjoyed a reciprocal relationship, sometimes denoting a sense of union. (4) "The Revelational Experience" was the least common type, and might be characterized as a cordial and confident union. Such experiences might be revelatory, a source of enlightenment or involve a commission.

These stages were offered as a developmental sequence. Glock and Stark found that 45 per cent of Protestants and 44 per cent of Catholics said that they had experienced "a feeling that [they] were somehow in the Presence of God".

We can see that the first experience is primary, the second is a conversion experience, while the third and fourth reflect the fifth use of religious experience we are interested in – mystical union. We saw earlier that James also sometimes saw this as a developmental sequence.

GREELEY: BIPOLAR, CONVERSION
AND MYSTICAL EXPERIENCES

In a national survey in the United States in 1974, Greeley asked a sample population (N = 1,467) this question among many: "Have you ever felt as though you were close to a very powerful spiritual force that seemed to lift you out of yourself?" This is called the Greeley "key question". Analysis of the result showed that 35 per cent responded positively, which he reported in his book, *Ecstasy: A Way of Knowing* (1974: 140). Greeley offers his account of what he means by ecstasy as a way of knowing:

> It is not just in religious mysticism or sexual union or murky drug trips that man breaks out of the boundaries of ordinary time and place. The height of a mountain, the expanse of an ocean or lake, a fresh breeze at the end of a hot summer day, a child struggling to walk across

a lawn, a flash of lightening, the elaborated design of a Picasso painting, the feel of a sail-boat dancing over the waters, the sight of a human body – male or female – on the beach may all free us out of the limitations of our self hood and the dull monotonous grey-tinged life of ordinary space and daily durations. These are moments of intimations of ecstasy, potential triggers . . . the Spirit of God is out there . . . and he is also inside us enveloping us with fire and warmth . . . (Greeley 1974: 138)

The most important triggers of mystical experiences in his sample population (from 49 per cent down to 30 per cent positive responses) were: listening to music, attending church services, listening to a sermon, watching little children, reading the Bible, being alone in church, prayer, beauties of nature, moments of quiet reflection.[11] Feelings resulting from these experiences (from 55 per cent down to 32 per cent): a sense of profound peace, a sense that all things would turn out for the good, a need to help others, the conviction that love is the centre of everything, a sense of joy and laughter, a great increase in knowledge and understanding. The length of these experiences for 50 per cent of respondents was less than fifteen minutes, while 20 per cent reported that they lasted from a half to a few hours and 20 per cent for a day or more.

In his work Greeley does not offer separate accounts of primary religious experience, conversion and mystical experiences, but includes them all, undifferentiated, in his inclusive survey.

HARDY: BIPOLAR, CONVERSION AND MYSTICAL EXPERIENCES

Sir Alister Hardy offers a summary as well as comment on the work of the Religious Experience Research Unit (RERU) in Oxford, of which he was founder and first director in 1969. Over 4,000 responded to Hardy's key question, published in the national press in 1969/70: "Have you been aware of, or influenced by a presence or power, whether you call it God or not, which is different from your everyday self?" (Hay 1990: 56; Hardy 1979: 18, 20). His collected data, and analyses, include primary religious experiences, conversions and mystical experiences. In answering this question, each respondent effectively defined "religious experience" in their own terms.[12]

Robinson published the first detailed study from the data collected by Hardy. He discovered that 15 per cent of all respondents stated that their first experience occurred in childhood, though they also said that their experiences in adult life were equally important (Robinson 1977a: 169). Robinson argues that "the original vision" of childhood, as he called it, is a form of knowledge. It was often mystical, self-authenticating in that it is authoritative and also brought a person to an awareness of their true self. It is purposive and religious. Here is one such experience:

When I was about five I had the experience on which, in a sense, my life has been based. It has always remained real and true for me. Sitting in the garden one day I suddenly became conscious of a colony of ants in the grass, running rapidly and purposefully about their business. Pausing to watch them I studied the form of their activity, wondering how much of their own pattern they were able to see for themselves. All at once I knew that I was so large that, to them, I was invisible – except, perhaps, as a shadow over their lives. I was gigantic, huge – able at one glance to comprehend, at least to some extent, the work of the whole colony. I had the power to destroy or scatter it, and I was completely outside the sphere of their knowledge and understanding. They were part of the body of the earth. But they knew nothing of the earth except the tiny part of it which was their home.

Turning away from them to my surroundings, I saw there was a tree not far away, and the sun was shining. There were clouds, and blue sky that went on for ever and ever. And suddenly I was tiny – so little and weak and insignificant that it didn't really matter at all whether I existed or not. And yet, insignificant as I was, my mind was capable of understanding that the limitless world I could see was beyond my comprehension. I could know myself to be a minute part of it all. I could understand my lack of understanding.

A watcher would have to be incredibly big to see me and the world around me as I could see the ants and their world, I thought. Would he think me to be as unaware of his existence as I knew the ants were of mine? He would have to be vaster than the world and space, and beyond understanding and yet I could be aware of him – I was aware of him, in spite of my limitations. At the same time he was, and he was not, beyond my understanding.

Although my flash of comprehension was thrilling and transforming, I knew even then that in reality it was no more than a tiny glimmer. And yet, because there was this glimmer of understanding, the door of eternity was already open. My own part, however limited it might be, became in that moment a reality and must be included in the whole. In fact, the whole could not be complete without my own particular contribution. I was at the same time so insignificant as to be almost non-existent and so important that without me the whole could not reach fulfilment.

Every single person was a part of a Body, the purpose of which was as much beyond my comprehension now as I was beyond the comprehension of the ants. I was enchanted. Running indoors, delighted with my discovery, I announced happily, "We're like the ants, running about on a giant's tummy!" No one understood, but that was unimportant. I knew what I knew.

It was a lovely thing to have happened. All my life, in times of great pain or distress or failure, I have been able to look back and remember, quite sure that the present agony was not the whole picture and that my understanding of it was limited as were the ants in their comprehension of their part of the world I knew . . . (Robinson 1977a: 12–13)

This religious experience was sent to the RERU by a woman, aged fifty-five. We will have occasion to analyse it in some detail later. The RERU publications, including Hardy's book, quote many such reported experiences (Beardsworth 1977; Maxwell and Tschudin 1990). Robinson explored the growth of religious awareness (1978) and mature reflections on the nature of religious experience (1977b). Within this latter group of statements, Professor Peter Baelz expresses a philosophical insight, which relates to the Hardy question:

Even that phrase – experiencing a sense of presence – is in a way interpretative. If you feel somebody is there, you have already got the notion of somebody; you are already applying the category of a human being which you have got from ordinary experience to interpret this feeling of presence. (Robinson 1977b: 81).

Baelz also says:

if you interpret this power in personal terms, as purposive or caring, the kind of life you develop in relation to it may be . . . subtly different from what it would be if you interpreted it simply in impersonal language . . . If we find this personal relatedness as the peak which evolution has so far reached, and if we want to go on to extend this personal relating even to this power beyond, whatever it may be, so that it would be an "as if" relationship, "as if" personal, I don't see why one should not take the next step and say that this power is personal. (Robinson 1977b: 74–5)

Baelz makes one thing very clear. The claim to have a religious experience is expressed in the interpretive language of not just personal discourse, but also in religious discourse. For example, only God-talk gives the appropriate mode of expression to help the 55-year-old woman relate her religious experience quoted above. This is true of most of the other experiences quoted by the RERU.

Religious experience is similar in structure to every other kind of experience. Every kind of experience, whether empirical – either just sense experience or scientific investigation – aesthetic, moral or of personal relationships involves perception, interpretation and attributed significance. As has been said before, there are no plain facts; humans cannot have uninterpreted experience. So-called facts are the results of interpreted perceptions. Different kinds of experience require different modes of discourse related to their corresponding forms of knowledge by those involved in them, in order to give an adequate account of them.

In examining the results of Hardy's project, one needs to remember that the respondents were a self-selected group, so the results have limited, though important significance. The results were classified into ninety-two types, grouped into twelve categories. These can be clustered into six groups which emphasize the characteristics stated (the Hardy twelve categories are in brackets):

I   (11) triggers, which initiated the experience;
2   (1–5, 10) means of communication of experience;
3   (9) underlying beliefs;
4   (7) cognitive and affective elements;
5   (6, 12) behavioural changes or effects;
6   (8) developmental characteristics.

Various triggers, which initiated religious experiences, were identified by those reporting them. The five most important triggers were natural beauty, sacred places, sharing in religious worship, prayer or meditation, and music.

Experiences were communicated to their recipients in various ways, such as through sight, hearing, touch, smell, perception or through dreams.

Those who reported religious experiences believed that they came from beyond themselves, "out of the blue", from deep within themselves, or most frequently in answer to prayer.

The main elements in and effects of these religious experiences

Hardy analysed in detail 3,000 of the scripts he had received.

The experiences here described include all those feelings which are most generally associated with the spiritual side of man: the sense of joy, peace, security, awe, reverence, and wonder; the feelings of exaltation and ecstasy, of harmony and unity, of hope and fulfilment, the sense of timelessness, the sense of presence, the sense of purpose, and the sense of prayer answered in events. There is also the darker side: feelings of remorse and guilt, of fear and horror. (Hardy 1979: 51)

An examination of cognitive and affective elements (Group 4 above) reveals that the most frequently mentioned elements cited together indicate that these significant experiences were experiences of personal transformation. For example, many reported a sense of security, protection and peace (25 per cent); a sense of joy, happiness, wellbeing, exaltation, excitement and ecstasy (21 per cent); a sense of presence (20 per cent); a sense of certainty, clarity, enlightenment (19 per cent); a sense of guidance, vocation or inspiration (16 per cent); a sense of prayer answered (14 per cent); or a sense of purpose behind events (11 per cent) (Hardy 1979: ch. 4).

The main effects reported were a sense of purpose or new meaning to life, and included healing, guidance or comfort, which contained a range of the elements quoted above. As Hardy says, it is not always easy to distinguish between an actual experience and its consequences or effects.

Hardy reports on religious experiences of evil. Some had an appalling sense of fear (4 per cent); or felt a sense of remorse or sense of guilt (2.5 per cent). Of these examples 155 were studied by Jakobsen in 1999. The experiences seem to be of hostile facial expressions or emotion. This compares with Glock and Stark's finding in 1965 that 32 per cent of Protestant church members tested believed they had been tempted by the devil (Argyle 2000: 48).

Key developmental characteristics of the religious experiences reported were a sudden change to a new sense of awareness, conversion or "moment of truth". Sometimes the change to a new sense of awareness was gradual (Hardy 1979: 26–28). Further analysis of the data in the last category listed above (Group 6) supports a developmental model for the spiritual nature of persons. This indicates that 76 per cent of those who mentioned periods of development indicated that childhood and adolescence were the major periods of development. Middle age, surprisingly, was more important than old age. About 24 per cent reported a growth in gradual awareness, while 15 per cent cited particular experiences as important in this development. In response to the Hardy key question, 45 per cent specifically stated "the beyond" as initiating a relationship with or responding to them.

Hardy summarizes his findings:

> It seems to me that the main characteristics of man's religious and spiritual experiences are shown in his feelings for a transcendent reality which frequently manifest themselves in early childhood; a feeling that Something Other than the self can actually be sensed; a desire to personalize this presence into a deity and to have a private I–Thou relationship with it communicating through prayer. (Hardy 1979: 131)

HAY: BIPOLAR, CONVERSION AND MYSTICAL EXPERIENCES

Hay used the Hardy key question,[13] and in his book *Religious Experience Today* collected the results of a number of surveys related to this question. He reported positive responses to questions about "religious experience" in eleven national surveys with sample sizes ranging from 985 to 3518. Back and Bourque conducted three surveys between 1962 and 1967, giving improving results of 20 per cent, 32 per cent and 41 per cent. We have already noted Greeley's result of 35 per cent. Hay and Morisy in the UK found 36 per cent responded positively to the Hardy question and 31 per cent positively to the Greeley question. Two polls in the United States in 1978 produced results of 31 per cent and 35 per cent. The Hardy question applied in Australia in 1983 produced a 44 per cent result, while in the UK and the United States in 1985 it was 33 per cent and 43 per cent respectively. In 1987 Hay and Heald had a 48 per cent result in the UK (Hay 1990: 79). In sum, these surveys consistently show that a third of the population consistently report having had one or more religious experiences, with the figure creeping up to 50 per cent on occasions.

A very important finding emerged from Hay and Morisy's 1976 poll, when they discovered the effect of education on the results of the Hardy key question. The positive responses increased the longer students stayed in full-time education. Leavers at fifteen – 29 per cent; sixteen – 37 per cent; nineteen – 44 per cent and twenty-plus – 56 per cent (Hay 1990: 82). This highlights the fact that education gives everyone a greater grasp of language so that they more easily understand the question. It does not indicate that they are more religious! In the 1987 Gallup Poll survey, we saw that 48 per cent of the population reported having had a religious experience. If premonitions are included, the proportion of the population rises to two-thirds. Parallel to this work, Hay ran a series of in-depth interviews with a sample of the Nottingham population, where he found that over 60 per cent claimed to have had a religious experience (Hay 1990: 57).

SUMMARY

At the beginning of this chapter we identified seven uses of the term "religious experience".

1   bipolar experiences – between God and a person;
2   habits of religious behaviour described as "religious experience";
3   religious knowledge, described as "religious experience";
4   a conversion experience related to a religious community;
5   mystical union, one kind of religious experience;
6   the discipline of meditation and contemplation leading to mystical union with God;
7   all experience is interpreted as religious, and is called "religious experience".

The reader will be very aware that only three uses have been illustrated and discussed from the writings so far explored in this chapter. They include the first, the primary religious experience – the person who on one or more occasions has experienced the Presence of God, which is a bipolar experience. The example given was that of the 55-year-old woman who reported her experience at the age of five. One example of the fourth use of the term "religious experience" related to conversion experiences was that of John Wesley.

No example has been given of the second use of "religious experience" – the habits of worship and the attendant religious way of life – as it has not been a major area of research interest. Places of worship are attended by many very worthwhile people who are committed to a life of faith, belief in God, the social activity of worship and living out the associated way of life faithfully, and often self-sacrificially. A few practising church-going Christians, interested in my project, say they have never had a personal religious experience like those reported by James, Hardy and Hay.

No example has been given of the third use of "religious experience" – religious experience is what you know – which related to the learning about a faith experienced by a child or an adult in a religious community. In the UK every person has been in this situation, if not in a religious community, then at least in a school religious education lesson. Such learning in early childhood is fraught with verbal misunderstandings of religious symbolism or is taken literally. For example, two small children, both of whom had religious parents, as adults smile at their youthful misunderstandings, one saying:

> As a small child I used to say my prayers following what Mummy said, 'Gobless Daddy, and Gobless Mummy and Gobless brothers and sisters and aunts and uncles and cousins and everyone else in the world, and Gobless me.' It was some years later that I worked out who Gobless was!

Another adult, already quoted in Chapter 4, said:

> I was taught in Sunday School that God was light. One day I told Dad that I had seen God in the hall. So he took me by the hand to show him, and I pointed triumphantly at the frosted window which framed a centre of diffused light. Dad asked me if he could open the window, so I said, 'Yes.' When he opened it we saw the street light lit up outside, as it was dark outside. And Dad said that I was right, as God was like light, and is the Light who helps us to see what is right and good.

The development of human religious understanding has been explored by many research workers, often using Piagetian and other techniques, and does not need comment here.[14] The term "religious experience" when used in this third way, is an intellectual activity with differing degrees of emotional commitment.

One example has been given of the fourth use of "religious experience" – conversion – that of John Wesley. Many more are contained in the RERU archives.

The fifth use of "religious experience" – mysticism – was quoted from James's classic account of mystical experience.

The sixth use of "religious experience" as indicating that mystical experience was sometimes the result of disciplined meditation and contemplation is a focus of Underhill's book *Mysticism*, mentioned earlier.

No mention has yet been made of the seventh use of the term "religious experience" when it is used to describe all experience, which is interpreted from a religious perspective. Many talk of "religious experience" as in one sense referring to a religious outlook on life. This is clearly one legitimate use of the term, which can be well illustrated from religious biography.

In the *Handbook of Religious Experience*, the editor "sought authorities who were willing to confront from within their own expertise what is known about religious experience. Religion was defined in a Jamesian fashion as some sense of the transcendent ... Experience was suggested to be an encompassing phenomenon broader than mere behavior, affect or cognition" (Hood 1995: 4).

This Jamesian definition is however immediately expanded by the six authors who wrote on their Faith Tradition, whether Judaism, Catholicism, Protestantism, Islam, Buddhism or Hinduism, largely using "religious experience" to mean the whole of life interpreted from a religious perspective, which I have called a "religious outlook on life", the seventh meaning of the term, which we are now considering.

One example from this group of contributors will illustrate the situation. The author of the entry on Protestantism does start by considering religious experience in a Jamesian fashion, discussing James's two types of religious experience, conversion and mysticism as in James. The author then analyses "religious experience in the Protestant tradition", discussing the nature of faith, the place of the Bible, church membership, conversion, the place of symbol, religious knowledge and personal morality. In this section of his chapter, "religious experience" has changed its meaning from an individual religious experience in time – the first use of the term, with which we are concerned – to the seventh use of the term, meaning a "religious outlook on life", which is outside our area of study.

We will look at one example of writers from within their own specialisms using "religious experience" to refer both to individual religious experiences and to mean the whole of life interpreted from a religious perspective. Tamminen and Nurmi (1995: 270) state:

> The development of religious experience refers to an underlying sequence of age related changes that is typical for the human life span. Religion is here understood on the one hand as a personal relation towards the ultimate value base of action (in Christianity towards God), on the other hand ... as an organized movement with doctrine and acting as a reference group for the individual's self understanding.

Such double usage is confusing.

However, before we leave discussion of the seven uses of "religious experience", I include an example of a personal religious experience taken from an autobiographical account of the distinguished Emeritus Professor of Religious Education at the University of Birmingham.

John Hull, born in Australia in 1935, has throughout his life had eye problems which grew worse until he totally lost his sight in 1983 at the age of forty-eight. All his lecture folders he could no longer read. He created an operating system to overcome this handicap with the use of tape-recorded lecture notes. On 22 December 1985 he wrote:

> Providence means "looking ahead" and refers to the idea of God leading you along a path. But meaning is conferred after the event ... Faith is a creative act. It is through faith that we transform the accidental events of our lives into signs of destiny. Meaning is conferred when chance is transformed through a rebirth of images ... The most important thing in life is not happiness but meaning.

On 21st April 1986 he wrote:

In recent weeks the thought has been in my mind that blindness could be a gift . . . I resist the thought, for if blindness is a gift, I would have to accept it. I have said to myself that I would never accept it. Yet I find the thought keeps coming back to me, and arousing my curiosity. Could there be a strange way in which blindness is a dark paradoxical gift? Does it offer a way of life, a purification, an economy? Is it really like a kind of painful purging through a death? Am I to expect that I shall enter into a new, more concentrated phase of life because of this gift?

On 28th April 1986 he wrote:

On Sunday 27th April I went with my colleague (Michael Grimmitt) to Mass in Notre Dame Cathedral in Montreal. The service was entirely in French, but it caught my attention. Although I hardly understood a word, I was rapt through the whole service. The organ is one of the most famous in North America, and it certainly was a powerful and beautiful sound. I found myself thinking again about blindness as a gift. As the service proceeded and as the whole place and my mind were filled with that wonderful music, I found myself saying, "I accept the gift. I accept the gift." I was filled with a profound sense of worship. I felt I was in the very presence of God, that the giver of life had drawn near to inspect his handiwork. He had drawn near as one who hardly dares to look at the result of his work. If I hardly dared approach him, he hardly dared approach me. "It's all right," I was saying to Him. "There's no need to wait . . . Go on, you can go now, everything's fine."

The use of "religious experience" here is the first use, a bipolar experience – a person in the presence of God. Later Hull does outline his religious outlook, our seventh use of "religious experience" in a postscript to his account of *Touching the Rock*.

There are many worlds and many states, but God is a transworld reality.
God is the God of every world, and the Lord of every state:
God is many and yet one, and in God there are many worlds yet one.
God does not abolish darkness; God is the Lord of both light and darkness.
If in God's light we see light, then in God's darkness we see darkness.
If a journey into light is a journey into God, then a journey into darkness is a
    journey into God. This is why I go on journeying, not through, but into.[15]

In this study we have focused on using "religious experience" to refer to (1) a bipolar personal experience of the presence of God, (4) a religious conversion experience or (5) a mystical unitive experience.

We have briefly explored seven uses of the term "religious experience", the primary personal experience, often alone – a person in the presence of God, a bipolar experience; the habit of communal worship accompanied by faithfully following its attendant way of life; intellectual understanding; a conversion experience – often related to a community; mystical union; disciplined meditation leading to mystical union; and the whole of life interpreted from a religious point of view – a religious outlook.

We have discovered that on average 35 per cent of the populations of the United States and UK have had an individual religious experience. However, the greater the educational opportunities people have had, the more likely they are able to understand questions about religious experience, so that of those who have been in full-time education after the age of twenty years, 56–60 per cent claim to have had a religious experience.

Professor Baelz underpinned the view that religious experience, like all other kinds of experience, is interpreted experience. He suggests that "the concept of God has a double parentage. In part it is derived from reflection on the world and on man; but in part it is derived from within experience itself" (Baelz 1968: 98). He also says that "religious experience is man's 'normal' experience in so far as it does justice to that fundamental relationship to the other in which he is in fact involved" (114; cf. Hick 1989: 158–9).

Religious experiences, in the first, fourth and fifth senses (the personal bipolar experience, the conversion experience and mystical union), are none the less authentic if they properly interpret what an individual has encountered.

We have seen that religious experiences, understood in the first, fourth and fifth senses used here, have a structure. They have *triggers*, like ants and a blue sky, or blindness (in the two religious experiences quoted earlier), which provoke reflection and lead to a religious experience, *situations* which provide a context for it to happen, a distinctive set of religious *emotions*, particular *cognitive insights* and psychological and behavioural *consequences* which can be evaluated. Are the results good or life enhancing for the individual and for others affected by that person? We will need to explore these issues more fully later.

James claims that these three kinds of personal religious experience (a bipolar experience, a conversion experience and an experience of mystical union) have complete authority, truth value and trustworthiness to the individual. While, on the one hand, I agree with James, on the other hand, Lash pointed out earlier that the individual has grown up in a society in which language, culture, relationships, religious and other institutions have greatly affected their personal experiences. Also, the experience and its consequences affect the community in which that person lives, and which legitimately makes a judgement on that experience. We know that conversion experiences only have validity if the relevant religious community accept them as such (Bouquet 1968: 69–80).

Those religious experiences which transcend religious community boundaries may well be authoritative to the person having them, but in the end they need to be given credence by a wider society in order to discover if that society finds them life enhancing. That endorsement may not come until after the death of the person who had the original religious experience – worked out in their life's work.

These criteria accord with those found in all areas of knowledge, e.g. science, the arts and moral discourse. In them the validity of scientific truth claims, the value of artistic creation, or the quality of moral judgements are finally affirmed by the scientific, artistic and moral communities respectively. These communities, concerned with truth in different areas and forms of knowledge, are the democratic guarantors of the evolving values in all sections of the community, local, national and international.

In each area of inspirational experience, whether in religion, science, the arts or in moral judgement there can often be a long interval before the relevant community accepts, or finally rejects, claims made by a religious leader, by the advocate of a scientific hypothesis, a radical musical composition or the "new" insistence on a moral principle.

Examples include the time it takes to accept the vision of Jesus of Nazareth and the time it takes to reject the worship of Neptune; or the time it takes to reject or accept the work of a scientist into the corpus of accepted scientific truth; or the time it takes to accept or reject the work of a composer into the repertoire of the great orchestras of the world; or the time it takes to accept the equality of women and men in all areas of society, local, national and international.

One cannot allow the charge of "rampant individualism" to be levelled against the autonomous authority of a religious experience for the individual who has it. A religious experience has to be tested by two criteria. The first is individualistic – the experience gives one delight. But the second criterion, that the religious experience leads to good consequential fruits, is judged not only by the individual, but also by the community.

From now on, our central concern will be with the first, fourth and fifth uses of "religious experience" we have briefly explored in this chapter.

NOTES

1   James (1968: 249): "I was taken to a camp meeting, mother and religious friends seeking for my conversion. My emotional nature was stirred to its depths; confessions of depravity and pleading with God for salvation from sin made me oblivious to all surroundings. I plead for mercy and have a vivid realization of forgiveness and renewal of my nature. When rising from my knees, I exclaimed 'Old things have passed away, all things have become new.' It was like entering another world, a new state of existence. Natural objects were glorified, my spiritual vision was so clarified that I saw beauty in every material object in the universe, the woods were vocal with heavenly music; my soul exulted in the love of God, and I wanted everyone to share in my joy."

2   The number of subjects was 192 (120 female and 72 male) (Starbuck 1899).

3   For this reason, it has been described here as a *secondary religious experience*. I reject Clark's view that this is primary for the reason given above, for it shares the same social context of a worship service which Clark does call secondary experience or behaviour. I agree with his view of tertiary religious experience or behaviour as describing what children learn at a cognitive level, often socially in church (Clark 1967: 23–4).

4   "When we think certain states of mind superior to others, is it ever because of what we know concerning their organic antecedents? No! it is always for two entirely different reasons. It is either because we take an immediate delight in them; or else it is because we believe them to bring us good consequential fruits for life" (James 1968: 37).

5   "By their fruits ye shall know them, not by their roots" (James 1968: 41).

6   "In its systematic variety, it is an abstract way of conceiving things as good" (James 1968: 101). Movements which are subsumed within this model of religious development include New England Transcendentalism (106) and Christian Science (118); "pessimism leads to weakness, optimism leads to power" (119).

7   (1) "Mystical states, when well developed, usually are, and have the right to be, absolutely authoritative over the individuals to whom they come . . . the mystic is, in short, invulnerable . . . " (James 1968: 407–8)

      (2) "No authority emanates from them which should make it a duty for those who stand outside of them to accept their revelations uncritically" (407).

      (3) "They break down the authority of the non-mystical or rational consciousness, based upon the understanding and the senses alone. They show it to be only one kind of consciousness . . . mystical states merely add a supersensuous meaning to the ordinary outward data of consciousness . . . " (407, 411).

8 "1. That the visible world is part of a more spiritual universe from which it draws its chief significance;

   2. That union or harmonious relation with that higher universe is our true end;

   3. That prayer or inner communion with the spirit thereof – be that spirit 'God' or 'law' – is process wherein work is really done, and spiritual energy flows in and produces effects, psychological or material within the real world."

Of the two models of religion which James explores, he clearly regards the model of the sick soul as one "in which all religions appear to meet" (1968: 464).

9 An "overbelief" is an ultimate belief which is a fundamental intuition of an individual in all areas of understanding. It is an individual judgement made in a particular form of knowledge, like religious, aesthetic or personal knowledge, which cannot be proved by quoting evidence, e.g. "That picture is beautiful"; "I felt God next to me"; "I love her"; "She is gracious." One can give reasons for a statement, but the reasons do not establish the validity of a form of knowledge. They only offer reasons to support a statement within a form of knowledge.

10 "Confining ourselves to what is common and generic, we have the fact that the conscious person is continuous with a wider self through which saving experiences come" (James 1968: 490).

11 Examples taken from Greeley's work can be found in Hay (1990: 70–1).

12 An insight from Professor Adrian Thatcher.

13 "Have you been aware of, or influenced by a presence or power, whether you call it God or not, which is different from your everyday self?" (Hay 1990: 56).

14 Consult the *British Journal of Religious Education*. An example of a specific work is Miles (1971).

15 Hull (1991). He makes it clear that he understands how he became blind; it was no one's fault. I have only quoted extracts of his search for meaning. Place this experience besides those related towards the end of Chapter 3.

REFERENCES

Argyle, M. (2000), *Psychology and Religion*, London and New York: Routledge.
Baelz, P. R. (1968), *Christian Theology and Metaphysics*, London: Epworth Press.
Beardsworth, T. (1977), *Living the Questions*, Religious Experience Research Unit RERU Westminster College, Oxford.
Bouquet, A. C. (1968), *Religious Experience*, Cambridge: Heffer.
Clark, W. H. (1967), *The Psychology of Religion* (1958), New York: Macmillan.
Donovan, P. (1979), *Interpreting Religious Experience*, London: Sheldon Press.
Flew, A. (1979), *A Dictionary of Philosophy*, London: Pan Books.
Glock, C. Y. and Stark, R. (1965), *Religion and Society in Tension*, Chicago: Rand McNally.
Greeley, A. M. (1974), *Ecstasy: A Way of Knowing*, Englewood Cliffs, NJ: Prentice-Hall.
Happold, F. C. (1975), *Mysticism*, Harmondsworth: Penguin.
Hardy, A. (1979), *The Spiritual Nature of Man*, Oxford: Clarendon Press.
Hay, D. (1990), *Religious Experience Today*, London: Mowbrays.
Hick, J. (1989), *An Interpretation of Religion*, London: Macmillan.
Hood Jr, R. W. (ed.), (1995), *Handbook of Religious Experience*, Birmingham, AL: Religious Education Press.
Hull, J. (1991), *Touching the Rock*, London: Arrow Books.
James, W. (1968), *The Varieties of Religious Experience* (1902), London: Fontana.
Lash, N. (1988), *Easter in Ordinary*, London: SCM Press.
Maxwell, M. and Tschudin, V. (1990), *Seeing the Invisible*, London: Arkana.
Meadow, M. J. and Kahoe, R. D. (1984), *Psychology of Religion*, New York: Harper & Row.

Meissner, W. W. (1984), *Psychoanalysis and Religious Experience*, New Haven: Yale University Press.

Miles, G. B. (1971), "A Study of Logical Thinking and Moral Judgements in GCE Bible Knowledge Candidates", MEd dissertation, University of Leeds.

Parker, P. L. (1906), *John Wesley's Journal (abridged)*, London: Pitman.

Robinson, E. (1977a), *The Original Vision*, Oxford: Religious Experience Unit.

Robinson, E. (1977b), *This Time Bound Ladder*, Oxford: Religious Experience Unit.

Robinson, E. (1978), *Living the Questions*, Oxford: Religious Experience Unit.

Smart N. (1968), *Secular Education and the Logic of Religion*, London: Faber and Faber.

Starbuck, E. D. (1899), *The Psychology of Religion*, New York: Scribner's.

Tamminen, K. and Nurmi, K. E. (1995), "Developmental Theory and Religious Experience", in R. W. Hood Jr (ed.), (1995), ch. 13.

Vardy, P. and Grosch, P. (1994), *The Puzzle of Ethics*, London: Fount.

Watts, F. (2002), *Theology and Psychology*, Aldershot: Ashgate.

Wulff, D. M. (1991), *Psychology of Religion*, New York: Wiley.

11 Religious and Mystical Experience

In Chapter 9 we explored different uses of "transcendence", not least because Michael Paffard, an agnostic humanist, so described the experiences of nature mysticism reported to him by his sixth form and undergraduate students, some of which were reproduced in Chapter 1.

Certain of these experiences were described as, and therefore interpreted as, religious. Some were described as experiences of nature mysticism, and we could not tell from the texts whether the underlying metaphysic, i.e. the underlying ultimate assumptions, were humanist or religious. Some were described as, and therefore interpreted as, Humanist.

Our brief investigation of 'spirituality' in Chapter 9 suggested that spirituality can be understood within religious perspectives, but that humanists also have a legitimate claim to talk of spirituality from their interpretive stance.

In Chapter 10 we explored those writers who studied transcendental experiences which they interpreted as religious experience, either as primary personal religious experiences, conversions or mystical experiences. Sometimes the boundaries of these different kinds of experience overlap and merge.

Now we turn to a group of humanist writers who claim that the "religious experiences" claimed by religious communities are very important human experiences and are too important to be monopolized by religious communities. They therefore seek to reclaim them as humanist experiences. Some of these writers belong to the humanist psychology movement we discussed in Chapter 3, where they made their contribution to the discussion of the topic of "what constitutes a human being".

NATURE MYSTICISM: A HUMANIST METAPHYSICAL STANCE

Leuba: *The Psychological Study of Religion*

The first major writer of the twentieth century who regarded mystical experience as part of the natural order was J. H. Leuba; he claimed that they were just human experiences. In *The Psychological Study of Religion*, written in 1912, he claimed there was no supernatural or theistic dimension to such experiences. In other words, it is not necessary to make the knowledge claim that there is a divine origin of the universe, or that God is the source of all being. He said, "I cannot persuade myself that divine personal beings . . . or the Christian Father, have more than a subjective existence." The profound emotions, fear, awe, reverence are not only religious; indeed, Leuba feels that "sacredness" should only be used for "the sacredness of life" (1912: 10).

In his theoretical examination of the mystical states, he accepts that they lead towards an inner reconciliation or union, but regards this as a natural phenomenon. He mocks James's attempt to suggest a hypothesis of divine action (Leuba 1912: 274). In regarding mystical states as natural phenomena, he said that "the only invulnerable thing in the 'union with the infinite' is a consciousness which does not reach beyond itself" (240).

The Psychological Study of Religious Mysticism was his second work, and is interesting because he argues that the mystical ecstasy in Wordsworth and Tennyson was a natural phenomenon. However, for all his naturalism his criticism was largely of assured doctrine. In the end he did not deny "the reality of any and every kind of Unseen" (Leuba 1925: 240). Mystical experiences, understood religiously, distracted people from recognizing that they were "the lawful workings of our psycho-physiological organism" (316). It must immediately be noted that this statement is a piece of physiological reductionism. In other words, Leuba is saying that a mystical experience is nothing but a *biological* process, and is confusingly suggesting that a religious statement should be a biological statement, without recognizing that both kinds of statement are legitimate and that the metaphysical interpretation could be theistic or atheistic.

There are two points in Leuba's thought which need comment at this stage. The first is his suggestion that mystical experience is just the production of psycho-physiological processes. As has just been said above, we can now see that this was a piece of reductionist thinking. He claimed that mystical experience is *nothing but* a psycho-physiological process, but in fact we now know, following our exploration of different kinds of knowledge, that mystical experience, like any other human experience, can be understood at a physiological level, a psychological level and a mystical level. And the mystical level can be interpreted in a humanist or a religious way.

Different kinds of knowledge are always needed to study human phenomena at different levels of understanding – which may include physical, biological, physiological, psychological, sociological, personal, aesthetic, moral and religious levels of understanding, as and when appropriate.

The second point in Leuba's thought which needs comment is his own unresolved tension. In his first book he rejects all religious claims, arguing that all such claims are natural phenomena. But in the second he does not deny "the reality of . . . every kind of Unseen". Perhaps he was rejecting dogma in his first book, and accepting some form of unseen Presence in the second book, or perhaps he was just agnostic.

Huxley: *The Doors of Perception*

This 1954 book needs to be placed in context. Some of Huxley's books focused on evil, extreme forms of perversity or stupidity, imbalance in life and the worship of the machine. One work in this group is *Brave New World* (1939). His corrective to this was his interest in an eclectic mysticism in works in 1936, 1941, 1944, followed by *The Perennial Philosophy* (1945). In this book he recognized "a divine Reality", found in various forms in all religions, with an ethic which places man's final end "in the knowledge of the immanent and transcendent Ground of all being" (Huxley (2004: vii). It is an anthology of and "An Interpretation of Great Mystics East and West". "Even an agnostic . . . can read this book with joy. It is a masterpiece of all anthologies" (*New York Times*).

In 1954 Huxley recorded his own experiences under the drug mescalin. It gave him a heightened sense of colour, so that he saw flowers in such a way that he felt he transcended time and space. He admitted his will suffered a profound change for the worse. Huxley suggested that such experiences were comparable to genuine mystical experience, giving him Wordsworthian insights. Huxley's idea of "release from ego" however, meant release from boredom (1954: 49). Such release was not life-fulfilling: he admits that the experiences were morally enervating (18–19). Clearly these experiences, on his own admission, fail the test that James suggests: they did not lead to good fruits, or an enhanced quality of life. The major effect of his book was to provoke Zaehner to write his, which we will explore later.

Laski: *Ecstasy*

Marghanita Laski was a humanist who wanted "to know what ecstasy felt like to other people" (1961: 1; 1980), so she designed a questionnaire asking, "Do you know a sensation of transcendent ecstasy? How would you describe it?" (1980: 9). It was given to a group of sixty-three of her middle-class friends. She was particularly concerned that such experiences, which were had by humanists as well as religious people, were marginalized by humanists because they were always described in religious language. She also collected one group of thirty texts from literary sources and one of twenty-four texts from religious sources (Laski 1980: 399–438).

She herself insisted that such experiences as she and her humanist friends had were purely natural, but because the only available language to describe them was "religious", she found she had to use religious language analogically to describe them. She found that these experiences were triggered by four main stimuli: nature, sexual love (not intercourse), art and religion.[1]

She classified and analysed the experiences she had collected, and found that they all expressed very similar feelings: 45 per cent reported feelings of gain, and 20 per cent feelings of loss. She defined ecstasy as "a range of experiences characterized by being joyful, transitory, unexpected, rare, valued, and extraordinary to the point of seeming as if derived from a praeternal source (i.e. outside of nature)" (Laski 1980: 5, 43). An experience was classified by her as an ecstasy if it possessed two of the following feelings: "unity, eternity, heaven; new life, satisfaction, joy, salvation, perfection, glory; contact; new or mystical knowledge; and at least one of the following feelings: loss of difference, time, place; of worldliness, desire, sorrow, sin; of self . . . up-feelings; inside feelings; feelings of calm, peace" (Laski 1980: 42).

She distinguished two main types of ecstasy, those which involved contemplation – especially of nature – and those which were very intense. She distinguished three kinds of intense experiences: first, those involving a sense of purification leading to a feeling of loving kindness to all; second, those which resulted from knowledge gained and led to creativity; third, those which led to "union-ecstasy", either with an aspect of nature or with another person.

As mentioned earlier, Laski both denied a religious monopoly on ecstatic experiences and criticized humanists for ignoring such an important area of experience. Finally, she believed that Wordsworth once thought that the source of his experience was the human mind: he later proposed a religious source in response to a rebuke from Blake (Laski 1980: 374).

Laski insisted that her twofold classification was unique, but we will see later that it is very similar to that of Stace, who identifies "extrovert" and "introvert" mystical states.

Maslow: *Religions, Values and Peak Experiences*

Abraham Maslow (1908–70) was the shy, solitary child of uneducated Russian Jews who had emigrated to New York. An atheist who identified with his free-thinking father, he was opposed to belief in God throughout his life, adopting a Freudian outlook and describing religious believers as "the childish looking for a big Daddy in the sky" (Wulff 1991: 602). Initially he was impressed by *behaviourism*, the belief that humans, like all animals, were responsive agents and that their destiny was determined by social or environmental stimuli, hence the term S–R (stimulus–response) psychology.

Maslow was concerned to release humans from external controls and to make them less predictable, i.e. to increase their freedom. However, there was a mystical strand in his personality, which found expression in music, art and nature. He recognized that Freud's patients were in some way inadequate or even neurotic, and so needed psychoanalysis. Maslow therefore sought out instead mature persons in order to study "the greatness of the human species". He sought to develop a broader humanist psychology.[2]

In his studies he discerned five levels of human need in hierarchical order, beginning from the bottom with *physiological* needs. These include food, water and sleep.

The next level of need is *safety*, which includes security, protection, structure, order and sequence of events. These needs can dominate the lives of the poor, the socially deprived, those of very limited energy or intelligence. In healthy adults from secure families, the safety needs may only appear in relation to a person's outlook on life or their work, in terms of their needs for guiding principles in life, whether based on science as in scientism, or a religious faith like Christianity, or a world-view like humanism.

The third level of need is concerned with a *sense of belonging and the need to receive and give love in family and other social relationships.*

The fourth level is concerned with *self-esteem*. An individual needs to develop social and technical skills, first to win the approval of those around him, i.e. *the respect of others*, and then freely to produce work or levels of performance of a standard which develops their own *self-respect*. Maslow calls these four levels the "lower" or "deficiency" needs.

The fifth and highest level is concerned with the need for *self-actualization*. Often called *metaneeds*, the individual is inwardly driven to achieve greater standards in one or more autonomously chosen areas, such as work, relationships and/or service to others. Despite the diverse areas of work such persons undertook, Maslow found some qualities in common amongst them. A self-actualizer would tend to have a creative and spontaneous personality which was non-authoritarian and democratic, acute perception and acceptance of reality, a creative detachment from the encircling culture, a few deep personal relationships, clear ethical standards consistently applied. Interestingly, it was "fairly common" for "self-actualizers to report mystical experiences".

He explored the topic of mystical experience in his 1964 book *Religions, Values and Peak Experiences*. Influenced by James and Laski, among others, he outlined three key hypotheses or arguments. He first argued that "organized religion, the churches . . . become the major enemies of the religious experience [because] most people lose or forget the subjectively religious experience and redefine Religion as a set of habits, behaviours, dogmas, forms which at the extreme become entirely legalistic and bureaucratic" (Maslow 1970: viii).

He was concerned to replace the approach to religion as a subject for reason to explore, which he felt organized religion had adopted, with an approach through the exploration of individuals' uplifting spiritual experiences. He distinguished two kinds of experience, which he called "peak experience" and "plateau experience". The peak experience was a mountain top experience, the feeling of being on top of the world. Because it was full of emotion, the peak experience surprised the individual, stopping them in their tracks. An experience of intense nature mysticism, such as Wordsworth experienced is a typical humanist or natural experience. But Maslow also accepted that it is quite natural for people to have experiences which are religious, e.g. the call experience of Moses, Jeremiah, Jesus or Paul.

Plateau experience is more like a consistent high residual level of satisfactory experience, which contains beliefs worked out through hard work, taking account of peak experiences, the higher spiritual human values, which are then formulated into an integrated view of life.

These states of mind are relevant to transcendental and transpersonal aspects of life, as Maslow regards man as having "a higher and transcendent nature" that is part of his biological endowment. It follows, then, that religious questions, religious yearnings, religious needs are perfectly respectable scientifically because they are deeply rooted in human nature. Indeed humanist psychologists would think any human without these needs and questions was sick or not fully human.

Central to his argument, Maslow focuses on the core religious or transcendent experience, which is a peak experience. Particular private peak experiences have been present at the very beginning of every known "high religion". So great World Religions like Buddhism, Christianity, Hinduism, Islam, Judaism and Sikhism are built up from the codification and communication of the original mystical experience or revelation. Maslow regards these revelations and mystical illuminations as peak experiences, ecstasies or transcendental experiences.

The second argument of his book, is the assertion that "it is very likely, indeed almost certain that these older reports, phrased in terms of supernatural revelations, were in fact, perfectly natural, human peak experiences of the kind that can easily be examined today". The reader needs to note that Maslow is here offering a reductionist hypothesis, namely that religious experiences are not experiences of individuals in communion with the Divine, but are *nothing but* human experiences of aspects of the natural world which transcend their own narrow or individual outlook on life. So, for example, such a naturalistic transcendent experience could be an experience of being in harmony with nature, or being in harmony with a family, a community or another human grouping. This view is shared by all humanist writers, who affirm that all experiences of whatever kind can only be of aspects of the natural world as we know it or discover it.

However, Maslow was also attempting to do something else when he called these ecstatic experiences peak experiences. He was attempting, as Marghanita Laski did before him, to reclaim ecstatic experiences from the monopoly of religious people. He emphasized that such experiences, however interpreted, were uplifting and made life extremely worthwhile and purposeful, marked by feelings of personal wholeness, creativity and supported an individual's integration into the community.

Maslow's third hypothesis, which underpins his claim of peak experience for the whole human family, is as follows:

> to the extent that all mystical or peak experiences are the same in their essence and have always been the same, all religions are the same in their essence and have always been the same. They should therefore come to agree in principle on teaching that which is common to all of them . . . This something common . . . we may call the "core-religious experience" or the "transcendental experience". (Maslow 1970: 18–20)

This argument is viewed differently from different religious perspectives. Some theologians take the view that their own, and possibly every religion, is unique and so cannot be "reduced" in this way to one kind of religious experience which has common structural and faith elements offering hope or illumination (Katz 1978: 22–3). Other theologians claim that religious experiences, which individuals have, do share common structural and faith elements (e.g. James, Otto, Stace and Smart).

Maslow's Concept of Self-Actualization   Maslow described the four basic levels of need, outlined above, as "lower" or "deficiency" needs. These levels of awareness are called "D-cognition" and focus on "D-values", which is shorthand for deficiency-cognition and deficiency-values.

We have also mentioned previously Maslow's fifth and highest level of need in his hierarchy of needs. This is the internal drive towards what he called self-actualization. In this level, through peak experiences, the individual is lifted above the modern disease of valuelessness. This disease is sometimes described as anomie or listlessness, amorality, anhedonia, which means joylessness, rootlessness, emptiness and hopelessness (Maslow 1970: 82). This self-actualizing level is concerned with the world as it is in itself, as an integrated whole (59) and as it can become, focused on what is *ideal* in human maturity, art, mathematics, science, religion, the environment and society (91).

There are intrinsic values associated with viewing the world in this way, such as truth, goodness, beauty, wholeness, the complementarity of opposites, uniqueness, perfection, completion, justice, order, simplicity, richness, playfulness and self-sufficiency (94). The self-actualizer is internally driven, not least because they view life with such emotions as awe, love, adoration, worship, humility, a feeling of smallness plus godlikeness (94).

Maslow calls this distinct form of knowing cognition of Being, B-knowledge or illumination knowledge (80). It is distinguished by the ability to view the world as it is itself, without regard for human use or profit. Likewise, human life can be viewed in a detached way, with a dimension of sadness for human weaknesses and the fact that life is lived within its own mortal span, in which there is time for quiet and persistent idealistic striving.

From the language used to describe self-actualization, we can see that Maslow was concerned that persons are freed and given the opportunity to become their "true selves". Let us drop his technical language and replace the goal of "self-actualization" with the goal of "maturity in wholeness". He recognizes that level four persons never reach it though they make a valuable contribution to society.

This last points needs expanding. Maslow recognized that level four persons, concerned with *self-esteem* achieved by gaining both the respect of others and their own self-respect, do not have peak experiences and B-cognition. But they do reach fulfilment in the world, using it for good economic, commercial or political reasons. They may be excellent captains of industry, leaders in different social institutions or sections of society, including members of parliament. They achieve self-esteem because of their own pursuit of excellence in their own realm of interest, driven by their own quest for self-respect, defined on their own terms.

Research inspired by Maslow   The Berkeley Religious Consciousness Project randomly selected 1,000 persons living in the San Francisco Bay area in the 1970s and found that three kinds of experience were reported: 50 per cent said that they had had a feeling of contact "with something holy or sacred"; 82 per cent had been deeply moved by an experience of nature; and 39 per cent claimed that they had a feeling of harmony with the universe.

These finding confirmed Maslow's assertion that these feelings were not unusual. Those who had any of these kinds of experience "in a deep and lasting way" tended to be able, organized and self-confident. They spent time meditating and reflected on the purpose of life. Those touched by the sacred had a clear purpose in life.

All three groups felt that their lives were very meaningful. These and other relevant research findings support Maslow's claim that such experiences happen to self-actualizing persons, i.e. persons who are driven by internal aims that are pragmatic or idealistic.[3]

TRANSPERSONAL PSYCHOLOGY

Maslow's work struck a chord in psychological circles in the United States and led to him founding the Association of Humanist Psychology with its mouthpiece the *Journal of Humanist Psychology*. Maslow and other associates soon became aware that they were interested not just in "humanist" matters, but in those involving what Tillich called "ultimate meaning", which involved exploring questions about the nature of the cosmos, life, its source and purpose.

Transpersonal psychology has often been viewed as the fourth psychological methodology, following behaviorism, psychoanalysis and the humanist psychology expounded by Allport and colleagues, which will be explored later. The founding editor of the journal outlined its purpose as focus on the empirical study of

> transpersonal process, values and states, unitive consciousness, meta-needs, peak experiences, ecstasy, mystical experience, being, essence, bliss, awe, wonder, transcendence of self, spirit, sacralization of everyday life, oneness, cosmic awareness, individual and species wide synergy, the theories and practices of meditation, spiritual paths, compassion, transpersonal co-operation, transpersonal realization and actualization; and related concepts, experiences and activities. (Wulff 1991: 613)

The editor also said that all such data can be interpreted from any perspective, naturalist, theistic or any other. This approach has attempted to incorporate psychological insights from Eastern World Religions on the one hand and holistic views from the sciences on the other (e.g. the work of Capra). This includes the attempt to include hierarchical levels of consciousness with an impetus towards integrating personal wholeness and Spirit. Wulff notes that such an outlook, being multidisciplinary, transcends the limits of psychology and should be called transpersonal theory.

Overview

The humanist critiques of "religious experience" as a mode of religious knowing have legitimacy in a number of areas.

First, it is now clear after our exploration of different kinds of knowledge that all kinds of knowledge are based on interpretation. It is therefore open to anyone to interpret their "transcendental promptings", and the unique quality of their experiences, either in religious or humanist language, whichever "rings true" to the experiencer.

There are other important points which need addressing and collecting together here.

Taking Christianity as an example, there is some justification in Maslow's first hypothesis, that people involved in "organized religion, the churches . . . lose or forget the subjective religious experience and redefine Religion as a set of habits, behaviours, dogmas, forms which [can] become entirely legalistic and bureaucratic". We will attend to this criticism later.

Maslow's second hypothesis, that revelations of the past are purely naturalistic happenings, can be rejected simply because it does not take account of the differences between personal knowledge with its mode of knowing and religious knowledge with its mode of personal knowledge of God, which religious people regard as the only adequate language to give even an approximate account of their experience.

Maslow's third hypothesis, that mystical experiences in different religions have a common core, but are expressed through different cultural stories or conceptual frameworks, has much plausibility and will be addressed fully in the next chapter.

Let us now address Maslow's first critical hypothesis, outlined above, in conjunction with his ideal, which he calls Self-Actualization, but which, avoiding technical jargon, we can call "Maturity in Wholeness". I agree with Maslow that there is cause for concern expressed in his criticism. The religious language used in Churches is always in danger of becoming outdated technical jargon, remote from the language used to express the life experiences of the "man or woman in the street".

For example, if one takes an interested group of young people from a secondary school "religious education class", who have no first-hand experience of religious services, into a "good traditional church service", the experience is as alien and incomprehensible to them as is the first experience of a group of religious studies undergraduates going into a Sikh Gurdwara, Hindu Mandir or Muslim Mosque, even after explanations before and after the experiences. Almost no existential connections are made between the liturgies observed and the day-to-day experiences of the visiting observers. Maslow is right; it is necessary to find contemporary imagery from the vernacular to close the gap which has widened between initial religious experiences in a religion and the outdated imagery which has expressed them well for many centuries.

Maslow is also right in drawing attention to the difference between the original experience, life and teaching of Jesus, the interpretations of the New Testament writers who were the first "spin doctors" expounding Christianity, and the later doctrinal accretions of the later centuries. The latter should be seen as interpretations for their own time rather than as binding statements of truth for all time. The fossilization of religious doctrine is one factor among many sociological factors which may account for the decline in Church membership.

Some may protest that my analysis fails to do justice to the present scene. They may point to the renewed surge of support in some sectors of conservative (e.g. fundamentalist) Christianity. But verbal learning of traditional doctrine, which cannot be translated into existentially significant language, is the stuff of the extrinsic religious orientation described in Chapter 3 in the section on humanist Psychology. Verbal learning is bound up with all its attendant disadvantages of conservatism, prejudice and exclusivism, as its members retreat into a thought-to-be-safe ghetto of psychological separatism and emotional security in a "threatening world of change".

Part of the paradoxical nature of religion is that some people can, on the one hand, retreat into an exclusive, conservative doctrinal system, embraced through verbal learning without existential understanding, which logically entails prejudiced and exclusive behaviour, but, on the other hand, live exemplary inclusive Christian lives. They follow their heart and deep tacit learning rather than the verbal learning they speak. Verbal learning is a stage all children grow through, with some remaining at that stage as adults, or regressing to it. Here is an example. An elegant, eloquent, liberal Jewish lady was giving a talk on the nature of her faith to a Methodist Local Preachers Meeting. At question time, an eighty-year-old local preacher told her that she could not be saved unless she believed in the Lord Jesus Christ. The subsequent discussion left this gracious gentleman, a good Christian neighbour to anyone in need, completely baffled. He tacitly accepted, without verbal understanding, the appropriateness and integrity of the speaker's faith. At the end of the meeting he went up to her to say goodbye and thank you. He said, as he shook hands with her, "I don't understand how you can be saved, but I think you are a very nice lady."

This raises a very important pastoral concern which needs exploring carefully. Our culture inherits two major influences relevant to our present discussion.[4] Our major religious influences come from the Middle East background of Judaeo-Christianity with its genius for expressing its deepest beliefs and philosophical problems through verbal stories, e.g. the Creation stories in Genesis. Our major source for philosophical thinking is the Greek inheritance.

The advantage of the stories approach is that people of all intellectual abilities can hear and accept them at levels of understanding either within their grasp, or according to the degree of intellectual curiosity and work they engage in with them. These stories, or verbal moving-pictures, can meet the needs of a whole community.

However, these stories are in constant need of reinterpretation in the vernacular language of the time. This is the situation which the churches need to continue to face if they are to meet the spiritual needs of persons with an intrinsic religious orientation. These are the people who are deserting the churches: they find old-fashioned literalisms and obsolete Jewish practices used as cultic analogies, like "Christ as a sacrifice for sin", unacceptable.

The humanists have two problems to face, but first let us be clear that there is some common ground between Christians and those humanists like Maslow who aspire to B-cognition, which I have translated as "Maturity in Wholeness". This is also the goal of a Christian with an intrinsic,[5] i.e. an internally digested, understanding of their faith. They express this goal in terms of "We will be mature, just as Christ is, and we will be completely like him" (Ephesians 3:13 CEV).

While there is common ground concerning many of the values shared, the routes to them are very different. The Christian route, through verbal pictures, uses language to express a "relationship"[6] with God which is most easily "seen" in the life and teaching of Jesus of Nazareth, which is contagiously attractive.

The humanist route of Maslow, and for many humanists, is through the Greek philosophical tradition of articulating philosophically their goals and values. "Humanists accept that they must set their own purposes, be they small and individual or large, long-term and collective."[7] Camus has given us a picture of a humanist saint in his portrait of Tarrou, a character in his novel *The Plague* who lived a life without "hope's solace" (Camus 1960: 237). This was Camus's way of describing the comfort of religious hope, which claims that death is not the end of human life, as there is a life hereafter.

NOTES

1   Art 21 per cent; nature 18 per cent; sexual love (not intercourse) 17 per cent; religion 10 per cent (Laski 1980: 486).

2   He was associated with the humanistic psychology movement (see Chapter 3), from which he emerged as the founder of the transpersonal psychology movement mentioned in the "Transpersonal Psychology" section below.

3   For a fuller summary see Wulff (1991: 606–7).

4   Our third cultural inheritance is Roman Law, which isnot relevant to this discussion.

5   An "intrinsic religious orientation" was described in detail in Chapter 3, p. 45. An intrinsic Christian *understands* their beliefs and how they relate to everyday life: they are not a parrot. The person with an extrinsic religious orientation is one which they do not *understand* and cannot explain existentially, i.e. in everyday language.

6   Of course a person can literally only have a relationship with another person. The "home" of the term is personal discourse. But it is impossible to describe a religious experience for many without using "relationship" in a similar but different sense, as a term within religious discourse.

7   A personal communication from Dr Charles Bailey.

REFERENCES

Camus, A. (1960), *The Plague*, Harmondsworth: Penguin Books.

Huxley, A. (1954), *The Doors of Perception*, London: Chatto & Windus.

Huxley, A. (2004), *The Perennial Philosophy* (1945), New York: Perennial.

Katz, S. T. (1978), "Language, Epistemology and Mysticism", in S. T. Katz (ed.), (1978), *Mysticism and Philosophical Analysis*, London: Sheldon Press.

Laski, M. (1961), *Ecstasy*, London: Cresset Press.

Laski, M. (1980), *Everyday Ecstasy*, London: Thames & Hudson.

Leuba, J. H. (1912), *A Psychological Study of Religion*, New York: Macmillan.

Leuba, J. H. (1925), *The Psychology of Religious Mysticism*, New York: Kegan Paul.

Maslow, A. H. (1970), *Religions, Values and Peak Experiences* (1964), New York: Viking Press.

Wulff, D. M. (1991), *Psychology of Religion: Classic and Contemporary Views*, New York: Wiley.

12 | Religious and Mystical Experience

The Model Builders

We have already explored some theoretical analyses of religious and mystical experiences in Chapter 10, in which we looked at some key empirical studies of our subject. But no empirical study can have meaning without a theoretical structure in which to present the findings. Each of the studies reported offered theoretical understanding of one or more of the three kinds of religious experience we are interested in as regards our present exploration: primary religious experiences, conversions and mystical experiences.

The same is true of the humanist studies we explored in Chapter 11: they too were theoretical studies offering structures to give shape to humanist understanding of our subject matter. There we read of Huxley's exploration of drugs as a trigger for religious experience, which disturbed Zaehner so much that he wrote his book in response.

FIVE MODEL BUILDERS: JAMES, OTTO, ZAEHNER, STACE, SMART

Authorities in this field most frequently quote five significant "model builders" who have developed models for understanding the kinds of experience we are studying.

We have already briefly explored, in Chapter 10, the first set of three models outlined by James – of an initial personal religious experience which spiritually enriches the individual, and which can develop into what James called the "religion of the healthy minded". James's second model is "the about turn" of religious conversion, from sinfulness to holiness, i.e. from self-centredness to other-centredness, which could lead toward religious maturity. James's third model is of mysticism, which can lead to saintliness, which is another way of talking about religious maturity.

We now need to describe the work of four other model builders: Otto, Zaehner, Stace and Smart; then we can try to integrate the major insights of these five model builders.

Otto: *Idea of the Holy*

Rudolph Otto (1959) was concerned with primary religious experiences and offers us a valuable model to enrich our understanding of them. He was a theologian and wrote

214

The Idea of the Holy (1917). His book was mentioned earlier when we needed an illustration of an *essentialist* or *perennialist* definition of religion, i.e. a definition that was concerned to try to list the key elements of all religions. Otto knew James's book but was concerned to build on ideas from Schleiermacher and Kant. Schleiermacher , in 1821, had tried to tempt the educated classes back to religion, which he said was found where "the whole soul is dissolved in the immediate feeling of the Infinite and Eternal" (1988: 16). He argued that religion was based on intuition and feeling, not on dogma, and that its highest experience was this sensation of union with the Infinite.

Kant argued that morals were not dependent on religion or anything else: they were independent or autonomous. The only good thing in the world is a good will. This is the basis of an independent principle called the Categorical Imperative: "Moral duties are categorical because they should be followed for the sake of duty only, simply because they are duties and not for any other reason" (Vardy and Grosch 1994: 68). If you ask why, the only answer is because it is your duty: there is no other answer or reason. Hence it is a categorical imperative. Hypothetical imperatives, however, do have a reason. For example, "If you want to win an Olympic medal, you must train." Why has this been mentioned here? Because for Otto "holy" contains two elements, a non-rational or feeling element which he called "the numinous", a word he derived from *numen*, a Latin word meaning the divinity or sacred power in a sacred place or object, and a rational element: "the moral".

"Holy" then, meant "the numinous" plus "the moral" for Otto. The feeling element was developed from Schleiermacher. Otto speaks of a unique numinous feeling, which is both felt and is at the same time objective and outside oneself. The rational element was taken from Kant and modified (Otto rejects Kant's use of the term "moral"), for he regards it as dependent on the "numinous". "Numinous" is the basic concept, for there "is no religion in which it does not live as the real innermost core" (Otto 1959: 20). Otto makes it quite clear that "holy" originally related to the numinous: " . . . if the ethical element was present at all, at any rate it was not original and never constituted the whole meaning of the word" (20).

The numinous does not need to be defended by appealing to any thing or reason outside of itself. This feeling leads to a sense of dependence resulting from feeling "as nothing" in the presence of a Sacred Mystery.[1] There is only one way to understand it. A person can be guided towards the idea, with the help of discussion, until he reaches "the point at which 'the numinous' in him perforce begins to stir, to start into life and consciousness" (Otto 1959:21).

We cannot describe a person fully by listing a head, a body, two arms and two legs, because a person is alive. But the description is a start! So also, we cannot fully describe the numinous as the *mysterium tremendum et fascinans* – the tremendous and fascinating mystery. But we can make a start by seeing what Otto means by this. Otto asks us to consider the deepest and most basic feeling we have when thinking about religious emotion. Now although we cannot fully describe a person by listing the six parts of their body, we can use that as a starting point. So we need to look at the five elements which constitute the "numinous".

Tremendum contains three elements and *mysterium* contains two elements. The elements in *tremendum* are "awefulness", "overpoweringness" and "energy". The elements in *mysterium* are the "Wholly Other" and "fascination".

In *Tremendum*, "Awefulness" contains a number of emotions that a divine presence would evoke: fear, dread, awe, moral inadequacy and fear of consequences.

"Overpoweringness" is the feeling one has in the presence of the majesty and greatness of an all-powerful lord or king, which evokes a feeling of nothingness and humility. "Energy" or "urgency" Otto describes as vitality and passionate activity.

Mysterium evokes blank wonder, astonishment, stupor in the presence of the "Wholly Other". Through positive feelings the "Wholly Other" is described as "transcendent" or "supernatural".

"Fascination" contains two emotions in tension, a sense of awe and dread, already noted, but now in tension with ineffable joy, frequently experienced in conversion experiences when the person is freed from guilt and sin. Quoting one of a number of examples from James's book, he likens the experience to one of "joy and exaltation . . . It was like the effect of some great orchestra, when all the separate notes have melted into one swelling harmony that leaves the listener conscious of nothing save that his soul is being wafted upwards and almost bursting with emotion" (Otto 1959: 51–2).

Otto concludes that the numinous is a manifestation of the Divine, which is "the highest, strongest, best, loveliest, and dearest that man can think of . . . God is not merely the ground and superlative of all that can be thought; He is in Himself a subject on His own account and in Himself" (53).

Otto's work is of comparable stature to that of James. His main claim is that the numinous evokes a *unique* sacred mental state (Otto 1959: 21). This means that it cannot be reduced to anything else.[2]

This indicates that there are different kinds of "ways of knowing". For example, I can say that "this male is beautiful" or "this male is honest", and each claim is "of its own kind". The first is aesthetic, the second moral. If you wish to agree or disagree, you must answer in "the same language game", i.e. use the same *kind* of language. If I reply to the first claim, "this male is beautiful", by saying "No, he is 80 kilograms", I have mixed up aesthetic and empirical kinds of language, and the conversation becomes nonsensical. If I reply to the second claim, "This man is honest", by saying "this male is asleep", I have mixed up a moral statement with an empirical statement. This exchange is similarly nonsense. None of these statements can be reduced into different kinds of statement. This is what Otto is claiming for the numinous: it cannot be reduced to anything other than itself.

We have already noted the criticism that this *essentialist* or *perennialist* definition of religion isolated the numinous from all religions. This can be countered by showing that the great religions of the world all contain this dimension, which is expressed in symbols found within each World Religion, as Otto takes pains to show.[3]

Zaehner: *Mysticism Sacred and Profane*

In 1957, in his study of comparative mysticism, Zaehner published his response to Huxley's view of religious experience. His central thesis was to suggest that there are three forms of mysticism. First, there is nature mysticism, which he called *pan-en-hen-ism* – "all-in-one-ism". In his account of nature mysticism, Zaehner includes Wordsworth, citing "Tintern Abbey". To have such an experience of nature means a person "has enjoyed a sense of communion or 'at-one-ment' with a reality transcending himself" (Zaehner 1973: 35).

The second form of mysticism is "monism" (one-ism) in which "the individual soul is substantially and essentially identical with the unqualified Absolute" (Zaehner 1973: 28). The individual and reality or God are One.

"Third, there is the normal type of Christian mystical experience, theism (Godism), in which the soul feels itself to be united with God by love" (Zaehner 1973: 29). Zaehner admits that this theistic experience is based on the theological premise that the individual soul is created by God, and that it has the capacity to be united with God. He concluded that "What goes by the name of mysticism, so far from being an identical expression of the self-same Universal Spirit, falls into three distinct categories" (198), nature mysticism and "two distinct and mutually opposed types of mysticism – the monist and theistic" (214). This view is heavily criticized by Smart, as we will see later. But treating work in chronological order, we must consider the work of Stace before that of Smart.

Stace: *Mysticism and Philosophy*

Stace was a civil servant who, in his spare time, wrote *Mysticism and Philosophy*, which was published in 1960. This significant philosophical analysis of the nature of mysticism, written three years after Zaehner's work, created renewed interest and began the modern study of mysticism.

Stace suggested as his model two related forms of mysticism, which he called extrovert mysticism and introvert mysticism.

He lists the common characteristics of extrovert mystical states of mind as follows:

1   The unifying vision, which is expressed in the formula "All is One". The "One" may be perceived through the physical senses in or through the multiplicity of objects.
2   A person grasps the "One", which is described as life, consciousness or living Presence, as the inner subjectivity of all things. Nothing is dead.
3   A person has a sense of the objectivity of this reality.
4   A person experiences associated feelings like blessedness, joy, happiness and satisfaction.
5   A person feels that what is apprehended is holy, sacred or divine. These qualities lead one to interpret the experience as being an experience of God.
6   The experience is full of paradoxes, e.g. God is closer than the air you breathe and farther away than the most distant planet.
7   The experience is ineffable: it cannot be adequately put into words. (Stace 1973: 79)

Not all characteristics are present in every case of extrovert mysticism. Stace particularly mentions "nature mysticism", which he thinks is one kind of extrovert mysticism. A sense of "presence" could develop into an extrovert mystical experience (Stace 1973: 80).

Quoting Wordsworth's "Tintern Abbey", Stace goes on to note that "there are underground connections between the mystical and the aesthetic (whether in poetry or other forms of art) which are at present obscure and unexplained" (1973: 81; cf. Armstrong 1987).

Basically extrovert mysticism looks outward through the physical senses into the external world and finds the "One" there. If one looks at glass, wood and stone, conceptual understanding identifies three objects, but if one passes beyond sensori-intellectual consciousness into mystical consciousness, then one sees these things as "all one". In extrovert mysticism, the distinction between three things does not disappear but they are at-one, and the experiencer is at-one with them.

Introvert mysticism is characterized by an absence of all sensation through the physical senses, and by the exclusion from consciousness of all sensual images, i.e. "abstract thoughts, reasoning processes, volitions, and other particular mental contents". This leaves a complete emptiness – a void which the mystic describes as "a state of pure consciousness – 'pure' in the sense that it is not the consciousness of any empirical content" (Stace 1973: 86). It is called the Void or the One or the Infinite. The paradox is that there should be a positive experience without positive content. An important element in the Void is that "the mystic gets rid of the empirical ego whereupon the pure ego, normally hidden, emerges into the light. The empirical ego is the stream of consciousness. The pure ego is the unity which holds the manifold streams together. This undifferentiated unity is the essence of the introvertive mystical experience" (87).

Stace concludes his analysis of introvert mysticism by saying:

> We may now fairly confidently assert that there is a clear unanimity of evidence from Christian, Islamic, Jewish, Mahayana Buddhist, and Hindu sources . . . that there is a definite type of mystical experience, the same as in all these cultures, religions, periods, and social conditions, which . . . [has] the following characteristics:
>
> 1   The Unitary Consciousness, from which all . . . sensuous or conceptual or empirical content has been excluded, so that there remains only a void or empty unity . . .
> 2   Being, non spatial and non temporal . . .
> 3   Sense of objectivity or reality.
> 4   Feelings of blessedness, joy, peace, happiness etc.
> 5   Feeling that what is apprehended is holy, sacred, or divine . . .
> 6   Paradoxicality.
> 7   Alleged by mystics to be ineffable. (Stace 1973: 111)

We can now examine the characteristics common to all mystical states, extrovert and introvert alike, including nature mysticism, in the table opposite.

Ignoring borderline experiences, which may well fall between the two, characteristics 3–7 in these distinct forms of mysticism are held in common and "are therefore universal common characteristics of mysticism in all cultures, ages, religions and civilizations of the world" (Stace 1973: 131–2). "The extrovertive mystic perceives the universal life of the world, while the introvertive reaches up to the realization of a universal consciousness of mind" (133).

Finally, we must look briefly at Stace's account of the content of mystical experience. Introvert mysticism is, as previously mentioned, concerned with the realization of the unity of the pure ego, which "over-reaches all individuals" (1973: 196). It is real, so it is trans-subjective. This means that the content of all mystical experiences, of both kinds, is not subjective or objective, but is of the trans-subjective Universal Self (197, 203).

Common Characteristics of Mystical Experiences

Extrovertive

1. The Unifying Vision –
   One, all things are One.
2. The more concrete apprehension
   of the One as an inner subjectivity,
   or life in all things.

Introvertive

1. The Unitary Consciousness;
   the Void, pure consciousness.
2. Non spatial, non temporal.
3. Sense of objectivity or reality.
4. Blessedness, joy, peace, etc.
5. Feeling of holy, sacred, or divine.
6. Paradoxicality.
7. Ineffable.

I find that this is a very abstract and complicated idea for most people, which can be simplified and translated into more familiar concepts for, for example, a Christian. Mystical union takes place with the Divine. God is the ideal "person". Humans are created in the image of God, giving them an opportunity to become "Divine-like". If Jesus of Nazareth is the closest model Christians have of the Divine, then this goal can be rephrased and expressed in familiar terms, already quoted once before: Christians can aim to "be mature, just as Christ is, and we will be completely like him" (Ephesians 3:13 CEV). If persons ever altruistically mature like this, they will become part of the Universal Self like Christ in God. In other words, all can share in the universal consciousness of mind. For theists, this consciousness is in and of God. For humanists, it is a natural phenomenon.

In order to meet the needs and forms of expression acceptable to their communities, other faith stances have formulated their own altruistic goals, and seen how far their accounts of mystical experience can relate to "what counts" as being part of "the Universal Self".

These two models need some discussion. Introvert mysticism seems to make sense of much reported experience which fits this model well. However, the model of extrovert mysticism has two aspects to it – the individual and the presence of God, either experienced "directly" or mediated through a person, a view, a piece of music or other medium. Structurally, this is similar to the model of religious experience outlined by Otto and briefly summarized earlier.

Smart: *Dimensions of the Sacred*

Professor Ninian Smart has made a long and distinguished series of contributions to the development of the phenomenology of religion in general which includes the discussion of the nature of the kinds of religious experience we are studying. In 1966, he examined Zaehner's distinction between panenhenic (all-in-one- ism), monistic (one-ism), and theistic (God-ism) mysticism. Accepting nature mysticism (panenhenic mysticism), he questioned the distinction between monistic and theistic mysticism. He demonstrated that the distinction between monistic mysticism – "realizing the eternal oneness of one's own soul" – and theistic mysticism – "the mysticism of the love of God" was based on a theological and not a typological distinction. A theological distinction involves a difference in *interpretation.* A typological distinction involves a difference in the *kind* of experience. Smart showed that the *structure* of a monistic and a theistic experience was the same, only the *interpretation* of the experience was different (1966).

For over thirty years Smart has developed his six, and now seven, dimensional model of religion, noted earlier in this chapter: experiential, mythical, ritual, doctrinal, social, ethical, material, and their political effects, outlined in Chapter 9. These dimensions eloquently articulate what a 'religious outlook' entails, with its corresponding 'religious attitude'. In the first articulation of this model in 1968, after he had described the first three dimensions – doctrine, myth, and ethics – he wrote: "These three dimensions can be said to represent the teachings of a religion . . . they express the perspective in which the adherent views the world and himself" (Smart 1968, 16). These three dimensions express the 'religious outlook', which finds expression through the other dimensions and the consequential way of life. Smart feels that we cannot understand "religion without paying attention to the inner life of those who are involved in the life of the dimensions we have so far considered" (18).

Smart, however, makes a clear distinction between religious experiences which have changed the course of 'religious history' and those which reflect the embracing of a religious tradition, outlook and way of life. "Indeed many of the seminal moments of *religious history* (my italics) have involved religious experiences of a dramatic kind: the Enlightenment of the Buddha, the vision of Isaiah in the Temple, the conversion of St Paul, the prophesy of Mohammad – we could scarcely explain the directions *religious history* has taken without referring to such moments" (18). He continues, "At a humbler level there is the testimony of countless religious folk who believe themselves to have had moments of illumination, conversion, vision, a sense of presence and so on. We may refer to this aspect of religion as the *experiential dimension*" (18). Smart is offering a 'top-down' view – 'religions form persons', where as others adopt a 'bottom-up' view 'persons form religions', though no one really accepts this second view without including that there is a Transcendent dimension which impinges on human consciousness to evoke a religious response, hence the development of religious language. However Smart's first group above should be seen as those who changed the course of *world* history, because they transcended the religious ideas they inherited, and the same is true to a lesser extend of the second group.

In 1973, Smart confirmed the clear distinction between two kinds of religious experience. Mysticism, in which the contemplative goal is mystical union with the divine, is one kind or form of religious experience. Nature mysticism is of the same form. It can be interpreted in different ways. If the individual, nature and God 'merge' into One, then that is an example of religious mysticism. If the individual merges with nature alone, he or she may interpret that experience as a purely 'natural experience'

– a humanist's position. The other form, or kind, is the primary personal experience. That is the religious experience in which the individual 'meets' or feels in the Presence of God. In this form, the 'numinous force', first identified and described by Otto above, evokes worship from the believer (Smart 1973:126; 1978: 13).

Smart made an interesting suggestion in 1973. "A religious system is somewhat like a collage" (149). This needs explaining. The collage children make on the wall of a classroom in a primary school may show a scene of Postman Pat! In a primary school in the USA the scene may be of a bear in the Rockies. It is constructed on a large painted background executed by the teacher, on which bits of cloth, milk bottle tops made of aluminium, bits of twigs and other kinds of materials are fastened. Indeed any kind of material bric-a-brac may be used to complete the picture. This collage consists of different *kinds* of material, but worked into an integrated whole which 'makes sense', and communicates in *one* form of knowledge – aesthetic knowledge.

Smart is saying that the religious form of knowledge can also have a collage-like structure. That means it can include experiences, myth (story), ritual, doctrine (teachings = interpretations), ethical and legal principles or statements. Social issues might incorporate the role of a priest, prophet, minister, imam, rabbi, mystic, or healer, and the role of organizations, like a synagogue, church, mandir or gurdwara. The material dimension is expressed in temples, churches, icons, statues, paintings. All these dimensions interact, and can have political, social and individual effects.

When religious language is used, it can contain all forms of discourse, but by its presence controls and therefore expresses the religious interpretative intentions or beliefs of the writer. So even religious language is often in the form of a linguistic collage, as we will explore later.

Smart makes another important point. He discusses the nature of understanding in relation to religious experience and notes that, for example, "psychological and theological accounts of Paul's conversion can be held together" (Smart 1978: 10). He sees that they are complementary and not contradictory. He implicitly recognizes the hierarchy in forms of knowledge which Peacocke outlined in Chapter 6. Any form of knowledge does not contradict another form of knowledge but complements it, by exploring a topic at a different level of understanding and/or from a different perspective. Any form of knowledge may include concepts and ideas from forms of knowledge lower in the hierarchy, as we noted also in Chapter 6.

In 1996, Smart put forward a preliminary theory that there are two main modes of experience in religion (1997: 167). The first is the 'numinous', the Other, outlined by Otto earlier, which Otto described as 'a tremendous and fascinating mystery' which evokes the sense of the 'sacred'. Otto suggested that 'holy' consists of a feeling element and a knowledge or cognitive element, the 'numinous' and the 'moral' respectively. The second mode "is the contemplative or mystical experience which does not postulate an outside Other and which feels the disappearance of the subject-object distinction" (167). Theistic mystics talk of an inter-personal mingling or union with the divine. Smart notes that in some religions there is no concept of the Other. He suggests that members of world faiths, without a concept of the Other, experience "a non-dual experience as inner feeling. There is a double initial contrast with the numinous: the latter experience is of an outside Other, the mystical of an inner non-other" (167).

Smart recognizes that Zaehner's 'panenhenic' (all-in-one-ism) category well describes an individual's feeling of union with the world or cosmos which is objectively there. Smart limits the term 'mystical' to relate to 'contemplative' experiences, and concludes that "So far, then, we have three varieties of experience: the numinous, the panenhenic and the contemplative" (Smart 1997: 170).

On the debate between those who defend the unique nature of religious and mystical experience within a particular world faith, like Katz (1978:22–3), and those who defend the idea of common experiential ground between World Religions, like Otto, Maslow and Stace, Smart suggests that there are relative rather than unique degrees of interpretation.

Summary of Smart's Model

The major contribution of Smart focuses on his three varieties of experience, which can be found in the various religious traditions in varying combinations. In some religious traditions some of these varieties may be given different degrees of significance or sometimes may seamlessly overlap. His numinous and contemplative modes of religious experience develop through three stages:

- *Panenhenic (all-in-one) experience* is the concept which Zaehner has used to describe "the experience of feeling absolutely at one with the world or cosmos . . . a kind of eyes-open fusion with reality which lies about the person" (Smart 1997: 170).
- Religious Experience
- Mystical Experience

Stage 1 *numinous experience* (revelation)	*contemplative* (mystical) *experience*
Stage 2 divine conversion (conversion)	luminous (mystical) conversion
Stage 3 divine disposition (ongoing faith)	luminous disposition (mystical union)
	(170, 195)

Smart suggests the term 'contemplative' can be divided into two groups, theistic and non-theistic, of which the latter is particularly important in relation to some Eastern Religions like Buddhism.

We are not here concerned with Smart's exploration of the relationship of the experiential with other dimensions of religions like the doctrinal or ritual dimensions. However, we do need to note that Smart, when he tests his theory of two religious modes of experience, suggests not only that religions often encourage both modes of his two pole theory, but that the theory allows three modes, the numinous, the contemplative or both (174). So it is possible to find that in a religious tradition, one kind or the other may be exclusive, one kind may be dominant, or both may be equally valued.

While accepting the panenhenic experience (all-in-one-mystical-union with nature and the cosmos) Smart points out the advantages of his theory of two modes of experience – 'numinous' and 'contemplative' over the idea of a single core theory advanced by Otto. However James makes a clear distinction between 'numinous' religious experiences and 'theistic mysticism' religious experiences. There is a further distinction, which is that the numinous experience is more likely to have a focal awareness on, for example, God, while theistic mysticism is more likely to have, for example, a tacit or subsidiary awareness of God. Smart suggests that the flexibility of his 'two modes of experience' theory recognizes the possibility of the varieties of experience, avoiding the notion of a single core to all religious experience. But Starbuck, James, Glock and Stark, Greeley, Hardy, Hay and Stace and others who defend both the theistic 'numinous' and 'nature mysticism' (panenhenic mysticism) are denying a 'single core' theory by using these two categories, as well as the categories of 'conversion' and 'theistic mysticism' all as *stage 1* varieties of experience. Their model-varieties derive from their empirical work to ascertain what kinds of religious experience and nature mystical experiences people report having. This is a 'bottom-up' approach to understanding religious experience.

Smart feels the flexibility of his model is extended by embracing an idea advanced by John Hick in 1983. If "the numinous experience provides a revelation of divinity . . . we can see the divine handiwork in the world around us . . . (so) . . . the presence of the Divine as Creator, in everything we encounter in life, helps to stimulate feelings of wonder and gratitude" (Smart 1997: 178). Hick describes this process of interpreting the natural world in a religious way as 'experiencing as' – 'experiencing (the world) as' created by God (Hick 1991: 140–2). But this is theistic mysticism – a tacit sense of a universal Presence, different from the direct confrontation, or focal awareness, of a numinous experience of the presence of God.

Smart suggests that "you can have the numinous experience; second you can have the vision of the divine; third you can have the continuing disposition to see the divine in the world (Smart 1997: 179). He hints just before this summary that the second stage "may be the point where a kind of conversion occurs." However this is to change the meaning of 'conversion', a *stage 1* experience, identified by most of the sources we reviewed in chapter 10, which is 'an about turn religious experience' into the developmental thinking which follows an initial religious or mystical experience (a 'stage 2' developmental phase) in contrast to the empirical evidence noted in chapter 10 that indicated that it was a 'stage 1' experience.

Smart also extends his view of mystical union. It can be linked, secondly, to the Creator through contemplation. This can provide a "framework for the luminous perception of the divine in everything". Thirdly, contemplation can be the basis for the attainment of pure consciousness for a disposition to see the world as an eternal network of linked events. These three stages can characterize the development of both theistic and non-theistic contemplation.

Smart takes a non-theistic Buddhist example. The three stages are: first, the high experience of nirvana, of liberation like the extinguishing of a candle, experienced as pure consciousness. This leads, second, to coming to see the Buddhas analysis of the nature of the world – 'experiencing (the world) as' proposed by the Buddhist view. Third, this leads to the continuing disposition to 'see', or interpret, the world in this way. Smart names the developmental stages of these experiences. "The first, dramatic phase we can call numinous experience, the second the divine conversion, the third the divine disposition. In the case of the contemplative triad, the first we can call the contemplative experience, the second the luminous conversion, the third the luminous disposition" (180).

Sometimes there will be a blending of the numinous and contemplative as when a "Christian mystic, for instance, may come to see the world about her not just as the handiwork of God (numinous, stage 1), but as glowing with the network of mystical illumination (contemplative stage 1)" (180).

Smart' 'top-down' view and stage development sequence gives an account of those whose religious experience occurs after they have experienced the doctrine, mythic and ethics dimensions of religion. However, from a 'bottom-up' perspective, the numinous and contemplative modes have two misleading stage 2 labels which question long-standing use of the concept 'conversion'. The use of 'conversion' by Starbuck, James, and many others, has established the meaning of conversion, which still stands in the Oxford Concise Dictionary to mean "a change from sinfulness to holiness". In other words it is "an about turn". Smart is not talking about 'an about turn' but a 'development' in his second stage. Smart agrees that the original numinous experience is an individual religious experience, in a manner similar to that described by Otto. But Smart does not acknowledge that a conversion experience, is an 'about turn' experience, regarded as a "stage 1" experience in sources mentioned in chapter 10, focused on a "bottom-up" approach.

Those taking a 'bottom-up' approach might feel it would be more appropriate to describe "Stage 2" in Smart's numinous and contemplative models as 'the divine confirmation' and 'the lumination confirmation' respectively, for the second stages are the locations in which the original vision of stage one is extended to embrace one's outlook on the world, into 'a religious outlook'. There is no change of direction here, as conversion, normally used, would imply, but the continuation of the direction indicated in a "stage 1" experience. In both the *numinous* and *contemplative* modes his "stage 2" represents a stage development towards a religious, non-religious or non-theistic outlook with its attendant disposition or attitude. These developmental models are important as they are designed to link up with Smart's other dimensions of the sacred. But this falls outside the brief of this study of religious experience.

However, it must be emphasized that these are models: real people use 'top-down' and 'bottom-up' approaches in diverse ways.

Comments on Smart's Model

In Chapter 10 seven uses of the term 'religious experience' were identified:

1   Bipolar experience between God and a person – personal religious experience of a numinous kind;
2   Habits of religious behaviour;
3   Religious knowledge;
4   Conversion;
5   Mystical union;
6   A tradition's spiritual discipline of contemplation leading to mystical union;
7   All experience interpreted as religious.

Here we repeat, for the sake of clarity of language use, only three of the above uses should be called 'religious experience' – numinous experience (1), conversion experience (4) and mystical experience (5). Smart has offered a 'top-down' account of the *development* of religious experience, which is outside the brief of our area of study. Our concern is to identify those *transcendental theistic and non-theistic experiences* a person has had which they can *only* describe by *using religious language*, where *God talk* is used. *Non-theistic religious* language is used by people who are religious like Buddhists. *Non-theistic language*, which is not religious, is used by people like agnostic or atheist Humanists. On this matter Smart has made an important contribution spotting an omission in previous work in the area to identify and articulate the mode of non-theistic religious language used in some Eastern World Faiths like Buddhism.

SUMMARY

The conclusion of this review, taking account of the five model builders we have considered, is that we focus our attention on models we have previously identified, to which we must now add Smart's "mystical non-theistic religious experience". Secondly, Zaehner's panenhenic model of mystical union needs to be divided into two models, non-theistic mystical *union* with nature and the cosmos; and theistic mystical *union* with nature and the cosmos. However not all mystical experiences need be *union* experiences, they can well be bipolar experiences, in which case we need to add two more modes of experience: non-theistic bipolar mystical experience with nature; theistic bipolar mystical experience with nature. Zaehner has argued that primary union experiences take place. James has shown that mystical union can be a development from previous initial theistic experiences. Smart observes that mystical union can *develop* from previous theistic and non-theistic experiences through contemplation. Thirdly, Stace's distinction between extrovertive and introvertive mystical experiences I take to be the same as bipolar mystical experiences and mystical union experiences respectively.

This gives us eight modes of transcendental experience:

Religious experience
- numinous experience with focal awareness
- conversion experience with focal awareness

Mystical experience
- bi-polar non-religious mystical experience with nature, e.g. an atheist
- bi-polar religious non-theistic mystical experience with nature e.g. a Buddhist
- bi-polar theistic mystical experience with nature, e.g. a Christian
- non-religious mystical union experience with the cosmos, e.g. an agnostic humanist
- theistic mystical union experience with God and the cosmos, e.g. a Jew
- non-theistic religious mystical union experience, e.g. a Buddhist experience of nirvana

These mystical experiences may have either focal or subsidiary awareness.

These models offer the sources for the evocation and necessary use of *religious language* in the form or mode of discourse used to articulate *religious knowledge* by human beings.

NOTES

1   Otto speaks "of a unique 'numinous' category of value and of a definitely 'numinous' state of mind . . . this mental state of mind is perfectly *sui generis* and irreducible to any other" (1959: 21). "Numinous is . . . felt as objective and outside the self" (25), which evokes "a numinous state of mind" that is accompanied by a "feeling of dependence", which he calls "creature consciousness" or "creature feeling". "It is the emotion of a creature submerged by its own nothingness in contrast to that which is above all creatures" (24). The numinous possesses the objective quality of "mystery" (*mysterium*), inspires awe (*tremendum*) and yet attracts (*fascinans*). Thus "*mysterium tremendum et fascinans*" was Otto's description of the numinous element of "the Holy". "*Tremendum*" was composed of three elements: first, awefulness, which included the emotions of fear, dread, awe (27–8); secondly, "overpoweringness" (*majestas*), which is the feeling of one's own submergence, of being into "dust and ashes and nothingness" (34); and thirdly, "energy" or "urgency", indicating vitality, passion, will, excitement (37). "*Mysterium*" contained two elements, the "wholly other", which may evoke stupor, wonder and numbness, and "fascination" a non-rational element with rational parallels of love, mercy, pity, comfort (39–40). Once confronted by the numinous "possession of and by the numen becomes an end in itself; it begins to be sought for its own sake" (47).

"The non rational apprehension of something transcendent, according to Otto, is then expressed through various symbols, or ideograms, as he calls them, ranging from the primitive symbolism contained in the idea of the Wholly Other, through mythological representations, to the formulations of theology . . . The sense of the holy or sacred . . . dimly and dumbly . . . points towards the Transcendent" (Otto 1959: 67, 74).

2   Otto calls it *Sui generis* (literally "of its own kind"). Scholars apply the term to irreducible states like the numinous, the moral, the aesthetic or the personal.

3   An important aspect of Otto's work was his concern with the experience of the numinous in eastern as well as western formulations of religion (e.g. 1959: 53, 145–6). While Otto discusses the autonomy of aesthetic judgement (165, 177), his discussion in no sense rules out the possibility, indeed the probability, that many may experience the numinous through nature: this is implicitly accepted in his reference to "the horror of Pan" (29). Otto's major work lay in this account of the numinous.

REFERENCES

Armstrong, K. (1987), *Tongues of Fire: An Anthology of Religious and Poetic Experience* (1985), Harmondsworth and New York: Penguin.

Fishbein, M. and Ajzen, I. (1975), *Belief, Attitude, Intention and Behavior*, Reading, MA: Addison Wesley.

Halloran, J. D. (1967), *Attitude Formation and Change*, Leicester: Leicester University Press.

Happold, F. C. (1975), *Mysticism: A Study and An Anthology* (1970), Harmondsworth: Penguin.

Hick, J. (1983), *Philosophy of Religion*, Englewood Cliffs, NJ.

Hick, J. (1991), *An Interpretation of Religion* (1989), London and New York: Macmillan.

Katz, S. T. (1978), "Language, Epistemology and Mysticism", in S. T. Katz (ed.), (1978), *Mysticism and Philosophical Analysis*, London: Sheldon Press.

Otto R. (1959), *The Idea of the Holy* (1917), London: Penguin.

Schleiermacher, F. D. E. (1988), *On Religion: Speeches to Its Cultural Despisers*, (1821), trans. R. Crouter, Cambridge: Cambridge University Press.

Smart, N. (1966), "Myth and Transcendence", *The Monist* 50 (4) 475–487.

Smart, N. (1968), *Secular Education and the Logic of Religion*, London: Faber & Faber.

Smart, N. (1973), *The Science of Religion and the Sociology of Knowledge*, Princeton, NJ: Princeton University Press.

Smart, N. (1978), "Understanding Religious Experience", in S. T. Katz (ed.), (1978), *Mysticism and Philosophical Analysis*, London: Sheldon Press.

Smart, N. (1997), *Dimensions of the Sacred* (1996), London: Fontana.

Stace, W. T. (1973), *Mysticism and Philosophy* (1960), London: Macmillan.

Vardy, P. and Grosch, P. (1994), *The Puzzle of Ethics*, London: Fount.

Zaehner, R. C. (1973), *Mysticism, Sacred and Profane* (1957), Oxford: Oxford University Press.

13 | Religious Experience and Interpretation

Religious experience, whether numinous or contemplative, can be interpreted in very different ways from those which we have sympathetically explored in Chapter 12, and from which we have selected models to aid our understanding of the different modes of religious experience, which can become fluid and merge in different ways.

In this chapter we will explore two different kinds of interpretation. The first kind will be from writers who do not accept the way we are using language in a "critical realistic way", i.e. *critical realism*, as when we described religious experiences in the seven "uses" outlined in the last chapter. The exploration of these critics is called "non-realist criticism of critical realism".

The second kind of "interpretation" we use, and will explore later, will refer to the complex task of levels of interpretation, which those of us who accept the different modes of religious experience need to take account of seriously. This exploration will follow under the heading of "Levels of Interpretation". This will be followed by an account of textual criticism.

Concerning "critical realism", let us be clear. We are using language in the form of symbols, images and verbal pictures to give an account of experiences and feelings which cannot be *precisely* described. We do this for the simple reason that we can never "know God as God is": we don't *know*. But we do feel that we know something of God, e.g. "the direction in which clues about and experiences of God point". To that extent we are critical realists: we are describing something real behind our experiences, even though we cannot help being vague or poetic about it.

As James has said, the experience is real for the experiencer and therefore to be trusted, or at least until the experiencer is persuaded that he is not a reincarnation of Napoleon, or until a religious community persuades him that his call and talents are not appropriate to being educated and trained to become a priest or minister. But even then, one may still trust the experience, or intuition, whether in a religious call, the pursuit of a scientific hypothesis, or the composition of a piece of music, and only find acceptance by the religious, scientific or musical community years after one has died.

Religious experience is like art, music or wine, in the sense that those who write about each of them are using words to describe that which is not experienced in words. How can words express the feelings and interpretation of a religious experience, a great work of art, musical composition or a glass of delightful wine? The art critic, the music critic and the wine critic have the same problems as the writer on religious experience: how to link language to a non-verbal experience.

Each of us is eccentric in some sense. I certainly am! I find some abstract art difficult to understand and so cannot make a link between a work of art and the language used about it by some art critics. I value music critics and appreciate what they say. Wine masters I rely on for my choice of good ordinary wines. But I cannot relate the language used in wine tasting to the wines. All these three types of critic use what I will call "flowery language" in their evaluations. I can only understand the music critic – that is me!

It is hardly surprising, then, that those who think religion is important to them as faith and practice, but who have never had a religious experience, reject the uses described by the five model makers quoted. I have said that I find the link between wine and "flowery language" unreal. Thus those religious people who have not had a religious experience either deny the reality of such claims or say that language is being used "in a different way".

In the 1970s and 1980s, in the UK, United States and Australia, 43–48 per cent of the population reported that they had had a "religious experience" (Hay 1990: 79, 82). This figure rose to 45–56 per cent for those who had received full-time education over the age of twenty (Hay 1990: 82). Their religious experiences and educational opportunities had given them an understanding of the links between experience and language so they were able to adopt a critical realist position about the nature of their religious experience.

NON-REALIST CRITICS OF CRITICAL REALISM

In 1972, Professor T. R. Miles, an active Quaker, wrote his book *Religious Experience*. He rejects dualism (1972: 5). He does not believe that the material world, which he accepts, co-exists with a non-material world, which he does not accept. He accepts that in a philosophical context, the word "materialism" stands for the assertion that "nothing exists except matter". "If a person has a moving religious experience, it is unnecessary to re-describe the situation by saying that he has received a visitation from a supernatural world" (2). One does not need to be "involved in discussions about 'materialism', 'dualism' or 'supernaturalism'" (5).

He argues that "the idea of a non-material Being is foreign to all biblical thought" (Miles 1972: 41). God is spirit, light and love. Miles continues: "When people claim to have experienced the presence of the risen Christ", this is not a literal reference to a 'non-material' Christ who has existed in a 'non-material world'", it is a myth or a parable. Nevertheless he accepts the compelling nature of "the demand that people should allow the spirit of Christ to work in their hearts, [and that it] carries many practical consequences" (44). It seems from this that Miles regards talk of "religious experience" as one way of expressing "religious commitment". One expresses one's response to cosmic questions through "silence qualified by parables" (50).

Perhaps the most influential non-realist is Revd Don Cupitt, former Dean of Emmanuel College, Cambridge. Since the 1970s, properly aware of its falling membership, he has tried to rescue the Church from being irrelevant to so many.

How? He says we need to shed the *"eschatological* faith" of the Gospels and the *"dogmatic* faith" (of Creeds), and try to hear the voice of Jesus concerned only "to relate the hearer to God" (Cupitt 1979: 7, 97). This prophetic challenge still stands.

However, Cupitt's explorations took him away from critical realism towards non-realism (Cupitt 1980: 8). God was not "an immense cosmic or supercosmic Creator-Mind" (8). "God is a personal religious ideal, internal to the spiritual self" (Cupitt 1985: 136). Cupitt's linguistic idealism, his belief in the primacy of language to shape experience, becomes clear here. He continues: "all of religious language is still usable and the religious life still viable . . . Moreover the God-ideal on my view remains transcendent, authoritative and irreducible" (136).

Professor John Hick notes that, even though Cupitt uses non-realist language, his religious vocabulary is no different from that of a religious realist (Hick 1989: 201), or a critical realist, I would say. For example, "God indwells the believer, enlightening his understanding, kindling his affections and enabling his will" (Cupitt 1980: 5). "Love is pure and disinterested when it is expressive and non-objective, so that its object [God] is internal to it. When one loves in that way then one is in the love of God" (69). This point of Hick needs emphasizing. Cupitt's personal pattern of worship has not changed, only his concept of God, from realism to non-realism. Cupitt says:

> a Christian non-realist like me may often find himself dropping back into the old type of God consciousness, praying and worshipping because he wants to or because it helps. And why not? I actually think I love God more now that I know God is voluntary. I still pray and love God, even though I fully acknowledge that no God actually exists. (1998: 85)

But no critical realist thinks of God as an objective entity, "for now we see in a mirror dimly" (1 Corinthians 18:12 NRSV). Cupitt's great contribution has been to stimulate thinking about how to present the Christian faith in a culturally relevant way today and still be true to the life and teaching of Jesus.

Miles and Cupitt are two of a group of thinkers who interpret some of Wittgenstein's writings as suggesting that religious language has no reference outside of itself. Both imply that religious language does not relate to a "heavenly" form of life, being and existence. Rather, religious language is an independent mode of discourse which operates as a symbolic system to aid personal transformation in the here and now through commitment to a life of faith. "God" is a controlling concept at the centre of this form of language.

Miles and Cupitt, then, are non-realists. Religious language used in story can express emotion and evoke commitment to a way of life, but refers to nothing outside itself. Once committed one experiences the world differently.

In contrast, traditionalist philosophers like Paul Tillich hold that religious language, even though used in different ways – literal, symbolic, paradoxical, myth-ical – has an ultimate referend: God. God is 'Personal Being' to whom one can relate, and in whose work of continuous creation one can share. Keith Ward notes five models of divine revelation, two of which are "the inner experience model" and "the new awareness model", in which the emphasis is on the personal experience of the recip-ient (Ward 1994: 227). This is critical realism.

Franks Davis rejects the approach of non-realists, arguing that they are "parasitic upon the descriptive content" (1989: 8). Franks Davis gives as an example the asser-tion that, "'God so loved the world that he gave his only-begotten Son . . . ' [which] inspires a sense of security because of the belief that there really is a divine power of infinite love somehow at work in the world, in our lives" (8). Secondly, Franks Davis notes that if not cognitive, then these stories or religious experiences must be useful fictions. But why use misleading fictions when a mature person can find better aids for moral guidance and reassurance about cosmic issues?

Franks Davis suggests that the argument of the non-cognitivists has a further weak-ness, it is prescriptive. They are imposing or prescribing a limit on "an account of a religious experience" made in good faith by the experiencer of a "received experience" – claimed as real with real content. If they say that the report of a religious experience is just "a parable", they distort from the outset an examination of the reported reli-gious experiences of many people. The effects of a religious experience are no doubt a test of the value of religious experiences. But non-cognitivists limit the religious person's choice of how to understand the initial experience to either believing in a crude picture of a God who performs miracles and makes his will known in visions and voices, or demythologization, which means "turning into fiction" all the God-talk picture language. However, there is a middle way, the one adopted by critical realism in its use of models and pictures in science and religion.

It is worth noting that some philosophical idealists, like Professor R. B. Braithwaite (1955), Professor T. R. Miles and Revd Don Cupitt who have explored the use and significance of language in describing the cognitive status of religious belief share commitment to the Christian way of life. In the exploration of the interpretation of Christian experience, realists are at one end of the spectrum, idealists at the other end, with critical realists in between: the Christian way of life is shared.

Models and Pictures in Science and Religion

We must now briefly explore the nature of God-talk picture language. Let us start with the picture language used in science. When scientists are talking about the inter-action of gas molecules, they picture them as balls on a billiard table in order to illustrate some of the qualities of the interactions of gas molecules, but no one takes it literally. Similarly, in Genesis, it says, "Late in the evening a breeze began to blow, and the man and the woman heard the Lord God walking in the garden." If I ask twelve-year-old pupils in school whether God was wearing evening dress, they laugh me to scorn! They know it is picture language, but that it is reality-depicting. Likewise in a religious experience, as we discovered when examining the call of Moses.

Scientists and religious people all use models and picture stories to represent aspects or characteristics of the real world they are exploring, but these are not taken literally. These models or stories picture aspects of what is real. The people who use them are critical realists.

Arguments from the linguistic idealists are not convincing. Take the eighteen cows I see in the field from my window. On a hot summer's day they go to the corner of the field and lie down in the shade under the oak tree's spreading branches. In winter when a full westerly gale is blowing the rain horizontally from west to east, the cows lie down snugly on the lee side of a hedge to keep out of the rain. At dawn on a cold winter's morning, with a clear sky overhead and a frost on the ground, the cows are all up at the top of the field, enjoying the warmth of the sun as its rays slowly descend the slope of the field. Prelinguistic humans no doubt did the same. Is that not knowledge which you first have and then put into words when words are available? Humans are different! They can "know that they know".

Look back to Chapter 6 and read or remember what Polanyi said about tacit knowledge: "you know more than you can say". Remember that he reminded us that we know how to ride a bicycle, but cannot put it into words. Look at a couple in love, or remember your own experience of the first exchanges, which involved all the senses – sight, sound, smell, touch, taste, pulse. Language is secondary, as a French man courting an English girl will make clear to you! Even religions, such as the Greek and Roman, die when they cease to meet the "tests" of human experience.

However, there is *some* truth in the claims of the postmodern linguistic cognitivists. To say that reality is constructed by language is half true. Yes, models, theories and descriptions are human *constructions*, but they are, for the critical realist, accounts of the way the world is: they are critical accounts of reality. Language becomes increasingly important as one moves up the hierarchy of levels of knowledge, or deeper into the analysis of micro or macro bodies, or complex organization in the physical world. Language will often be very important, for without it there would be no verbal knowledge, but the knowledge claimed by a scientist will be that of a critical realist.

So, at sophisticated levels of enquiry, either of the natural or human worlds, and of the "world hereafter", knowledge is *constructed* by and *expressed* or communicated through language. But, and this is the difference between linguistic constructivists and critical realists, that which is offered as knowledge must *ring true* to human experience sooner or later.

Finally, Wittgenstein is quite clear that "what determines . . . our concepts . . . is the whole hurly burly of human actions" (Thiselton 1992: 540).

CRITICAL REALISM: LEVELS OF INTERPRETATION

In the study of religious and mystical experiences, the exploration of the relationship between experience and interpretation received a considerable step forward in the work of Peter Moore (1978), which has been developed further (Miles, G. B., 1983). This suggests that one can construct a matrix of eleven distinct "levels" involved in the relationship between experience and interpretation, and six "columns" or perspectives from which interpretations can be made (see table 13.1, opposite).[1]

If one is studying the text of a modern or recent religious experience, the elements involved in the linguistic expression of the experience are the physical triggers which set off the experience, e.g. the lady in Chapter 10 whose experience was triggered by ants, trees and a blue sky. There are also psychical factors involved, like feelings, thoughts, perhaps previous experiences. Language is involved, because how one interprets and describes the experience will depend on the vocabulary to which one has access.

If one is studying a religious experience found in Scripture, which has been revised over many hundreds of years by a considerable number of editors repre-

Level of Interpretation	Perspective of Subject (Individual) (a)	Perspective of third person inside tradition (b) ISI	Perspective of tradition's interpretation (corporate) (c) TI	Perspective of third person on a tradition's interpretation (d) 3T	Internal pan-religious perspective interpretation (e) IPR	External pan-religious interpretation (f) EPR
1. Raw experience	O	O	O	O	O	O
2. Reflexive interpretation	Experience 'described' by God-talk.	ISI	TI	O	O	O
3. Incorporated interpretation	Experience related to beliefs already held.	ISI	O	3T	IPR	
4. Retrospective interpretation	Experience reinterpreted later.	ISI	TI	O	O	O
5. Doctrinal interpretation	Experience related to theology or naturalism for interpretation.	ISI.	O	3T	IPR	EPR
6. Generalized interpretation	Many experiences related to theology or natualism.	ISI.	TI	3T	IPR	EPR
7.		Internal secondary interpretation (ISI) 2b, 3b, 4b, 5b, 6b				
8.			Tradition's interpretation (TI) 2c, 4c, 6c			
9.				Third person's interpretation of tradition (3T) 3d, 5d, 6d		
10.					Internal interpretation (IPR) 3e, 5e, 6e	EPR
11.						External interpretation (EPR) 5f, 6f, 10f

Table 13.1   Religious experience interpretation matrix: levels and perspectives

senting different communities, the task can be very complex. It is possible, when studying some of these experiences that eleven levels of interpretation may be distinguished.

"Raw" Experience

First, if we start with the experience related in Chapter 10 which began with the sight of ants, trees and a blue sky, we can say there was some raw sensory experience of what was seen. While I accept, for the sake of building an analytical model, that there is no such thing as "raw experience" or "uninterpreted experience", it is here theoretically postulated that an individual has some experience which is unaffected by previous beliefs, expectations or intentions. This will be called *raw experience*, though by the time it has been put into words, it is already at the second level of understanding. This first level can never be expressed (level 1a in the matrix).

Reflexive Interpretation

Secondly, concepts derived from reflection through reason were used in the "representation" of the experience or some aspects of it. So she had a sense of the smallness of the ants and her hugeness; then she looked at the sky and had a sense of her smallness and the largeness of the sky. This was her spontaneous description, which involved interpretation by selecting concepts to "describe" what has been seen and felt. This we can call *reflexive interpretation* (level 2a) (Luckman 1967: 43; Barbour 1974: 51–2; Moore 1978: 108; Schillerbeeckx 1980: 13; Franks Davis 1989: chs 1, 4, 6).

Incorporated Interpretation

Thirdly, she already had some religious ideas in her mind, beliefs which she had learnt at home, a place of worship, school, from the media or elsewhere. "A watcher" is such a word, which she quickly uses to describe "the bigness" looking down on her from the sky. This we will call an *incorporated interpretation* because it includes or incorporates a religious idea which she found necessary in order to give a faithful account of her experience as a five-year-old (level 3a).

Retrospective Interpretation

However, when I used this passage with classes of students in an open access sixth form college (Miles 1983), they detected a level of language use outside the understanding of a five-year-old. They found reference in paragraph five to a "Body" and recognized that the lady must have studied St Paul's writings in the New Testament either in school, in church or at home when she was in her teens. She tells us that she often thought of this experience, but in her teens or later she incorporated this Pauline language and so reinterpreted it in a way she found more "true" to her maturing understanding of the original experience. This we can call a *retrospective interpretation*, because it happened after the initial event when she was a five-year-old-girl. (level 4a).

Doctrinal Interpretation

Fifthly, if the experience is set within a doctrinal framework after the experience, then such an activity is described as *doctrinal interpretation*. This may be theological or naturalistic. The Pauline interpretation mentioned in the paragraph above shows that the experience has been interpreted at both levels 4 and 5 at the same time. In paragraph four of the ant experience, there is further evidence of mature religious thinking, perhaps added at some time between her twenties and fifties, indicating more retrospective theological interpretation (level 5a).

Generalized Interpretation

So far we have discerned five levels of interpretation involved by one person reflecting on a single religious experience. But the same person may be able to refer to many experiences, which together may be interpreted within a tradition, or doctrinal framework, in the form of a *generalized interpretation*, a sixth level of interpretation (level 6a).

Now let us turn to the powerful religious experience of Hosea (CEV). He marries Gomer (Hosea 1: 2–8; 3: 1–2) and has three children. She is unfaithful and goes off with a lover. Hosea buys her back for fifteen pieces of silver and 150 kilograms of grain – a sort of ransom.

Clearly all the six levels of interpretation already described are involved in this experience, and more, for Hosea takes all his experiences as the foundation of a model of how he believes God thinks of his chosen people Israel, whom he loves and wants to "win back" (Haughton 1972: 14–15). Similar imagined or observed experiences must have led Jesus to embrace the same model of a loving "waiting" God, leading to his Parable of the Prodigal Son, who "wasted all his money in wild living" (Luke 15: 11–32).

Secondary Interpretation

But a seventh level of interpretation is involved if a faithful worshipper interprets the experience of Hosea or the Prodigal Son as relevant to them – a *secondary interpretation*. This seventh level of interpretation is made by, for example, someone inside Judaism in the case of the story of Hosea, or inside Christianity in the case of interpreting the parable of the Prodigal Son. Either case would involve *secondary interpretation*. This provides a second column of interpretive perspectives in the analytic matrix (levels 2b, 3b, 4b, 5b, 6b). These are individual secondary interpretations of a third person from within the same tradition.

A Tradition's Interpretation

What further happens when a biblical editor collects together many traditions of Scripture, and places an interpretation on them characteristic of their own religious community? If you care to read a commentary on the first five books of the Bible, you will find reference to four editorial traditions known as J, E, D and P. They each offer a generalized religious interpretation of the experiences of the community they have collected, which are then interpreted and written down according to the religious beliefs of their own community at the time. This eighth level of interpretation can be called *the tradition's interpretation* (levels 2c, 4c, 6c). Level 6c would represent the activity of the systematic theologian attempting to incorporate the range of experiences and interpretations into a systematic theology, i.e. a coherent overview of a faith like Christianity or Judaism, or even of a tradition or movement within a World Faith, like the Roman Catholic Church or Baptist Church within Christianity.

External Interpretation

A ninth level of interpretation is involved if a religious experience is interpreted by a person who is outside the religious or cultural tradition of the experience under consideration. Such a statement can be called *an external interpretation* (levels 3d, 5d, 6d). This provides a fourth column or perspective in the analytic matrix.

Internal Pan-Religious Interpretation

A tenth level of interpretation is involved if a comparative study of religious experience and its interpretation is made by a person who is a sympathetic member of one of the religious traditions being studied. We have already explored work of this kind in, for example, Otto's view of the numinous, or Stace's view of the core religious experience. Each study has lead to a particular *internal pan-religious perspective*.

Other topics, such as Sacred Writings, Ethical Teachings, or Worship in a number of religious traditions, could be the focus of studies searching out the similarities and differences within the World Faiths, or areas for fruitful dialogue these Faiths may wish to explore. The series of books edited by Jean Holm with John Bowker provide some other fruitful topics and resources for this approach in eight World Faiths (Holm and Bowker 1994).

External Pan-Religious Interpretation

An eleventh level of study of World Faiths, by a writer from outside or alienated from the religious traditions, could provide a view which may be favourable but which is usually hostile. Hostile views are expressed by writers like Freud, Marx and Nietzsche. We will examine some of these later. But one brief illustration can be provided from two brief comments from Freud. He thought favourably of religion before science emerged, as cement that held society together. Now, in the light of science, he decided that religion was a universal neurosis and should be replaced by scientism. He hoped that his therapies would liberate repressed energies and free the world of both religion and unnecessary restraints (levels 5f, 6f, 10f).

The first ten levels are not necessarily distinct and may cluster differently about any experience. Developing understandings of concepts and their status and localization within philosophical or religious language use prevents any fossilization or reification of these levels of interpretation. They are outlined to alert the reader to the complex levels of interpretation and the different perspectives from which religious experience can be understood.

INTERPRETING TRANSCENDENTAL EXPERIENCE

In Chapter 1 we discovered the work of Michael Paffard, who was investigating experiences of nature mysticism. He was trying to find out if the students (aged 16–18) to whom he taught Wordsworth's poetry in sixth forms in an English School[2] and University students (aged 18+) actually had had any direct experience of nature mysticism. He found that some of those reporting such experiences described them as religious experiences. Paffard included both nature mysticism and religious experience in the inclusive category of transcendental experience.

Table 13.2 Kinds of psychological relationships in mystical, contemplative, ecstatic or transcendental experiences

Kind of Experience	RELATIONSHIP with the revered subject, nature human or divine	MYSTICAL UNION with the revered subject, nature human or divine
Natural experiences involving personal effort	'withdrawal ecstacies' (Laski) 'plateau experiences' (Maslow)	
Natural experiences received as personal gift	'nature mysticism' (Zaehner et al.)	'intensity ecstacies' (Laski) 'peak experiences' (Maslow)
Experiences which may be natural or religious gift	<------- 'nature mysticism' (Wordsworth et al.) -------> <------- 'transcendental experiences' (Paffard) ------->	
Religious experiences involving personal effort	religion of the 'healthy minded' (James) 'immanent divinity' (Emerson) 'way of transcendence' (Kee)	'Contemplation' (cf. Butler, Williams et al.★)
Religious experiences received as personal gift	<------- 'absolute dependence' (Schleiermacher) -------> 'religious experience' (James et al.) 'sick-soul conversion' (Starbuck, James) I–Thou (Buber) 'numinous' (Otto et al.) 'ecstacy' (Underhill) 'response experience' (Glock and Stark) 'ecstacy' (Greeley) 'signals of transcendence' (Berger, Ward) 'signals of God' (Deeks) 'sense of presence' (Hardy, Hay)	'mystical experiences' 'mystical experience' (Otto) 'unitive life' (Underhill) 'revelational experience' (Glock and Stark) monism, Christian mystical union (Zaener)

★ The survey of literature excludes reference to the "contemplative life". Cf. Butler (1967); Williams (1979); Turner (1995); Cupitt (1998).

In Chapter 3 we found that many people working in the field of humanist psychology described ecstatic experiences as transcendental experiences. In Chapter 9 we explored further some uses of the term transcendence.[3] We can now summarize this work, and our review of religious and mystical experience found in Chapters 10–12, in table 13.2 (above). This table also provides an opportunity to incorporate examples of other work which may seem to be outside our area of concern, but which can contribute to extending our understanding of Smart's categories, renamed as "the religious outlook" and "the mystical outlook".

These explorations have demonstrated that the meanings and use of transcendence always refer to some aspect of self-transcendence. These uses of transcendence may describe an attribute of God without, or a quality of experience that transcends self-consciousness in the form of God within.

Transcendence may be experienced as nature mysticism; as consciousness of the "ultimate depth" of life, or of the "Divine"; of the ultimate within the moral; as a way of life; in relationships; as "signals" within life; or as experience in another dimension. Such experiences may be described or interpreted as natural or religious.

THE PROCESS OF INTERPRETATION: HERMENEUTICS

We have all been involved in the task of interpreting both written texts, like novels, Scripture or business reports, for example, and the spoken statements we hear from all kinds of people, such as neighbours, actors, preachers and business executives. The development of new insights and techniques for the interpretation of communicative acts through speech or texts, identified at the end of the nineteenth century but vigorously developed since the 1970s, is the textual critical method called hermeneutics. Hermeneutics is simply the theory of the interpretation of communicative acts in speech or texts. Understanding involves interpretation. It was initially applied to sacred, and then legal, texts. In principle and practice these methods are, or can be, used on any text such as in reading and understanding a novel or letter, as well as a point of view you are listening to. Different scholars offer different and frequently incomplete accounts of their views of the nature of the study of texts and their underlying assumptions.

What follows is an attempt to outline a hermeneutic method, without discussion of underlying philosophical assumptions, which will be mentioned at the end. The method of study is governed by principles, with different principles being important to different people. These include the following:

- ✧ Interest: The text of a document becomes interesting to a reader as it grasps their attention, otherwise why bother trying to *understand* it fully?
- ✧ Text: The text of the Bible, and any other sacred writings, should be treated like any other text.
- ✧ The author's intentions: Some say the reader should try to recapture the original intent of the text, indeed "re-experience the creative act of the original author" (Stiver 1998: 88). As this is difficult, if not impossible, it is ignored by some scholars.

✧ Grammar: The reader should be guided by grammatical interpretation. This means attending carefully to the meanings of words, sentences, stories, narratives and different literary forms like myth, analogy and metaphor as well as literal statements.

The process of interpretation involves three stages:

Stage 1: Initial Understanding: The Hermeneutical Circle

An initial understanding of a text is achieved through examining the circular interaction between the meaning of words, sentences, paragraphs, i.e. between parts of the text and the whole text. There is "a truth" in the text, as there is a truth in a work of art, that is not purely subjective or a matter of taste, which controls the *meaning* of the text. This the reader tries to discover in order to *understand* the meaning of the text. This understood meaning is sometimes called "*the world in front of the text*". The term "hermeneutic circle" is just technical shorthand for this interaction, just as "over" is technical shorthand for "six balls" in cricket. This initial achieved understanding of the text grasps the reader's attention.

Intuition   Understanding the text involves the use of intuition and the creative imagination in using all the skills available.

Stage 2: Explanation, Critical Evaluation and Interpretation

Once an *initial understanding of a text* has been gained, a text needs *explaining*. What caused the text to be written? This is a historical task. What did the text mean at the time it was written, and how does the text disclose "the world behind the text"? Where did the text come from? The answer comes through historical and literary criticism, which has developed to a great extent in the last two centuries. How was the text made? The answer involves analysis of the text's structure. How is the text unravelled? By using all the methods and stages involved in analysing the text, including language use and the process of interpretation (Thiselton 1992: 489).

Language Use and the Process of Interpretation   Our exploration earlier of the uses of language in the twentieth century (Chapters 2, 3 and 4) began with the claim of the logical positivists that meaningful language, which could be regarded as true, was largely literal. They used scientific knowledge as the model for all kinds of knowledge.

In the 1960s philosophical analysis replaced logical positivism, indicating that there were many uses of language and forms of knowledge. While we have looked earlier at different forms of knowledge and the hierarchy of forms of knowledge (Chapters 4 and 6), we have only looked at some of the uses of language.

In order to explore some other aspects of the method of interpreting texts called hermeneutics, we need to recall that, by the medieval period (CE 1200–1350), there were three widely known ways of understanding the use of religious language. Some said that religious language is *analogical*, expressing religious insights through pictures, as in "God is the Good Shepherd" (Psalm 23). Some said that religious language is *literal*, i.e. univocal or having one meaning, as in "God is almighty". Some said that religious language helps people *evoke an experience of the Divine* while they travel along one or more of four paths through life, on their spiritual journey. The four paths are: "sharing God's delight in creation"; "letting go and sinking into God"; "creating with God from within oneself"; and/or "transforming life through compassion and social justice". Some recognized and accepted any or all of these paths. Explanation also involves the hermeneutics of suspicion.

Hermeneutics of Suspicion   Knowing these medieval ways of understanding religious knowledge, and welcoming modern critical methods of studying sacred texts, like the study of historical sources, literary forms and sources, one modern scholar, Paul Ricoeur, became concerned. He realized that unconscious motives or prior beliefs would colour or prejudice a writer's ways of thinking and ways of recording events, and likewise a reader's response. This can arouse suspicion.[4]

Sometimes a reader might read a religious text and say that on the surface it is perhaps talking about heavenly reward: "How great your reward will be in heaven if you work hard at your unpleasant occupation now." But deep down the text is *avoiding* describing the human situation now, which should be changed by social action now, not projecting "satisfaction" into the future in "another world". Hence Marx called religion the "opium of the masses" because religion can be seen to distract attention from legitimately pressing for social and economic betterment.

Perhaps theology is really disguised anthropology, say some. If so, understanding involves being aware of both the surface meaning and a hidden meaning. The reader needs to be *suspicious* of the presence of unconscious influences or personal prejudices or beliefs which will skew the way an author writes his narrative: you, the reader, must watch out for these influences. Likewise, prejudice or unconscious motives can influence how a reader interprets the text that they are reading.

Another example is the phrase "God forgives him who repents", which may be a disguised ecclesiastical power claim meaning "Forgive him who submits to the priest" (Thiselton 1998: 105). Is this how the Church exercises power?

Stage 3a: Post-critical Understanding [5]

Interpretation involves three stages: *interpretation*, which seeks the "world in front of the text" (stage 1); *explanation and critical evaluation and interpretation*, which seeks the "world behind the text" (stage 2); and then the reader brings their own pre-understandings – their outlook on life – to the text to *post-critically understand* it (stage 3a).

Let me say this again in a different way. The first stage involves seeking the meaning of the text – "the world in front of the text"; this is called the "horizon" of the text. This understanding is modified in the light of the *critical analysis, explanations and interpretations* (stage 2). In reading a text, the text's "horizon", mingles with the pre-understandings of the reader, called in shorthand, "the reader's horizon". This mingling of horizons leads to *the reader's post-critical understanding* of the text (stage

3a), which is open to revision later, perhaps leading to a fresh understanding. While an understanding need not be final, it obviously establishes what the reader *knows*.

If this reader's understanding becomes shared by members of a community, like a church, shared knowledge becomes "the shared teachings basis", or doctrinal basis of the church. Likewise, if the scientific interpretations of an individual, or their literary or musical interpretations, are shared by the respective scientific, literary or musical members of those communities, such insights are added to the bank of knowledge in that society.

Exploring a text involves three stages. We have explored the two processes of, first, *initial understanding,* and then, second, *explanation, critical evaluation and interpretation,* which then leads to, third, *a post-critical understanding.* The second half of the third stage, which we have not yet mentioned, is *application.*

Stage 3b: Application

Even now, the reader's interpretation is incomplete if they just consider the influences on the writer, whether unconscious, ideological (his beliefs and prejudices), historical or literary. The reader must search for the fullest meaning or meanings of the text, particularly "the ones which are most bound by the presence of the sacred" (Stiver 1998: 103). The reader reaches their understanding of the text through the intermingling of the horizon of the text and their own horizon, which is affected by their own prejudices and prior beliefs and unconscious influences. This is the reader's full, post-critical understanding of the text.

However, the reader's post-critical *understanding of the text* is still not complete until it is *applied.* In Christian terms, this can be seen in the work of those who indicate that language is frequently performative or evocative, building on the work of John Austin.[6] For example, language mediates the authentic existence of Jesus, and his use of the metaphor of the Kingdom of God and the parables, which are extended metaphors. While myth creates pictures of what the world is like, parables subvert one's world, for they "function to disturb and unseat conventional understanding" (Stiver 1998: 129). Their application amounts to an invitation to follow in the steps of the Master. The reader may find it is a summons to do just that, and if so it will change their orientation and way of life.

Just as we saw that *the hermeneutical circle* involves establishing the meaning of a text by attending to the part–whole interactions within a text, so also *the hermeneutical arc* (Stiver 1998: 129), which summarizes the three stages of reaching *an initial understanding* of a text, then an *explanation and critical evaluation and interpretation of the text* is completed by the *post-critical understanding of the text,* in which the horizons of the text and the reader merge. The result of this should lead to the *application* of this *post-critical understanding* to one's life. However, because that critical understanding is not final, it can lead to a hermeneutical spiral, in that this interpretation, in the light of further study and reflection by oneself or another, can lead to another interpretation, and so on.

TEXTUAL CRITICISM

Philosophical Claims

The central claim, which has possessed some philosophical theologians, centres on the nature of being, i.e. on *what is* and *therefore can count as knowledge*. It comes from George Hans Gadamer's major work, *Truth and Method*. I have been constantly defending the traditional view of knowledge embraced by critical realists, which is "we have an experience, and then put it into words," or "we discover something and try to explain it."

Against this traditional view Gadamer said, "It is not that the understanding is subsequently put into words, rather the way understanding occurs – whether in the case of a text or a dialogue with another person who raises an issue with us – is the coming into language of the thing itself" (Gadamer 1991: 378–9; Silver 1998: 95). His famous statement is, "Being that can be understood is language." This is the source of what informs the thinking of some modern philosophers and theologians like T. R. Miles. It is an idealist's doctrine which, as an absolute statement of "the nature of being", is just wrong, for reasons I gave above in my comments on T. R. Miles's thinking. Being is not language. Understanding is not language. Being that can be understood is usually understood *through the use of* language: yes! But there are many non-linguistic ways of *understanding being* for humans (Argyle 1967). Simply, this doctrine will not stand the tests of making sense of human experience.

This doctrine of Gadamer does contain much truth in that most knowledge can only be formulated by language. But language is not being. It is a tool for those who enjoy *being*, such as living vibrant persons. For example, in the city of Pompeii language is nothing more than dead words on dead pages in the libraries of people who were exterminated by the volcano, Vesuvius (Thiselton 1992: 125). Language is the *main means of communication* between human beings, and the main means of creating much knowledge.

Language is but one vehicle of communication between human *beings*: the most important vehicle carrying knowledge to and fro. Another important argument against Gadamer's doctrine is that personal knowledge and religious knowledge are both different kinds as well as being "higher" forms of knowledge than semiotics, the name for the form of knowledge concerned with the study of words, signs and symbols.

However, language, which has developed in the last half a million years, is an indispensable tool for the knowledge and understanding which has developed in parallel interaction with it over the same period. This, in turn, has led to the transformation of the nature of being human.[7]

Common Practice

The complexity of understanding a text, outlined above, may look pretentiously academic. If you reflect on what you do when you read a long letter from a friend, you may find that this picture of textual criticism tells you what, in fact, you do intuitively. Perhaps this is the first time you have had it put into words.

Let us start putting into words your reactions to your friend Mary's long letter! As you read it, here are some thoughts that may come into your mind: "Oh come off it! Your criticism of our head teacher, George, was influenced by your socialist beliefs and your disillusionment with George after you were overlooked in the last round of promotions!" "Yes, I know what you are saying about your deputy in your department, but reading between the lines I can see what you really mean!"

"Oh, and why did you choose that form of words to say that? Cynical aren't you?"

"Oh no! Fred did not mean you to think that you were special to him when he gave you that bunch of flowers, because yesterday I saw him with . . . !"

Each kind of critical skill listed above you can, and do, employ on letters or other texts you read!

Take another example – reading a novel. The novelist writes in such a way that there are "spaces" in her text, so that you complete an understanding of the text through the use of your intuition or creative imagination. Take Daphne du Maurier's novel *Rebecca*. Had you noticed that the heroine has no name so that every female reader can see herself as the heroine? Did I not mention earlier that my father and I each fell in love with a girl in a novel we read? The two occasions were thirty years apart! How's that for application! You can now see what is meant by the horizon of the text mingling with the horizon of the reader for a full understanding of the text to emerge.

Falling in love with a heroine in a novel helped me discover something of the qualities of the kind of real girl I could hope to meet some day. Admiring particular qualities in some "older boys" (17–18 year-olds) in school,[8] and in some teachers in school and university, helped me work out the kind of person I would like to become. The same process was at work when I admired people like Martin Luther King and Nelson Mandela. How much more is this process at work when I look at, and get to know, Jesus of Nazareth and find his contagious attraction irresistible. There comes a moment when, reflecting on the nature of God as love, one "sees" "the human face of God" in the life of that extraordinary man, Jesus of Nazareth. And then one goes to church and repeats the now "dead-for-me" language of the Nicene Creed. So what should one do?

I have mentioned these brief autobiographical comments in the last paragraph only to show you what *you* do: you read the text of the lives of people you are attracted to, exploring the horizon of their text and mingling it with your own horizon. You apply "textual criticism" both to written texts and to the living texts of people you encounter in different ways.

One further comment needs to be made on this process, concerning the pre-understanding "horizon" each "reader" brings to the stage of post-critical understanding of a text or person. Each of us is "eccentric" in some way or other. I have already mentioned how I find it difficult to reconcile the taste of wine with the language used by the knowledgeable wine experts to "describe" different wines.

We all have psychological sets of "pre-understandings" from the influences of our previous life relationships and experiences, and our receptive and creative responses.

SUMMARY

To summarize, critical textual criticism, called hermeneutics, involves three stages and many critical skills.

It involves reading the text to *understand* it, by seeing how the whole relates to the parts within in order to discover the "world in front of the text", sometimes called the horizon of the text.

It includes reading the text to *explain* it, by taking account of historical, literary, ideological, unconscious and other factors which influence the way it is written in order to discover the "world behind the text". Then the reader uses these insights to make a fuller *interpretation of the text* in order to have a better view of "the world in front of the text".

The reader then interacts with the text, thus engaging in an interaction of the text's and their own horizon to achieve a *post-critical understanding*.

Finally, the reader *applies* the text to their philosophy of life, religious or humanistic, changing their orientation, or confirming its direction.

These considerations of this and previous chapters will inform us when we examine some examples of religious experiences in detail in the next chapter.

NOTES

1   The matrix notes that interpretation may be religious or naturalistic, though discussion in the text will focus on our main concern, religious experience and interpretation.
2   Now forms eleven and twelve in schools in the UK.
3   Hick (1991). This is an excellent overview of religion from the perspective of transcendence, as the title and contents indicate. See also Hick (2004).
4   Thiselton (1997: 531). "A fine balance is achieved by Ricoeur between the need for a hermeneutic of suspicion which 'destroys idols' and a hermeneutic of retrieval which seeks to recover the creative power of symbol, metaphor and narrative" (534). This important point must not be lost, but cannot be pursued here.
5   Technically called the third element in the *hermeneutical arc*, a term used to describe "three moments": an initial moment of *understanding* followed by a second moment of *explanation and criticism*, and then a third moment of "post-critical understanding" which must lead to and include *application*.
6   See e.g. Stiver (1998: 80–2) for a brief account.
7   There are three schools of thought concerning the relationship of language and thought: "language and thought are the same"; "thought is dependent on or caused by language"; or "language is dependent on and reflects thought". By implication, thought in the third school is based on experience. In 1921 Wittgenstein gave support to the second view espoused by Gadamer and idealist philosophers and theologians mentioned when he said, "The limits of my language mean the limits of my world." The last view, critical realism, which is defended by Polanyi, Piaget and Vygotsky, is the view I have supported here. (Cf. Gross and McIlveen 1998: part III, esp. ch. 37.)
8   I enjoyed attending Ashville College, Harrogate, then a boys' school (evacuated to the Hydro Hotel, Bowness-on-Windermere, during the war) and now a school for girls and boys.

REFERENCES

Argyle, M. (1967), *The Psychology of Interpersonal Behaviour*, London: Penguin.

Barbour, I. G. (1974), *Myths, Models and Paradigms*, London: SCM Press.

Braithwaite, R. B. (1955), *An Empiricist's View of the Nature of Religious Belief*, Cambridge: Cambridge University Press.

Butler, D. C. (1967), *Western Mysticism*, London: Constable.

Cupitt, D. (1979), *Jesus and the Gospel of God*, London: Lutterworth.

Cupitt, D. (1980), *Taking Leave of God*, London: SCM Press.

Cupitt, D. (1985), *Only Human*, London: SCM Press.

Cupitt, D. (1998), *After God: The Future of Religion*, London: Phoenix.

Franks Davis, C. F. (1989), *The Evidential Force of Religious Experience*, Oxford: Clarendon Press.

Gadamer, H. G. (1991), *Truth and Method*, New York: Crossroads.

Gross, R. and McIlveen, R. (1998), *Psychology: A New Introduction*, London: Hodder & Stoughton.

Haughton, R. (1972), *The Knife Edge of Experience*, London: Darton, Longman & Todd.

Hay, D. (1990), *Religious Experience Today*, London: Mowbrays.

Hick, J. (1991), *An Interpretation of Religion: Human Responses to the Transcendent* (1989), London: Macmillan.

Hick, J. (2004), *The Fifth Dimension*, 2nd edn, Oxford: One World.

Holm, J. (ed.) with Bowker, J. (1994), Themes in Religious Studies: *Human Nature and Destiny*; *Worship*; *Making Moral Decisions*; *Myth and History*; *Attitudes to Nature*; *Picturing God*; *Sacred Writings*; *Women in Religion*; *Rites of Passage*; and *Sacred Place*, London and New York: Pinter Publishers, ten volumes.

Luckman, T. (1967), *The Invisible Religion*, New York: Collier-Macmillan.

Miles, G. B. (1983), "A Critical and Experimental Study of Adolescents' Attitudes to and Understanding of Transcendental Experience", PhD dissertation, School of Education, University of Leeds.

Miles, T. R. (1972), *Religious Experience*, London: Macmillan.

Moore, P. (1978), "Mystical Experience, Mystical Doctrinal, Mystical Technique", in S. T. Katz (ed.), (1978), *Mysticism and Philosophical Analysis*, London: Sheldon Press.

Schillebeeckx, E. (1980), *Interim Report on the Books Jesus and Christ*, London: SCM Press.

Stiver, D. R. (1998), *The Philosophy of Religious Language*, Oxford: Blackwell.

Thiselton, A. C. (1992), *New Horizons in Hermeneutics*, London: Harper Row.

Thiselton, A. C. (1997). "Theology and Hermeneutics", in D. F. Ford (ed.), (1997), *The Modern Theologians*, 2nd edn, Oxford: Blackwell.

Thiselton, A. C. (1998), "Biblical Studies in Theoretical Hermeneutics", in J. Barton (ed.), (1998), *The Cambridge Companion to Biblical Interpretation*, Cambridge: Cambridge University Press.

Turner, D. (1995), *The Darkness of God*, Cambridge: Cambridge University Press.

Ward, K. (1994), *Religion and Revelation*, Oxford: Clarendon.

Williams, R. (1979), *The Wound of Knowledge*, London: Darton, Longman & Todd.

14 | Religious Experience

The psychological community used to recognize four major movements or "forces", the first of which was behaviourism. However, "There are no behaviorists any more; they are extinct" (Argyle 2000: 7; cf. Gross and McIlveen 1998; Scott and Spencer 1998). The psychodynamic approach, the second force, includes the work of Freud, Jung and object relations theory. This was followed by humanistic psychology in the 1960s. The fourth force was transpersonal psychology, which was introduced by Maslow (1987). Freud and Jung still exercise much popular influence and will be explored in this chapter, as will object-relations (O-R) theory, which is of contemporary relevance.

FREUD, PSYCHOANALYSIS AND RELIGION

Freud and Human Nature

Before discussing the religious ideas of Freud, it will help us if we first look at Freud's outlook on life and his views on human nature.[1] Freud described himself as "a godless medical man and empiricist" (Shafranske 1995: 202). Reason was his god, and scientific work was the only route to knowledge of reality. Religion was a cultural neurosis which impeded this work and progress. Freud's picture of the mind was based on the mechanistic nineteenth century principles of physics. The mind contained energy, such as sex and aggression, but the nervous system was held in a state of inactivity until awakened by external stimuli or internal needs or drives to which the mind gave passive, compliant responses.[2]

Freud's Scientific Approach   A scientific approach is focused around concepts like cause and effect, mechanisms, entities, structures and processes. This approach portrays the human mind as a mechanism which reacts to external stimuli and internal physical needs, and its view of mind was based on the idea that the neural system needed to be held in a constant state of stimulation. Too much excitement meant that the extra stimuli were released down the easiest escape route, like dreaming.

The mind was pictured as having three elements. The *id* was the seat of basic and uncontrolled instincts, like the needs for food and sex, and curiosity, e.g. will there be more food over there? The *ego* is the essential "me". The *superego* is the conscience or moral dimension of human nature. Later Freud added the concept of *ego-ideal,* "the person I should become", to his model of the human mind. Basically the stimulus, the *id*, finds itself in conflict with the *superego,* a situation that is responded to by the *ego.* This model of the mind is a version of the stimulus–response (S–R) theory of human behaviour, a model of the behaviourists. The mind is entirely passive, concerned with need or drive reduction (cf. Meissner 1984: 190–1).

Human energy is portrayed as a closed system which needs to be kept in balance in the ways described. Thus, because I am or will be hungry, I search for or go and buy some food, and that restores the state of equilibrium in the body.

Freud's Humanities Approach Paradoxically Freud also focused on *meanings.* In his clinical approach to the interpretations of dreams, he was concerned not only with the meanings of them, but also other meanings of, for example, mistakes, jokes and slips of the tongue. So his clinical theory was expressed in terms of motivations, personal relationships and interpersonal interactions.

But this is an approach, not through the sciences, but through the humanities, with the focus on experiences described in terms of unconscious fantasy, wishes, dreams, feelings, reasons and meanings. In searching for deeper meanings of these experiences, the analyst makes use of functions like projection, introjection and repression.[3] Freud based his work on three assumptions. First, he assumed that the unconscious mind could be retrieved and understood by psychoanalysis. Secondly, he used analogy as a tool of interpretation. For example, he believed that the primitive mind = the child's mind = the neurotic's mind (cf. Shafranske 1995: 205, 209). Thirdly, he assumed that all human activity was concerned to meet the needs of human wishes, which he called wish-fulfilment, either through achievement or when they could not be achieved, frequently through illusory goals in the future.

This second approach, through the humanities, focused on human activity rather than on human passivity and foreshadowed *the object-relations model* of human behaviour.[4] The object-relations model of the human mind is that of an open system which emphasizes growth more than stability, creativity more than equilibrium, becoming more than existing. The human is viewed less as a responsive robot and more as a creative source of energy which does not just respond to stimuli, but is able to initiate changes in the environment, social relations and personal life plans. The object-relations model views the human as a source, not of reactive, but of transactional behaviour.

These two approaches, the scientific and the humanities approaches, influence Freud's ideas of religion, but before we examine them, it is important to notice that, while Freud was heavily influenced by the mechanistic science models of his day, he "took psychoanalysis out of the realm of science and planted it firmly in the realm of the humanities. The difference between these two disciplines is fundamental . . . science seeks answers in terms of causes, the humanities in terms of reasons" (Meissner 1984: 197).

Despite this important achievement, Freud was philosophically tied to the passive mechanistic view of the human being from his scientific approach. It was only much later, with the advent of modern philosophical analysis after Wittgenstein's later work, that we became clearer about the different kinds of knowledge humans can hold without there being any conflict between them, because they are each concerned with different kinds of understanding. But this is to anticipate a later stage of this story of the development of our understanding of religion.

Freud, Neurosis and Religion

Using his second approach focused on *meanings*, mentioned above, Freud uses *analogy* to discover these meanings. He says that religion is like a collective neurosis,[5] i.e. religion is a set of shared repressions. The obsessional neurosis of an individual resulting from repression means that the painful experience is kept repressed but finds an outlet through one or more *rituals*.

One example might be when the ritual of folding and placing clothes when going to bed has to be carried out in a precise way, and omitting it or varying the procedure at the required time causes untold anxiety and trouble to the person's conscience. Secondly, the ritual is carried out in private. Thirdly, it is carried out without understanding. Freud suggests that religious ritual is designed to contain and control repressed guilt feelings. The pious describe themselves as "miserable sinners", so Freud thinks that the rituals must be performed to avoid unconscious consequences.

Comments   However, this use of analogy collapses when examined:

- ✧   The neurotic engages in ritual without understanding; the pious can give reasons why they worship, naming the sources of their guilt and seeking pardon and a fresh start.
- ✧   The neurotic celebrates secretly in private; the pious openly in public worship.
- ✧   The neurotic wants no disturbance; the pious welcome the participation of others, even when the young, for example, may hope for modification of the ritual by agreement.
- ✧   The neurotic's ritual is cognitively meaningless to himself, while religious ritual is full of meaning to the participants.
- ✧   The neurotic cannot miss a ritual performance without great anxiety; the pious can miss a ritual for adequate reasons without having a conscience or any anxiety.
- ✧   For the neurotic the ritual is consciously isolated from life; the pious find worship intrinsically related to all life.
- ✧   For the neurotic the ritual prevents the emergence of an unknown repressed threat; the pious performs a ritual to ask for and receive forgiveness, thereby consciously removing a threat.
- ✧   Finally, as Pfister, Freud's Christian confidant and correspondent, said, Freud emphasizes the limited and pathological aspects of religious experience, "but ignores precisely the noblest utterances of religion" (cited by Meissner 1984: 90–1).

By this comment Pfister means that Freud takes as his source of evidence for religion the views of his immature neurotic patients while ignoring completely the mature writings of theologians like Augustine, Aquinas or Luther.

The eight points of comparison compare a neurotic with an enlightened and informed believer. They refute the analogy that "religion is like a collective neurosis": it is false. However, it must be conceded, as Pfister did, that it is possible for uninformed superstitious believers at the other end of the range of believers to use a ritual strongly focused on guilt and atonement to encourage "compulsive formations of a collective neurotic character" (cited by Wulff 1991: 316).

However, Freud's argument, as a generalization, cannot stand, not least because it is impossible to offer a theory of the nature of humankind or of religion based on the neurotic mind alone, which was the primary source of all Freud's data. His concern here to identify the psychological origins of the religion of all individuals in this way failed.

Freud, Society and Religion

In *Totem and Taboo* (1913) Freud uses two methods to search for the source of religion. He uses the method of searching for the origin of religion in the psychological nature of the individual used previously, but this time it was used in support of a second method through which he sought to find the origins of religion in the social behaviour of primitive man. Using some anthropological sources (e.g. Atkinson, McClellan, Reinach, Frazer and Robertson Smith), Freud accepted that there are two crimes which violate "the sacred law of blood" within a tribe, murder and incest.

Freud adopted the theory that a taboo prevented sexual relationships between members of the same tribe and murder within the tribe. This was enforced because, in the distant tribal memory, the male members of a tribe had revolted against the male tribal chief who had sole rights to sexual relationship with the tribe's females. The young men rose up and murdered the chief. They were overcome by guilt, one part of which acted as a taboo on murder within the tribe, and another denied them sexual relationships with the tribe's women.

These young men, having a collective mind on the matter, adopted a totem animal as a "stand in" for the slain tribal chief, which became the ancestral animal of the tribe. On occasions the totem animal was ritually killed for a tribal feast which both atoned for the guilt of the sons, joyfully celebrated the honour of the slain tribal chief, and ritually shared his power amongst them all.

Freud even traced the theory back to the organization of gorillas, using the evolutionary work of Darwin. However, to quote Freud, "as time went on the animal lost its sacred character, and the sacrifice lost its connection to the totem feast; it became a simple offering to the deity, an act of renunciation in favour of the god" (Freud 1913: 212).

Finally, Freud links the memory trace theory in the tribal memory of the origins of the totemic meal with the Oedipus complex, which he regarded as the psychological origin of religion.[6] Thus Freud's theory of the origin of religion was that it began with the ritual against murder and incest. He supported this theory with the anthropological and evolutionary theory of totem and taboo and the psychological theory of the Oedipus complex, one particular neurosis.

Freud, then, views the psychological constraints of the individual, in particular the *id*, as containing, and therefore influenced by, a collective hereditary memory of the beliefs and rituals of the past (cf. Shafranske 1995: 203).

Comments   There are fatal criticisms of this argument that totemism was the most primitive form of society and that totemism was the origin of religion. The oldest known forms of society know neither totemism or totemic practice. Totemism is not universal, and is unknown in three important racial clusters – the Indo-Europeans, the Hamito-Semites and the Ural-Altaics. Of the many hundreds of totemic races known, only four have anything like a totemic ritual meal. Cannibalism and promiscuity were unknown in pre-totemic tribes, so a parricidal meal would be impossible in the earliest primitive tribes.[7] So Freud's anthropological theory of the origin of religion collapses.

The Oedipus complex, which also underpinned Freud's thesis in his book, is based on Freud's analysis of a decade of his own "very considerable neurosis", which consisted of severe attacks of anxiety and extreme changes of mood. Analysing these experiences, he invented the Oedipus complex and discovered it in himself (Wulff 1991: 259–69). It fitted him, not least because he was as close in age to his mother as was his father at that time, the time leading up to his father's death.

This is not a sufficient basis to project his complex onto humankind, for the love-hate directed to a father is much more likely to be related to the universal wish of every child to move from dependence to independence.[8]

However, Freud is important in introducing the idea of the projection of a father-figure leading to the formation of the God-concept, which is both useful and illuminating. It has provided a basis for the exploration of the psychological need for "transitional space", which will be explored later.[9]

Jung rejected Freud's notion that sexuality was the central source of human energy, replacing it by the notion of "psychic energy" (Jung 1967: paras 271–8). One must therefore conclude that Freud's theory that religion was based on the Oedipus complex is refuted. Clearly, sin, guilt, the need for the restoration of relationships both between humans and between God and humans, as well as the nurture of good relationships, are central to religion and involve a much greater range of emotions and reasons for self-centredness than sexuality can encompass, even with Freud's broad definition of it.

Centrally, religion is concerned with the meaning of life to which the mythology of an Oedipus complex is not central. The Oedipus complex has no explanatory power when it comes to mature religion, though it may have some slight use in understanding the somewhat magical religion of neurotic persons.[10]

Magical religion holds the belief that the individual can get God to manipulate his life situation, and the magical or superstitious believer thinks that he can get God to change his situation and promote or restore his health, wealth and relationships just through his prayers.

God, as a Projection of a Father-figure

In *The Future of an Illusion*, written in 1927, Freud outlines the origins of religion in the individual. A child is afraid of his father's power, but needs his father's protection. When the growing individual realizes that he will always, psychologically, remain a child, he realizes "that he can never do without protection against strange superior powers, [so] he lends those powers the features belonging to the figure of his father; he creates for himself the gods whom he dreads, whom he seeks to propitiate, and whom he nevertheless entrusts with his own protection" (Freud 1927 in Dickson 1991: 204). In other words, the individual projects the qualities of his ideal parent, portrayed as a father, onto God.

Comments The mechanism of projection[11] has become well known since Freud popularized its use. It has its use as a method used to formulate and hold ideas, but one must not confuse the mechanism of concept formation with the truth or otherwise of the idea. To say that God is nothing but a projection of a father-figure is a reductionist statement, which Freud intends.

More importantly it is also a genetic fallacy, i.e. it confuses the mechanism of concept formation with questions of the truth or falsity of a concept which are of a different *kind* or *genus*. For example, the Football Association may ask its members to say which of two types of football is better for match play after a year's trial in which the kind of ball used in matches is alternated, with the players not knowing the material of construction. One type, marked A, is made of leather in Halifax, and the other, marked B, is made in Leeds of a man made material. Ball B is chosen. A journalist will correctly say that ball B was chosen because there was unanimous agreement that it performed better. It would be a genetic fallacy to say that the ball performed better because it came from Leeds, because that is to confuse quality with origins. However, we are greatly indebted to Freud for introducing the concept of projection, as we indicated earlier.

Religion as "An Illusion"

In his book Freud then considers the nature of evidence for religion. He rejects "teachings" because they are inherited with "proofs" that cannot be challenged. He suggests that doctrines will not stand the tests of reason, so they are therefore "illusions, fulfilments of the oldest, strongest, and most urgent wishes of mankind" (Freud 1927 in Dickson 1991: 212; cf. McGrath, 2004: 67).

Freud then defined "illusion" as a belief based on a wish which may or may not come true. Intuition is not a sufficient basis for a true illusion, Freud argues. Despite this definition, he continues his argument by assuming that all religious beliefs are delusions.[12] He dismisses these beliefs because he believes that they are formed as a result of infantile and neurotic origins, and so are unacceptable to rational adults.

Comments First, Freud's view was not based on an exhaustive survey of the relationship of religious belief and mental health, merely the data from his neurotic patients, which led to his view that "religion brings with it obsessional restrictions, exactly as an obsessional neurosis does" (Freud 1927 in Dickson 1991: 227). His association of "infantile and adult neuroses" and "religious belief" indicates that he regards the former as the content of the latter (227).

Secondly, if one analyses the extensive evidence, one can explore the relationship between those involved with religion and mental disorder, for the result is different depending on how an individual relates to his religion.

Batson and colleagues have explored a three-dimensional analysis of individual religion: as means, end and quest. The "means" dimension is the outlook on life which is based on the extrinsic religious orientation. The "end" dimension is the outlook on life which is based on the intrinsic religious orientation. The "quest" dimension is the outlook on life concerned with the open-ended exploration of religious and other explorations of the meaning of life.

If individuals adopt an *extrinsic religious orientation*,[13] it means they are persons who have turned to God without turning away from themselves. Self-centred still, they use religion for their own end, to provide personal security, status and self-justification. Beliefs may be handled literally and are not internalized, but evoke cognitive assent.

If one person adopts an *intrinsic religious orientation*, one no longer remains self-centred, but embraces one's faith and internalizes it, living the life of faith. One's faith, with characteristics outlined in Chapter 3, is: (a) well differentiated, i.e. one searches for harmonious relationships with others and with God; (b) it produces a consistent morality; (c) it is dynamic, transforming the individual without them becoming fanatical; (d) it is comprehensive and encourages tolerance; (e) it seeks to integrate all aspects of life, recognizing areas where life is determined and where one has freedom, noting where evil lurks; (f) it is heuristic, i.e. it is held as a working hypothesis, open to revision or expansion as one's faith matures.[14]

If one adopts the quest approach, "an individual who approaches religion in this way recognizes that he or she does not know and probably never will know the final truth about such matters . . . There may or may not be a clear belief in a transcendent reality, but there is a transcendent, religious aspect to the individual's life" (Batson *et al.* 1993: 166).

If one looks at eighty findings from forty-five different studies of the *extrinsic religious orientation,* forty-eight findings show a negative correlation with seven concepts of mental health[15] (thirty-one were neutral and one positive) (Batson *et al.* 1993: 261–86). As an extrinsic orientation has many similarities with authoritarian and even neurotic religion, it is not surprising that it seems to lend some support to Freud's thesis if reformulated thus: "an extrinsic religious orientation has some features in common with a neurotic religious orientation".

However, when one turns to ninety-three findings from forty-six different studies of the *intrinsic religious orientation,* forty-nine findings show a positive correlation with seven concepts of mental health (thirty neutral and fourteen negative), which lends support to the comment on Freud's 1907 paper on *Religious Actions and Neurotic Practices,* namely, that a mature or intrinsic religious orientation has almost nothing in common with neurosis. Rather, an intrinsic religious orientation is a positive support to mental health (Batson 1993: 287, table 8.7, 286).

Finally, the quest dimension is supported by twenty-four findings in twelve studies cited (Batson *et al.* 1993: 288). The evidence supports a positive relationship between the quest dimension and open-mindedness and flexibility. This again totally refutes Freud's idea of religion being a "collective neurosis".

Freud does modify his judgement on religion expressed in *The Future of an Illusion,* in which he expressed a negative view of it. Later, he still maintained that religion was not ultimately true, i.e. he denied that the Being of God existed, but nevertheless religion was socially and functionally true.[16] This accords with the apparently contradictory view, where he argues "on behalf of retaining the religious doctrinal system as the basis of education and of man's communal life" (Dickson 1991: 236). He gives his reasons: society needs illusions to hold it together, and these need to be taught until they are replaced by the illusions of a scientific world-view, which we call scientism – the belief that scientific knowledge is the highway to understanding the meaning of life and its purpose.

Freud's life mission was to replace religion with scientism as the new lifeway and lifework for all (Lee 1985: 7–8). A key reason for this is Freud's assessment of the function of religion in society. Because of the hardness and disappointments of life, humans may choose some alternative goals – powerful deflections like making light of our aims, substitute satisfactions like art or science, or intoxicating substances to drown our sorrows. Freud thinks that each person should face these choices alone, with the help of science.

Freud holds that religion restricts choice, imposing one path towards happiness and protection from suffering. By requiring belief in itself, it reduces adults to the status of infantile dependence, a mass delusion which is a collective neurosis. Freud thinks that religion at least saves each person from an individual neurosis.[17]

Science is "The Proper Religion"

Freud was concerned with "education to reality". Well aware that maturity was concerned with controlling human instinctual desires, he argued that: "my illusions are not, like religious ones, incapable of correction". He recognized "that man's intellectual life is powerless in comparison with his instinctual life . . . nevertheless . . . the voice of the intellect is a soft one . . . after a countless succession of rebuffs it succeeds". Freud's God was Logos, or Reason, and science was the way to personal and social salvation (Dickson 1991: 238–9).

Comments   Freud recognized the positive social role of religion in the past, which he respected because of its important function of holding society together. Scientism should now take over this role he believed. Freud's inadequate notion of religion was based on the religious views of his neurotic patients. He took no account of the great spiritual leaders and their insights, and the work of great theologians.

Freud took no notice of his great Christian friend and correspondent, Pfister, who more than answered his criticisms during a thirty-year friendship. Pfister's answers fell on deaf ears. Pfister turned down a professorial post in order to adopt Freudian psychoanalysis as a pastor at a Zurich church. He was enthusiastic and excited because he had found previously that he was unable to cure defects of the religious and ethical lives of his members in need. Pfister wrote: "In hundreds of instances which I had attacked in vain by methods grappling directly with the conscious – in vain because the religious ethical conflict was situated and did its work in the unconscious – I now achieved the cures I had so long desired."[18]

After the publication of *The Future of an Illusion*, Pfister wrote to Freud totally rejecting his attempt to replace religion by natural science, for it would undercut the noblest flowering of human nature and achievement as well as the realistic basis for morality. Pfister continues:

> Religion concerns itself with the meaning and value of life, with the singular drive of reason towards a universal view of the world encompassing both being and obligation, with the yearning for home and peace, with the impulse towards mystical union with the absolute, with the chains of guilt, with the thirst for freedom and grace, with the need for a love that is removed from the unbearable uncertainties of earthly life, and with innumerable other anxieties which, if not relieved, can choke and disquiet the soul but through religious harmony can lift the life of man towards radiant mountain heights with indescribably joyous vistas, can strengthen the heart, and by the imposition of heavy moral obligations in the spirit of love can elevate the value of existence. . . . It is no surprise that some of the greatest scientists thought of their work as service to God and that some of the greatest artists and poets laid down their laurel wreaths before the altar of God. (Quoted by Meissner 1984: 99)

Pfister's comment emphasizes the place of reason in working out a coherent universal view of the meaning of life, the pre-eminence of love and the lure towards mystical union with the absolute. The fount of love is the foundation for a fully creative life, making it possible to overcome the problems of guilt and take on heavy moral obligations.

This, however, makes clear that Pfister was an advocate of what is now called *the intrinsic religious orientation*. Freud was concerned with *an extrinsic religious orientation*.[19] There was thirty years' friendly exchange of correspondence and much mutual personal and professional respect between the founder of psychoanalysis and this Christian psychoanalyst, but Freud had no bridge across the divide between the soliloquies they each delivered.

Freud and Religious Experience

Discussion of Freud so far has been on his concept of religion, which is focused either on doctrine which has been passed down and which he wrongly assumes cannot be questioned because it is revealed, or ritual based on it. Freud's view of religion, in the outline and discussion of his views so far, fits the seventh concept of "religious experience" outlined in Chapter 10, namely, that "it describes all experience interpreted from a religious perspective".

However, Freud does discuss "religious experience" in the first sense discussed in Chapter 10, i.e. that it refers to a person's personal religious experience, when alone, between oneself and "Another". This occurs in the beginning of his *Civilisation and Its Discontents*, where Freud is attempting to make sense of a report of a friend criticizing *The Future of an Illusion*. The friend said that Freud had no idea that "the true source of religious sentiments . . . is a peculiar feeling . . . which he [the friend] would like to call a sensation of 'eternity', a feeling of something limitless, unbounded – as it were 'oceanic' . . . [which] is a source of religious energy . . . even if one rejects every belief and every illusion" (Batson *et al.* 1993: 251–2).

Freud admits he cannot discover this oceanic feeling in himself. He traces the feeling back to an early phase of ego-feeling and the unconscious, interpreting it as being nothing but a feeling of infantile helplessness. This is later linked to religion and labelled "oneness with the universe", which is an attempt at religious consolation so that the ego need not feel threatened by the dangers of the external world.

Comments   Once again, Freud reduces religious feeling to "nothing but" a feeling of "infantile helplessness" which provides a basis for religious consolation. He does not understand how "religious . . . feeling . . . [can be] a source of religious energy" for someone holding *an implicit religious orientation.* Because he regards religion as a neurosis, as *an extrinsic religious orientation,* he views it as an infantile emotional consolation crutch.

Freud, Moses and Religion

In 1939, Freud published *Moses and Monotheism,* in which he largely repeats the theories of his earlier books, but this time applied to Moses. Freud adopted a theory of Sellin, that Moses was murdered, a theory totally rejected by all Jewish and Christian scholars (Freud 1939 in Dickson 1990: 275; cf. Spinks 1963: 86; Watts and William 1988: 24). When first written in 1934, it was titled "The Man Moses: A Historical Novel". That would have been a better title, for it bears no relationship to biblical scholarship concerning the historicity of the death of Moses, or the relationship of Jesus to it.

The new point made by Freud in this book is: "If all that is left of the past are the incomplete and blurred memories which we call tradition, this offers the artist a particular attraction, for in that case he is free to fill in the memory according to the desires of his imagination . . . and according to his intentions" (Dickson 1991: 314).

Freud then introduces the psychological concept of "latency", a psychoanalytic term for the interval of about five years between infant sexuality and the development of normal sexuality at the time of puberty, which he then analogically transfers to the interval between the murder of Moses, when the memory was repressed, and the "tribal need" for a human sacrifice to atone for the crime once it emerged into the conscious mind. Then the remorse which it generated led to the need for a redeemer. Paul of Tarsus suggested that Jesus is both the sacrifice and the redeemer, says Freud (Dickson 1991: 330–1).

Freud's logic is that the death of the father (Moses) can only be atoned for by the death of a son (Jesus) even though Jesus was personally innocent of the crime. In other words, Jesus took the blame for and "paid for" the death of Moses.

Comments   This story of the tribal murder of Moses and the dream of a Messiah to resolve it was just a piece of creative thinking by Freud, and was without any historical foundation. The work was most likely the result of Freud's final attempt to resolve his conflict with his father. "A hero is someone who has the courage to rebel against his father and has in the end victoriously overcome him" (Dickson 1991: 248).

Secondly, Freud certainly identified himself with Moses. Freud too was betrayed by followers like Jung, and would never get to his "promised land", which would be the triumph of scientism over religion. The book can be viewed in part as a piece of self-analysis concerning his unresolved relationship with his father, and his own subsequent self-concept.

This work links all Freud's analogies together. His first analogy linked the repressed guilt of the neurotic with its rigorously attached but meaningless ritual inappropriately to the meaningful ritual of the religious believer.

His second analogy linked the Oedipus complex – the guilt felt by the son who wished to possess his mother leading to his jealousy of his father but also his fear of his father's power of revenge – to the guilt of all persons and their need to placate an angry God. There is no easy fit here as the strained link is guilt not sex. This Freud regarded as the psychological source of religion.

His third analogy linked the tribal horde murder atoned for in the patricidal meal with, for example, the Holy Communion service of Christians. This Freud regarded as the sociological source of religion, though he held that both the psychological and sociological sources were interrelated.

The fourth analogy linked the first three analogies to the murder of Moses, repressed until Jesus comes to atone for his death. In each case Freud hardens the analogy into a cause – a totally fallacious procedure. *Moses and Monotheism* has nothing to do with the history of Moses or Christ and their religious significance in the Jewish and Christian communities (cf. Meissner 1984: 49–56). It is just an interesting novel, as its original subtitle *The Man Moses – A Historical Novel* indicated, except that it was not even historical.

Freud: Summary

Freud viewed religious experience in two ways. Usually he treated religious experience as being all experience interpreted from a religious perspective and summarized by religious doctrines or beliefs. Religious beliefs were illusions based on wish-fulfilment, on belief in God, which was nothing but the projection of "father" into "God". He held that religious acts are similar to the habitual acts of an obsessional neurotic.

The origin of religion, said Freud, can be traced to two sources that are often inextricably linked in societies. The cultural source of religion was in totemism, the organization of the most primitive religion, in which there was a taboo on murder and incest. In Judaism, the memory of killing the tribal father, Moses, remained latent until invoked by Paul, Freud claiming that Jesus was both the sacrifice for the blood guilt and also the redeemer of the believing community.

The psychological source of religion was the Oedipus complex in which the boy coveted his mother, or the Electra complex in which the girl coveted her father. The boy wishes the father dead, and the girl wishes her mother dead, so that each can achieve their sexual possession. This led to the guilt an individual felt after wishing a parent dead as a result of sexually coveting the other parent for himself or herself.

Freud also explored religious experience, meaning the "personal religious experience" of an individual alone. He had never had it himself, and regarded such religious feeling as a feeling of "oneness with the universe", which was infantile.

Comments   Freud's concept of religion does bear a resemblance to an *extrinsic religious orientation*, i.e. when religion is used in a self-centred way to bolster an individual's personal security, status and self-justification and provide comfort and company. However, the concept of religion held by Freud was immature, based on wish-fulfilment, an authoritarian father-figure who imposed prohibitions, guilt, and rituals not understood but practised obsessively. These are characteristics of an extrinsic religious orientation.

The anthropological basis of totemism has collapsed, and the Oedipus concept emerged from a self-analysis, which is not necessarily universal, but the practice of a neurotic minority. "Projection" is a mechanism of concept formation, not a method of testing "truth". Freud's attempt to reduce religious ritual to a universal neurosis; God, a projection, to "just a father-image"; religious feeling to an infantile security wish; was nothing more than the psychological result of his predetermined philosophical aim to replace religion with scientism.

A central cause for his failure was that his attempt to reduce religion to its "material" constituent parts failed. This was because he had not realized that when he used personal discourse in the interpretation of dreams, in which he properly used concepts like wishes, feelings, motives, meanings, which involved interpretation, he, like most of his contemporaries, did not then have the philosophical understanding to realize that he had moved out of scientific discourse into personal discourse. Nor did he realize that religious understanding involves interpretation, relationships and meanings, and as such was closer to personal knowing than scientific knowing. Here, notice that he is not using his mechanistic scientific methodology and scientifc discourse, but the humanities methodology in which personal discourse is central.

The major criticism of Freud's view of religion is that, because it was based on data from neurotic patients, it is ignorant of the *implicit religious orientation*, which is concerned with developing good relationships with others and God, and the selfless service of others based on a rational faith of beliefs which can be and are subject to critical scrutiny. He might have discovered this if he had not had the predetermined philosophical aim in his work of wishing to replace religion with scientism, and had had a data bank of experiences of the mentally healthy from which he may have discovered the implicit religious orientation.

Freud made important contributions to the study of religion. His major significance, concerning our topic of religious experience, lies in the fact that he developed and introduced psychoanalytic theory to the study of religion. This is important because it provides a perspective from which one may seek to understand the place of religion in the psychological world of the individual and particularly the place of the unconscious, which builds on James's use of the subconscious.

He has issued appropriate warning about one form of religion, *the extrinsic religious orientation*, and how it may become a collective neurosis for this kind of group.

Pfister, Freud's Christian disciple, felt that Freudian psychoanalysis did help explain, and could help to undo, why neurotic fear, with an emphasis on guilt and atonement, was a major element in Christian ecclesiastical practice, "whereas Christ met fear with an attitude of understanding and love" (Wulff 1991: 316).

Freud's discovery of one method of forming the concept of God through the projection of the father image is a fruitful contribution in itself, and was a stimulus to later thinking.

Freud recognized that wish-fulfilment was a key element in the formation of religious "illusions", a wish which may or may not come true, even though he used the term to mean a "delusion", a false unshakable belief based on neither evidence or reason. However, his use of the term "illusion" has been important as a stimulus to much later thinking.

It is understandable, but unfortunate, that his mission to replace religion by scientism blinded his reason so that he could not even hear his friend Pfister outlining the nature of an *implicit religious orientation*. Had he done so he might have explored the concept of a mature religious outlook, even if he would never have shared it. Finally, one has to grasp the violent irrational emotional hostility Freud had to religion. Let Jung relate a conversation he had with Freud which clearly illustrates this:

> I can still recall vividly how Freud said to me, "My dear Jung, promise me never to abandon the sexual theory. That is the most essential thing of all. You see we must make a dogma of it, an unshakeable bulwark." He said that to me with great emotion, in the tone of a father saying, "And promise me this one thing, my dear son; that you will go to church every Sunday." In some astonishment I asked him, "A bulwark – against what?" To which he replied, "Against the black tide of mud" – and here he hesitated for a moment, then added – "of occultism." First of all it was the words "bulwark" and "dogma" that alarmed me; for a dogma, that is to say an indisputable confession of faith, is set up only when the aim is to suppress doubts once and for all. But that no longer has anything to do with scientific judgement; only with a personal power drive. This was the thing which struck at the heart of our friendship. I knew that I would never be able to accept such an attitude. What Freud seemed to mean by "occultism" was virtually everything that philosophy and religion . . . had learned about the psyche. (Jung 1973b: 173)

Standing back, how should one evaluate Freudian theory? Two groups think critical evaluation is unnecessary. Some think that Freud is hopelessly dogmatic and unscientific, an outdated product of nineteenth-century scientific determinism. The other group of practising psychoanalysts think that Freud's ideas are basically true and only need fine tuning.

We have seen that Freud's theories lie on the boundary between scientific methodology and the methodology of the humanities, between scientific discourse and personal discourse. Perhaps Freud's theories should be evaluated

> not by subjecting them to piecemeal testing by experiment . . . but in terms of criteria we would use to judge the adequacy of narrative: coherence, comprehensiveness, continuity, and the capacity to elicit an aesthetic response. Freud's theories are thus seen as guiding metaphors, subject to artistic truth, not as hypotheses to be tested according to historical truth. (Wulff 1991: 301)

Have any insights emerged from the testing of Freudian theory in relation to religious behaviour? The answer is many, of which one example may provoke reflection. Freudian theory assumes that the pattern of early relationships affects all later relationships. The way a person "relates to God" has been tested by three hypotheses using data from eleven societies. The results suggest:

I   If a child always gets an immediate nurturing response from mother or the attendant adult as soon as it indicates its need, by crying etc., that former child, now an adult, thinks that by performing compulsive rituals, God can be "compelled" to respond to a particular need.

2   If an older, reality-oriented child is taught to expect help only when one asks a parent, then supernatural help will be assumed to depend on propitiatory rituals, through which God is coaxed into action.

3   If parents help, even without being asked, later supernatural help will be assumed to be available without ritual action or obedience to certain demands (Wulff 1991: 305–6).

JUNG, ANALYTIC PSYCHOLOGY AND RELIGION

Before discussing *Psychology of Religion*, the book Jung wrote in 1938, it is necessary briefly to review the development in Jung's thought through three periods.[20]

In the first period (1902–12), his key work was *The Psychology of the Unconscious*, written in 1911–12. Jung distinguished two different kinds of thinking which led to two different kinds of truth. The first he called "directed thinking", which led to literal truth achieved by consciously directed thinking. Normal scientific investigation produces such empirical truth.

The second kind of thinking he called "fantasy thinking", which was unconscious and led to psychological truth.[21] He asserted that "metaphysical thinking" was superfluous for psychological enquiry. Jung was interested in this second kind of thinking. He asked two questions. The first was 'Why are there religious projections?' His answer was that they were psychic responses to unfulfilled wishes. In other words, "we imagine what we lack" (see Heisig 1979: 23). The second question was 'Where do religious projections come from?' His answer was that they came from a pool of archaic inclinations expressed by humans everywhere, which emerge as "images".

The patients he met in his consulting rooms constantly presented to him symbols which a study of religion showed were universal symbols for divine creativity, which he accepted as the libido regarded as the source of personal creative energy. To Jung, at this stage of his thinking, God was a fantasy picture of the child's libido, an archetypal image.

These religious symbols wean the child's libido from incestuous focusing on a parent, transferring it into power to create and forgive. Jung believed that the age of symbolic illusion should give way to the stage of moral autonomy: "belief should be replaced by understanding: then we would keep the creative beauty of the symbol, but still remain free from the depressing results of submission to belief . . . " (Jung 1977: 42–4). Psychoanalysis should free a person from their religious projections, helping them to recognize unconscious desires for what they were (Heisig 1979: 25).

In his second period (1912–37), we see a restatement of ideas expressed in his book *Psychological Types*, published in 1920, dressed up in a new scientific terminology. The god-image is a symbol of the accumulated energy of the unconscious which mediates between the ego and the unconscious as a means of embracing the Self. The concept of "Self" for Jung was "the ideal Self" which I wish and need to become. Self and God seem implicitly synonymous. However, Jung begins to see the god-image as having three elements. It is not only a projection of unfulfilled wishes, but is also a symbol of the Self, and is a symbol which transcends the Self.[22]

When Jung wrote *Psychology and Religion* in 1938, it marked the beginning of the final period of his thought in which he expresses his conviction that religion is an important ingredient in mental health.[23] Our concern here is with his discussion of religious experience.

As already noted, Jung accepted Otto's basic assumption that the "holy consisted of the 'moral' and the 'numinous'" (which we explored when discussing Otto earlier). Jung is mainly concerned with the numinous. Jung defines religion: " 'Religion' it might be said is the term that designates the attitude peculiar to a consciousness which has been altered by the numinosum" (1973a: 6).

This is one part of the experience of religion. The other part is faith, "the loyalty, trust, and confidence towards a definitely experienced numinous effect" (Jung 1973a: 6). The example Jung gives is of the conversion of Paul on the Damascus road.

Jung describes how an individual can experience the numinous:

> "[T]he numinosum" . . . seizes and controls the human subject, which is always rather its victim than its creator. The numinosum is an involuntary condition of the subject, whatever its cause may be . . . this condition [is] due to a cause external to the individual. The numinosum is either a quality of a visible object or the influence of an invisible presence causing a peculiar alteration of consciousness. (Jung 1973a: 4)

Jung notes that ritual performances are carried out for the sole purpose "of producing at will the effect of the numinosum by certain devices . . . such as invocation, incantation, sacrifice" (Jung 1973a: 5).

Jung's final verdict was that: "religious experience is absolute. It is indisputable." (Jung 1973a: 113). But this only means that Jung uses the word "true" to indicate psychological truth. For example, if I insist that something is true, for me it is a psychological truth in my mind. But in fact you may have evidence to show that it not true, *or* we may be talking about a *personal* truth. For example, Frank says that Mary is the best girl in the world, and George says that Alberta is the best girl in the world. Each is talking about his wife. Both those judgements are correctly regarded as personal truths in the mode of personal discourse. Religious experience, however, operates in the mode of religious knowledge, which is transpersonal. Jung is saying that for the person who claims to have had a religious experience, it is true for them. But a religious community needs criteria by which to recognize these claims as truth claims which have transpersonal validity. A religious community needs to be able to test the validity of these claims in the life of a person, e.g. by their fruits you will know them.

The psyche is thought to operate at four levels of consciousness:

1. The realm of heightened sense experience.
2. The level of personal being, in which one explores one's past; particularly one's feelings formed in the past, perhaps through reflecting on memories or by exploring the unconscious through psychoanalysis.
3. The level of social being, explored through the archetypal images, myths, symbols and rituals through which communal identity is expressed, and which emerge through what Jung calls the *collective unconscious*. Archetypes, as we have noted earlier, are well-displayed in fairy tales and dreams e.g. the villain, the victim and the hero.

4   The level of the mysterium is the highest level of consciousness, and dreams experienced are those of people committed to a spiritual path, suffused with a sense of the numinous.

Some dreams at level 3 can lead into the fourth level of consciousness, if they have a numinous feeling, with the search for an icon which illustrates the sacred in the ordinary, or the discovery of the god-image or Self, which is the symbol of personal integration, or individuation as Jung calls it.[24]

For Jung the ego is the centre of conscious experience. The "Self" is the centre of the whole personality, analogically corresponding to God. The purpose of the Self is to bring a person to integration and wholeness (Watts and Williams 1988: 96). In this sense the Christ-image is a symbol of the Self, the goal of spiritual and personal maturity.

Jung himself has said: "The seat of faith . . . is not consciousness but spontaneous religious experience, which brings the individual's faith into immediate relation with God." The only way to find it is to take "the first step towards *the unconscious, the only accessible source of religious experience*".[25]

In the final pages of his book *Psychology and Religion*, Jung says:

> No matter what the world thinks of religious experience, the one who has it possesses the great treasure of a thing that has provided him with a source of life . . . And if such experience helps to make your life healthier, more beautiful, more complete and more satisfactory to yourself and those you love, you may safely say: "This was the grace of God." (Jung 1973a: 113–14)

Comments   Jung says: "I am an empiricist and adhere to the phenomological point . . . I approach pyschological matters from a scientific and not from a philosophical standpoint . . . I refrain from any application of metaphysical or philosophical considerations" (1973a: 1–2).

He was an empiricist. He did collect views of individuals and regarded them as phenomena, and as "psychological facts". So truth for Jung was related to the simple claim that a person had made a statement – "The moon is made of green cheese." He was not concerned whether the statement was tested for any kind of truth, empirical, moral, aesthetic, personal or religious. He asserted that he did not approach matters from a philosophical or metaphysical position. That just indicates that Jung failed to understand that all positions or statements assume certain metaphysical and philosophical assumptions.

So, when he says something is true, he only means that it is pyschologically true. In other words someone reported that experience. Such a position means his work has limited value for anyone concerned with truth claims in the different domains of truth – within empirical, moral, aesthetic, personal, religious and other ways of knowing. He therefore unwittingly belittles religious truth claims made by theology asserting that religion is simply a matter of personal faith. He denies, and therefore does not understand, that metaphysics and philosophy are concerned with the assumptions and limits of every mode of knowing, or form of knowledge.

We need to explore Jung's methods of identifying the god-image in a patient:

✧ A patient may claim that he has experienced a god-image.

✧ The patient describes an experience which the analyst questions to gain more information, which may include clarification by the patient that it was an experience of a god-image.

✧ Jung classified an individual's "highest value" as a god-image.

✧ Jung classified unconscious phenomena in the patient by reference to motifs found in literature like myth, fantasy and folk-tales. The god-image was universally present among all humans past and present.

✧ Jung described experiences when described by a patient using terms which were polar opposites as evidence of the god-image.

✧ Some unconscious dream images alluded to by a patient, Jung would describe as a god-image.[26]

Even though Jung thought of the god-image as the unconscious projection of an archetype, it is difficult to defend more than the first two of the six cases mentioned above.

However, Jung goes beyond simple observation or labelling in his study of archaic images, and in exploring symbols within both fairy stories and religious stories. His development of psychological theories involving the construction of concepts like the "collective unconscious", "archetypes", "Self", "individuation", to name a few, indicates that he was constructively engaged in developing a mode of knowing – analytic psychology.

In this work Jung developed hypothetical structures of the mind and of personality development in psychological terms. He included a religious dimension in this work on the basis of what he would admit was subjective judgement based on intuition. As soon as the analytic terminology was exceeded by a concern for a person's total development, in the development of the "Self", then that presupposes a metaphysical, philosophical and theological stance of the individual. This needs to come from the individual and not be imposed by a Jungian therapist who intuitively inherits Jung's assumptions.

By 1938, Jung accepted that Otto's analysis of religion was correct, and that there was a core religious and sometimes mystical experience common to all mature formulations of the "Self". In making that assumption he exceeds the role of a psychotherapist and imports under that professional hat religious assumptions. The religious ideas need detaching and evaluating according to truth criteria within the domain of religious knowing.

Mature formulations of the "Self" may be "materialist", "humanist" or, like Jung, "religious". However, the incorporation of those philosophical and theological beliefs are outside the realm of analytic psychology. Any individual, however, in formulating his concept of "Self" will hold one set of ultimate assumptions, possibly one of the three mentioned, or another.[27]

Once Jung's unconscious conjunction of psychoanalysis and religion is separated, it allows for the conscious conjunction of psychoanalysis and an outlook on life after the patient has indicated their own outlook on life. Jung's personal testimony in favour of the numinous is just that, but it does not establish whether it is true: that is the task of theology.

However, Jung's positive regard for religion is important because he not only regards the unconscious as the vehicle of religious experience, but also of the positive and negative desires of the individual, and of archetypal images like those of the good or bad mother, or good or bad father.

Religion for Jung was the intermediary between the unconscious and conscious minds. He was able to use archetypal symbols, including that of the Self, to promote the integration of the good and shadow or dark sides of human nature, which he called individuation. Jung's analytic psychology's achievement was to promote health and wholeness of the individual with religion as a major resource in that process. That major psychological contribution to our understanding of the structure of the mind and human personality leaves the task open for theologians and religious thinkers to continue the task of uncovering the form and content of true religion.

OBJECT RELATIONS THEORY AND RELIGIOUS EXPERIENCE

The term "object relations" is universally used by psychologists working in this field because it has a quasi-scientific flavour, and they naturally use scientific discourse. It would be better called "relationship theory" because it is concerned with the structure of human relationships. For some reason they call the focus – a person – an object. Most of us understand an object to be inanimate. I prefer to call a person a person, and so avoid reductionist overtones. However, here we will be using their language, and will talk of "object relations theory".

Object relations theory (O-R theory) has grown out of classical Freudian theory, as its early pioneers started off their work from within psychoanalysis. "The concept of 'objects relations' in psychoanalytic writings of the last two generations means relations with significant others and their internal representations, starting with mother" (Beit-Hallahmi 1995: 255).

Broadly, O-R theory is concerned to study the nature and origins of interpersonal relations. It rejects some Freudian ideas, like innate instincts except object seeking, and the *id-ego-superego* structure of the personality, in favour of *ego* and *superego* development. While classical Freudian theory holds that the development of personality is based on drives, like sex and aggression, with an information-processing system to handle relationships, O-R theory holds that the main determinants of behaviour are introjection and projection, i.e. ideas are internalized or projected onto other people or objects. "O-R theory is . . . pessimistic . . . [because] the results of early learning will be irreversible" (Beit-Hallahmi 1995: 256–7). The reason for this focuses on the change of emphasis from classical Freudian views. Freudians held that the key stage of development was between three and six, while O-R theory holds that the key stage is from birth to two years. Freud focused on instincts, O-R theory focuses on the way the infant relates to the mother and other significant people.

How can O-R theory inform our understanding of religion and religious experience? First of all, research based on O-R theory, while accepting "projection" as a mechanism used in "relationship", replaces Freud's idea of God as being "a projected father figure" by the finding that the idea of God is based on "a projection of the ideal parent", usually the mother (Wulff 1991: 305).

A key contribution to O-R theory is Winnicott's idea of a "transitional object". A good mother holds together all the separate elements of an infant's fragmented life. The breast grants the child a "moment of illusion", i.e. that the child thinks it can create a breast whenever required. A good mother provides a "non-demanding" presence that allows an infant to develop a state of "going-on-being" – a state of being alone.

The transition from hallucinatory omnipotence, of creating a breast at will, to the world of objective reality is aided by a "transitional object". This has been thought to be a favoured blanket, teddy, doll or other object cherished by the infant, and over which the child has sovereign rights (Wulff 1991: 337).

This idea has been related to religion by Pruyser, who finds the idea of a "transitional object" helpful when considering visual arts, literature, music, science and religion, all of which he feels use the notion of "illusion". However, when Pruyser uses the term "illusion'" he means something different from Freud. Freud used "illusion" to mean "an idea which may come true", but then he inferred that religious ideas do not become true, so they are "delusions". Pruyser, before choosing to use the word "illusion", noted that the word came from the Latin word, *ludere* (to play). By "illusion" Pruyser means "the play of the imagination". "Illusions" then can become "ideals" to aspire towards, and perhaps achieve.

The "illusionist world" comes midway between the "autistic world", the infant's world, characterized by features like untutored fantasy and omnipotent thinking, and the "realistic world", characterized by reality testing, hard facts and logical connections.

The "illusionist world", which is the equivalent of the "transitional object", is characterized by playing, adventurous thinking, imagination, creativeness, inspired connections, images which can be expressed in words, symbols. These "transitional activities" replace the infant blanket or teddy. In this way of looking, God is seen as a "transcendent object" prefigured by the child's transitional object.

> It is in the illusionistic world, Pruyser concludes, that religion most appropriately finds itself, for "the transcendent, the holy, and mystery are not recognizable in the external world by plain realistic viewing and hearing, nor do they arise directly in the mind as pleasurable fictions. They arise from an intermediate zone of reality that is also an intermediate human activity – neither purely subjective nor purely objective." (Wulff 1991: 339–40, citing Pruyser 1977: 77–97)

Sheer physical survival means that the autistic world has to be replaced by the realistic world, but a thoroughgoing realism will inhibit the play of imagination and thus restrict the growth of human potentialities (Wulff 1991: 340, citing Pruyser 1983: 176).

O-R theory, as mentioned before, is a development from Freudian psychoanalysis. Concerning the nature of mystical experience, O-R theory supports Freud in regarding it as a regression to the infant state characterized by the "oceanic feeling" of union with mother. Some say this regression is a resource for "transformation in the direction of inner harmony and wholeness" (Wulff 1991: 363). Thomas Merton, an important celebrant of the mystical tradition, accepts that the Freudian view may characterize an early stage of mysticism for some. Nevertheless, he holds that unitive experience that is "merely regressive and narcissistic would be invalid religiously and mystically" (quoted by Wulff 1991: 363). Merton is clear that this psychoanalytic view does not describe the deeper and more genuine parts of mystical experience.

We must note some O-R observational studies on Winnicott's theory of the "transitional object" which question its significance:

1.   There is no consistent relation of the use of a transitional object either to early experience with mother, or to later relationships or creativity.

2.   The use of a transitional object is promoted by both mother and infant, and is not controlled by the infant.

3.   The blanket is probably not a transitional object but an extension of the child's self, and is usually given up by the age of three years.

4.   Less adequate mothers who allow a child to keep a transitional object longer risk symptoms of maladjustment in a child (Wulff 1991: 363, abbreviated).

Such reported studies question rather than undermine the work of Winnicott, and indirectly Pruyser, who we noted earlier argued that all creative activities occupied an "illusionist world" between the infant's "autistic world" of fantasy and the "realistic world". Religious ideas are formulated in this middle world in which the human imagination shapes meanings related to the realistic world. Like Pruyser, Watts and Williams, I still think this way of understanding the formation of religious images has some explanatory power (cf. Watts and Williams 1988: 37, 153).

However formed, these religious images and concepts are signposts for the critical realists which point towards the "Personal Other" whose existence is affirmed by each believer's religious intuition. For intuition, we have learned, is the bridge which links the reality of all perceptions to all forms of knowledge and the language we use to express those perceptions in words.

Finally we need to note two of the insights of O-R theory into the psychology of religion which bears on an understanding of religious experience. First, the God-image is related more to the mother than the father. Secondly, the God-image develops in relation to how one describes oneself and religious teachings which influence one. If one develops high self-esteem, this positively relates to images of a God who is kind and loving. If one develops low self-esteem this relates to images of a controlling, vindictive and impersonal God. Clearly the child's upbringing and educational and social experiences are major factors in whether a child develops an intrinsic, an extrinsic, or a quest religious orientation.

NOTES

1   For a brief, valuable overview of Freud and other psychological approaches to religious experience, see Watts and Williams (1988: chs 2 and 3).

2   Freud attempted to outline this scientific view of the mind in his *Project for a Scientific Psychology* (cf. Meissner 1984: 191).

3   "Projection": the unconscious transfer of one's own feelings or impressions onto an external object or another person, e.g. unconscious transfer of guilt or feeling of inferiority onto another person, or transfer picture of parent onto "God".

"Conscious": the sum total of everything of which we are aware.

"Pre-consciousness": the reservoir of all we can remember or recall from the memory.

"Unconscious": the deep well of the mind whose feelings, pre-concepts and processes, such as the primal instincts or drives of human nature which are prior to, and unknown to, those of the conscious mind.

"Introjection": the unconscious transfer of new or learnt ideas into one's own mind.

"Repression": the unconscious exclusion of painful thoughts or experiences from conscious awareness.

4   Meissner (1984: 193). This theory is explored later in the chapter.

5   "Neurosis": painful experiences unacceptable to the conscious mind are pushed down or "repressed" into the "unconscious". They can emerge into the conscious in the form of a neurosis expressed in neurotic behaviour which is not intelligible to the conscious mind of the patient (Freud 1907).

6   The "Oedipus complex" is a largely unconscious complex in which a son develops a sexual attachment to his mother but finds he is jealous of his father, thus producing a guilt complex. A daughter develops a similar complex, the Electra complex, in which she wishes to possess her father sexually, is jealous of her mother and thus feels guilt. Freud regarded such feelings of both love and hate towards the father or mother as natural in a family.

7   Father Wilhelm Schmidt in 1935 cited by Spinks (1963: 83); cf. Wulff (1991: 278).

8   While Freudians defend it, Shafranske notes that anthropological studies have not supported the universality of the Oedipus complex. It therefore follows that Freud cannot claim that all religious experiences bear the marks of an Oedipus complex. However, there may be rare occasions when a God-representation is a displacement from an Oedipal father complex (Shafranske 1995: 221, 223, 224). However, the tension between a child/parent does exist on independence/discipline and this is more likely to provide a universal basis for projection theory when a child does need the qualities of a father later, and so projects them onto a God-image. This is much better explained by Rizzuto, who argues that the formation of the image of God is formed within the transitional space which develops in the developing psychical life of the individual child facing the tension between needing to become independent of one's father, but needing a substitute which can soothe, comfort but also provide inspiration and courage (cf. Shafranske 1995: 224).

The problem with Freudian theory, especially that concerning the Oedipus complex, is that it cannot be disproved, nor can any other mythology. So the only test for Freudian theory is its usefulness. Clearly much Freudian theory does provide a useful and illuminating mythology of the mind, particularly through the use of such concepts as id, ego, superego, ego-ideal, the unconscious, repression, neurosis, projection and introjection.

9   The term "transitional space" used by Rizzuto (1979: 46) builds on the concept of a "transitional object". "Transition object" is a term introduced by Winnicott. The mother is the first transitional object, helping the baby to relate subjective awareness with increasing exposure to the external reality introduced to the child since birth. The mother also uses other transitional objects to wean a child off her breast, which gives food, warmth, security and constant presence, by giving a child perhaps a dummy, blanket or teddy, which provides in time an acceptable substitute "*illusion*" and *object* for security for at least limited periods of time (Winnicott 1988: 101–7).

10 It may be relevant to a very few people who may be clinically diagnosed with an Oedipus complex which is related to their religious views, but that does not provide a basis for a universal theory.

11 "Projection": taking a concept attached to one object and transferring it to, or projecting it onto, another, e.g. father's (ideal) attributes projected onto God.

12 Freud (1927 in Dickson 1991: 227). He talks of religion = "wishful illusions" + "a disavowal of reality", which was his definition of delusion (213, 237).

13 These concepts of "extrinsic religious orientation" and "intrinsic religious orientation" have been described earlier in Chapter 3 in the section on "Humanistic Psychology's Models of Being Human".

14 Cf. Batson *et al.* (1993: 161, 261, 287); Allport (1960: ch. 3): Note that the Allport terms "intrinsic" and "extrinsic" parallel Fromm's terms "authoritarian" and "humanistic" (see Fromm 1971: 34–5).

15 The seven dimensions of mental health are: absence of illness; appropriate social behaviour; freedom from worry and guilt; personal competence and control; self-acceptance, self-actualization; unification and organization; and open-mindedness and flexibility.

16 In his "Postscript" to his *Autobiographical Study* (1935a in Dickson 1991: 182), he says: "In *Future of an Illusion* I expressed an essentially negative evaluation of religion. Later, I found a formula which did better justice to it: while granting that its power lies in the truth it contains, I showed that that truth was not a material but a historical truth."

17 "Religion restricts this play of choice and adaptation, since it imposes equally on everyone its own path to the acquisition of happiness and protection from suffering. Its technique consists in depressing the value of life and distorting the picture of the real world in a delusional manner – which presupposes an intimidation of the intelligence. At this price by forcibly fixing them in a state of psychical infantilism and by drawing them into a mass delusion, religion succeeds in sparing many people an individual neurosis." (Dickson 1991: 273)

18 Cited by Meissner (1984: 74). Chapter 4 is an excellent summary of the Pfister–Freud relationship.

19 The "quest orientation" is of course very important but not central to this argument, though it has much more in common with the intrinsic than the extrinsic orientation (see Batson *et al.* 1993: 166, 295–6).

20 Hostie (1957); Heisig (1979). The following account draws on the work of these two scholars, particularly Heisig.

21 "Unconscious fantasy thinking" emerges through "the collective unconscious", a concept which Jung introduced. The "unconscious" is a term Freud used, which Jung called the "personal unconscious" as it refers to acquisitions determined by but out of the conscious reach of one individual. Beneath the unconscious lies the "collective unconscious", which is a deposit of thousands of years of the human species, and is the source of prototype human drives and needs through which "archetypes" or archetypal images ascend. "Archetypes" are not memories. Jung called archetypes "forms without content". Archetypes are "a kind of readiness to produce over and over again the same or similar mythical ideas. They are the ruling powers, the gods, images of the dominant laws and principles, and of typical, regularly occurring events in the soul's cycle of experience" (Jung 1912).

These are found in dreams, fantasies, fairy tales, folk tales and myths. Some archetypes are the *father*, the *mother*, the *child*, the *persona* – the mask we wear to conceal our true nature or to make an impression, the *shadow* – the negative side of personality which is the unconscious. When the persona has integrated the shadow, it then accepts the archetype that personifies the soul, which is of the opposite sex, called either the *anima* or *animus*. A very important archetype is the *self*. The Self is the mature synthesis of the unconscious mediated through archetypal images, and the conscious, expressed in the ego. Christ is the supreme archetypal image of the Self. The *child* archetype can take many forms, such as a child hero or a child god. The focus of child archetypes is on the future. Another archetype is the *wise old man* who may appear as king, prophet, grandfather, professor or other figure of authority. (For a good account of Jung's picture of the human psyche see Wulff (1991: 422–3), which is a summary source used above.)

22 Heisig notes that "the significance that the God concept had lost when seen as a projection of unfulfilled wishes has begun to be restored through his growing concern with archetypal symbolics" (1979: 39).

23 "Religion . . . is a carefully scrupulous observation of what Otto termed the 'numinosum' and that the unconscious mind exercises an authentic religious function" (Jung 1973a: 4).

24 "Individuation": the integrating of all aspects of a personality into a mature formation – the collective unconscious, the unconscious, the shadow, and the ego through the development of the Self, partly through recognizing its form in, for example, the God-image or Christ as the image of the Self.

25 Jung (1957: 100–1), quoted by Halligan (1995: 231–53). I am indebted to this article for many insights in the second part of this digest on Jung.

26 This summarizes key points made by Heisig (1979: 123–9). For comments on Jung's psychological theory see Heisig (1979: 130–1), Wulff (1991: 431–2) and Halligan (1995: 248–9).

27 Cf. appraisals of Jung's views of religion in Halligan (1995: 248–9), Wulff (1991: 456–7) and Clarke and Byrne (1993: 187–8).

REFERENCES

Allport, G. W. (1969), *The Individual and His Religion* (1950), Toronto: Collier-Macmillan.

Argyle, M. (2000), *Psychology and Religion*, London and New York: Routledge.

Batson, D. *et al.* (1993), *Religion and the Individual*, Oxford and New York: Oxford University Press.

Beit-Hallahmi, B. (1995), "Object Relations Theory and Religious Experience", in R. W. Hood Jr. (ed.), (1995), ch. 12.

Clarke, P. B. and Byrne, P. (1993), *Religion Defined and Explained*, London: St Martin's Press.

Dickson, A. (ed.), (1990), *The Origins of Religion* (1985), Penguin Freud Library, vol. 13, London: Penguin.

Dickson, A. (ed.), (1991), *The Penguin Freud Library*, vol. 12, *Civilisation, Society and Religion*, London: Penguin.

Freud, S. (1907), *Obsessive Actions and Religious Practices*.

Freud, S. (1927), *The Future of an Illusion*, in A. Dickson (ed.), (1991).

Freud, S. (1939), "Moses and Monotheism", in A. Dickson (ed.), (1991).

Freud, S. (1913), "Totem and Taboo", in A. Dickson (ed.), (1990).

Freud, S. (1935a), "Postscript" to *Autobiographical Study*, in A. Dickson (ed.), (1991).

Gross, R. and McIlveen, R. (1998), *Psychology: A New Introduction*, London: Hodder & Stoughton.

Halligan, F. R. (1995), "Jungian Theory and Religious Experience", in R. W. Hood Jr. (ed.), (1995), ch. 11.

Heisig, J. W. (1979), *Imago Dei: A Study of C. G. Jung's Psychology of Religion*, Cranbury, NJ: Associated Universities Press.

Hood Jr, R. W. (ed.), (1995), *Handbook of Religious Experience*, Birmingham, AL: Religious Education Press.

Hostie, R. (1957), *Religion and the Psychology of Jung*, London: Sheed & Ward.

Jung, C. G. (1957), *The Undiscovered Self* (1933), trans. R. F. Hull, New York: New American Library.

Jung, C. G. (1967), "The Theory of Psychoanalysis" (1913), in H. Read, M. Fordham, and G. Adler (eds), *The Collected Works of C. G. Jung*, vol. 4, Princeton, NJ: Princeton University Press.

Jung, C. G. (1973a), *Psychology and Religion* (1938), New Haven: Yale University Press.

Jung, C. G. (1973b), *Memories, Dreams, Reflections* (1963), London: Fontana.

Jung, C. (1977), *Psychology of the Unconscious* (1911), in H. Read, M. Fordham, and G. Adler, (eds), *The Collected Works of C. G. Jung*, vol. 7, London: Routledge & Kegan Paul.

Lee, J. M. (ed.), (1985), *The Spirituality of the Religious Educator*, Birmingham, AL: Religious Education Press.

Maslow, A. (1987), *Motivation and Personality*, 3rd edn, New York: Harper & Row.

McGrath, A. (2004), *The Twilight of Atheism*, London: Rider.

Meissner, W. W. (1984), *Psychoanalysis and Religious Experience*, New Haven: Yale University Press.

Pruyser, P. W. (1977), "The Seamy Side of Current Religious Beliefs", *Bulletin of the Menninger Clinic* 35, 77–97.

Pruyser, P. W. (1983), *The Play of the Imagination: Towards a Psychoanalysis of Culture*, New York: International Universities Press.

Rizzuto, A. M. (1979), *The Birth of the Living God*, Chicago: University of Chicago Press.

Scott, P. and Spencer, C. (1998), *Psychology: A Contemporary Introduction*, Malden, MA, and Oxford: Blackwell.

Shafranske, E. P. (1995), "Freudian Theory and Religious Experience", in R. W. Hood Jr. (ed.), (1995), ch. 10.

Spinks, G. S. (1963), *Psychology and Religion*, London: Methuen.

Watts, F. and Williams, M. (1988), *The Psychology of Religious Knowing*, Cambridge: Cambridge University Press.

Winnicott, D. W. (1988), *Human Nature*, London: Free Association Books.

Wulff, D. (1991), *Psychology of Religion: Classic and Contemporary Views*, New York: Wiley.

15 | Religious Experience

The approach of depth psychology has been explored in the previous chapter at considerable length because of the enormous influence of Freud and Jung on the general reader. The present importance of object-relations theory received briefer treatment.

Psychology in the twenty-first century can be divided into four main areas: biopsychology, cognitive psychology, developmental psychology and social psychology (Gross and McIlveen 1998; Scott and Spencer 1998). These approaches are starting points which converge, with other areas of psychology, in order to understand the many aspects of the psychology of personhood (Gross and McIlveen 1998; Scott and Spencer 1998). Biopsychology includes considerations of the emotions (see Chapter 7). Cognitive psychology includes the consideration of perception, attention, memory, learning, language and thought, including problem solving (see Chapter 2). Developmental psychology includes the consideration of social and cognitive development from childhood onwards (see Chapter 4, Piaget). Social psychology covers social cognition. It focuses on how individuals *attribute* causes of behaviour to specific agents or situations in ways which are often superficial and biased; hence the focus on *attribution theories*. The study of social relationships, influence and behaviour also come into this domain.

EMOTIONS AND RELIGIOUS EXPERIENCE

Biopsychology includes the treatment of the emotions (see Chapter 7, "Emotional Knowledge").

Some psychologists use a two-factor theory, like Schachter's theory (1962) of the emotions (Gross and McIlveen 1998: 155), which suggests that emotion must confirm a religious experience. In this kind of theory an emotion is assumed to be composed of a physiological state of arousal, like being frightened, awestruck or inspired. This is the first factor. The second factor is the cognitive or knowledge appraisal of the situation. In practice, the cognitive appraisal comes first and takes place very quickly, causing an arousal of particular emotions. For this reason it is sometimes called a "cognitive-arousal theory". Sometimes the interpretation takes the form of an "attribution", which means that the believer brings to the experience ideas, beliefs and conceptual frameworks which they know and which are usually learnt from the surrounding culture. The meaning of the experience, then, could be "attributed" to a particular person or beliefs selected to explain the experience (Beit-Hallahmi and Argyle 1997: p. 95). Three other factors may affect emotional arousal. A particular psychological state may occur through, for example, prayer, fasting, meditation or reflection. The personality or religious upbringing of an individual may be contextual factors (Argyle 2000: 73).

Some psychologists talk about what they call *affective theory*. "Affect", is often distinguished from cognition (knowledge), volition (willing) and emotion. "Affect" is similar in meaning to emotion but broader, including feelings, moods and emotions. Attitude describes it most satisfactorily, if it is being taken to mean an emotional disposition.

Professor Hill is concerned with the structure and function of an emotional disposition in relation to religious experience. He regrets that there are no overarching theories of affect in research on religious experience, due to the theories of rival psychological camps like the cognitivists, who disparage emotion (Hill 1995: 353–5).

Biosocial theories minimize the importance of cognition, focusing on our biological processes, which give evidence for a set of innate emotions. But there are difficulties if the importance of the cognitive or knowing element is reduced. For example, if the causes of fear could be listed, how would that relate to "fear" in religious teachings in relation to a concept or picture of a punishing God, and to a picture or concept of a loving God?

However, the manner of appraisal can be affected by different factors, for example: (a) if the person appraising is the person engaged in the action or is another person; (b) if the action is approved or disapproved; and (c) whether the action affects the narrator's well-being. For example, if a friend is hurt by another's action, the affect or attitude will be different if I cause the hurt (guilt) or if another causes the hurt (anger).[1] A religious example: after a personal religious experience, is the subject of the experience going to decide to respond in a way pleasing or displeasing to God?

Religious coping is a way of handling worry and stress in life. St Paul made this clear to Christians in Phillipi: "Don't worry about anything, but pray about everything" (Philippians 4:6). People turn to religion for comfort when threatened by situations in life. For example, religion may help one cope with bereavement.

One way of dealing with the absence of any consensus concerning theories about the nature of emotion is to have a list of emotion-attitudes with defined characteristics. That would be provided from a list of atttributes of an affect or emotion-attitude. This could be established by "watching" a series of "exemplary" incidents which produced responses. Then a group of readers or observers might agree on the identification of an emotion and its description – a proposal put forward by Russell in 1991. For example, the script of a prototype for "anger" might be "harmful intention by offender, victim glares and scowls, victim feels tension and agitation, victim desires retribution, victim strikes at offender" (Hill 1995: 369).

Hill concludes his review of affective theory and religious experience, with a briefly summarized reference back to William James, who said in 1902, "You see now . . . why I have seemed so bent on rehabilitating the element of feeling in religion subordinating its intellectual part" (James 1902: 492) (Hill 1995: 371).

Affective theory, then, largely focuses on the judgement that cognitive appraisal of a situation releases the appropriate mix of emotions which generate relevant energy and disposition to act in response to any kind of experience, including a religious experience.

All situations need *cognitive appraisal,* i.e. one must assess what cause or causes create the stress. This illustrates the fact that talk of emotion is impossible without trying to describe what one knows. Basically Hill is siding with William James in a determination to emphasize the primacy of emotion over cognition when talking about religious experience (Hill 1995: 360–71).

COGNITIVE APPROACHES TO RELIGIOUS EXPERIENCE

One way of studying religious experience emerges from a model Wallas outlined in 1926. This model was explored by Vinacke (1952). Vinacke showed how there were often four stages involved in creative thinking which were equally apparent for those working in mathematics, the physical sciences, composing music, indeed in all creative work in the arts and sciences and in all other creative activities. The four stages in the model are:

1 *preparation* – the unsuccessful attempt to solve a problem by using old ways of thinking;
2 *incubation* – give up the attempt to solve the problem, so it sinks into the unconscious;
3 *illumination* – the sudden emergence of a new way of looking and a new structure of thinking which reorganizes the elements in the "problem", leading to a new solution, e.g. Archimedes' 'eureka'!
4 *verification* – test the value of the new solution: does it work?

This 1926 model of Wallas was then successfully applied to religious experience and religious conversion by Batson in 1971 (Batson *et al.* 1993: ch. 4). Professor Batson suggests that the sequence includes:

✧ *existential crisis*
✧ *self surrender*
✧ *new vision*
✧ *new life.*

This model illuminates "what happens psychologically during dramatic, reality-transforming religious experiences" (Batson *et al.* 1993: 115). Religious experiences which seem to emerge totally out of the blue have to be put into words for the experience to be understood by the recipient as well as to tell others. The language used has been previously learned from the community in which that person lives. Likewise, examples of religious experience, ritual and behaviour within the faith traditions known to the individual are usually available as models which can be adopted or adapted to "describe" their own religious experience. However, this model is subject to important feminist criticism for it offers a very male perspective.

There are a number of different feminist approaches to religion. We will explore one, which discusses religious experience. In examining ways of talking about religious experience Professor Mary Jo Neitz suggests that models used by men reflect the dominant male culture. While women and men seek and find experiences of otherness, or receive these experiences unexpectedly, "the meaning of religious experience has been different for men and women reflecting both differences in personality structure and the constraints of gendered life opportunities" (Neitz 1995: 522).

Neitz (1995: 522, citing Batson *et al.* 1993: 115) singles out the four-stage model of religious experience which involves:

- ✧ existential crisis
- ✧ self-surrender
- ✧ new vision, involving submission to new discipline
- ✧ new life.

Neitz comments: "The pre-existing condition for men in the narrative is autonomy . . ." (1995: 523). Would a feminist position start from "autonomy"?

It is suggested that girls and women do not have the experience of exercising their will, in other words, do not have the freedom to chose their opportunities in the same way as boys and men do, e.g. in having similar freedoms when growing up and choosing careers. If true, girls and women therefore start off with more constraints than boys and men. If females have not had the experience to exercise their will, that inhibits female spiritual development.

One strand of feminist theory suggests that while the gender identity model embraced by boys is "autonomy", the identity model embraced by girls is "the-self-as-connected-to-others". The female model is "connectedness" or "relatedness".

Females and males each have a different starting point:

- Males — less constraints — autonomy,
- Females — more constraints — connectedness with others

So Neitz finds the model offered by Batson and colleagues above as male gender oriented.

So perhaps women would find different spiritual paths appropriate to develop their "whole selves". Perhaps positive religious experiences for women might entail exercising the will rather than more self-surrender and more submission plus more discipline. I would argue that the admission of women to ordained ministry or spiritual leadership needs to be universally embraced, following the lead in, for example, some Christian churches, some reformed synagogues and some Sikh gurdwaras. Other examples of constraints on women are economic inequality and violence in the home, neither of which is tolerable.

For a male, a religious experience inviting a response to God's presence evokes different kinds of response from different kinds of males. Those with an extrinsic religious orientation may well respond in the style of the "classical model" of submission, obedience and subservience. That, however, is to adopt a heteronomous view of a relationship with God which is immature. To be unthinkingly under "the law of another" (nomos, i.e. law + hetero, i.e. of another = heteronomous) is, as we saw earlier, to be like a child not eating sweets before lunch because Mummy says so rather than for an autonomous reason (auto, i.e. self + nomos, i.e. law = autonomous), such as it will spoil your appetite (Piaget 1932).

Those reacting to a religious experience with an intrinsic religious orientation will respond as to a friend with better insights (cf. Exodus 33:11, "the Lord would speak to Moses face to face just like a friend"). Because the insights will "ring true" at a deep inner level, the responder will "make them her or his own" and be in a reciprocal partnership as a junior partner – an autonomous response. Perhaps it is this intrinsic religious orientation which many females are now seeking. Is this not their search for a satisfying relationship – a connectedness with the Other, freely and autonomously responding to the offer of an interdependent relationship?

The main problem seems to be with the articulation of the model of religious experience from Batson, Schoenrade and Ventis (1993: 115). It is a small change which could be made, for they expound the difference between the implicit and extrinsic religious orientations eloquently with research support. Should not the model of an intrinsic religious orientation focus less on autonomy and more on reciprocity, less on self-surrender and more on mature relationships, less on submission and more on internalizing ideals which "ring true"? I greatly prefer the vision of new life which Neitz outlines.

Neitz properly argues that women do not want to deny themselves but to affirm themselves. Aspiring to an intrinsic religious orientation, or a quest orientation, is surely seeking to fulfil oneself. If I understand Neitz, I think she is arguing in favour, not of autonomy, but of a reciprocal goal of shared responsibility. And that surely would be the more mature goal for men as well.

The feminine exercise of will has been focused on the search for connectedness, which involves reciprocal power sharing, or democracy, which is a finer ideal than the more individualistic personal autonomy sought by many men. This is not an argument for a uni-sex goal. Men and women have complementary physical, mental and spiritual characteristics which they can bring to any shared goals. It is an argument for removing the "extra constraints" women have in comparison to men.

Cognitive Theory and Religious Experience

In her review of cognitive theory and religious experience, Beverly McAllister notes that James said, "No account of the universe in its totality can be final which leaves out these other [non-rational/mystical] forms of consciousness . . ." (1995: 313, citing James 1902: 303). She notes that cognitive psychologists, with their concern about the personal structure of self-knowledge, cannot just focus on behaviour, motives or attitudes of persons but must also study mental operations, i.e. how people think. This largely theoretical review does point towards some practical insights.

McAllister mentions the work of Watts and Williams (1988). One of their concerns is to explore "what contribution (religious experiences) make to religious knowing" (23). In particular, Watts and Williams are concerned to explore how people who have a religious experience struggle to find the right words to express it. Watts and Williams find it is a process rather like that which occurs when psychotherapy patients explore their strong feelings, which, when they succeed in putting them into words, give them powerful insights into themselves (23).

After exploring "psychotherapeutic insight" as an analogue or model "of a sort of 'knowing' which consists neither of simple intellectual propositions nor simply of emotional feelings" (Watts and Williams 1988: 59), Watts and Williams explore the similarities between aesthetic knowing and empathic knowing, and the nature of perception, feeling and cognitive interpretation involved. This is followed by an exploration of meditation and perception and the elements involved in moral "knowing".

A number of similarities were found between aesthetic and moral "knowing" which are illustrated in St Paul's conception of love outlined in 1 Corinthians 13.

The characteristics in these forms of knowing include conceptual or propositional knowledge which is formulated after a number of processes are involved. The perceiver genuinely examines the data before or within them with careful attentiveness, looking at the evidence from a variety of perspectives, avoiding stereotypic interpretations as each perception is particular and evokes different aesthetic and/or moral feelings. This is similar to Simone Weil's "concept of attention to love of the neighbour" (Watts and Williams 1988: 69). Iris Murdoch rejects accounts of morality which emphasize the role of the will in making moral decisions:

> Her alternative account of morality, emphasizing its perceptual features, explicitly borrows from Simone Weil the concept of "attention" . . . the same "just and loving gaze directed upon an individual reality" as Simone Weil describes in the religious life. Once a person has learned through "attention" to see things as they really are, it will be obvious what needs to be done. It is accurate perception, not correct resolution, that she regards as the central moral task. (Watts and Williams 1988: 70)

Watts and Williams then explore further their comparison of knowledge gained through psychotherapy sessions, and knowledge gained from religious experience. They focus on the *grounds* of personal knowledge and insight which emerges from conversations with the psychiatrist or psychotherapist. Patients can refer to a powerful emotional experience without being able to identify the basic emotion or its cause. Watts and Williams compare this to "the path of 'unknowing' in Christian contemplation as reflecting these facts about the difficulty of articulating raw insights" (1988: 73).

Watts and Williams reach two conclusions. First, religious insight that has little experiential base is similar to intellectual insight in psychotherapy. Secondly, where the insights in psychotherapy and in religion are based on experiences which are initially unsymbolized – i.e. they are experiences which are difficult to put into words – this leads to *effective* insights. That is, in both psychotherapy and in religious experience, insights through the second way of knowing lead to marked consequences. Then, "an insight into the nature of God and a passionate love of God are bound up together" (Watts and Williams 1988: 74).

Watts and Williams not only explore the structure of the nature of religious knowledge derived from specific religious experience, but also explore the inclusive nature of religious knowledge: "Religious knowing involves . . . coming to know the religious dimension of the everyday world . . . It is with an interactional reality, not with a separate religious reality, that religious knowledge is concerned" (1988: 151).

DEVELOPMENTAL APPROACHES TO RELIGIOUS EXPERIENCE

Attachment Theory and Religious Experience

What is attachment? Attachment is a very close personal relationship between two people, of which the first is between mother and child. It endures over time, and significant separation from the other causes considerable stress. Examples of attachments are two very close friends, husband and wife, man or woman and God.

Professor Kirkpatrick, in his review of attachment theory and religious experience, draws attention to the fact that psychoanalytic study "has long been estranged from research-oriented fields of social, developmental personality and even clinical psychology" (1995: 446). These disciplines form the area of empirical psychology focused on proposing hypotheses of "religious experience formation" and then testing them through collecting empirical data, like analysing written biographical accounts, opinion polls, questionnaires and interviews.

Commenting on the estrangement between psychoanalytic approaches and empirical approaches Kirkpatrick says: "This is an unfortunate state of affairs because the strengths of each are the weaknesses of the other" (1995: 446). To oversimplify, depth psychology is strong on theory and weak on empirical data, and empirical approaches are strong on objective behavior measures and weak on theory. Kirkpatrick and Hood recognize that even the main empirical model of "intrinsic and extrinsic religious orientations" "offers little more than a descriptive typology (or classification) and fails to provide much in the way of an explanatory framework for understanding the emotional, cognitive and behavioral variability of religious experience" (Kirkpatrick 1995: 447).

Kirkpatrick offers a solution to this problem by using Bowlby's "attachment theory" as a bridge over the gulf between psychoanalytic theory and empirical approaches. "A psychiatrist, trained in the British object relations school of psychoanalysis, John Bowlby (1969) set out explicitly to construct a new theory to replace psychoanalysis in view of what he regards as a variety of conceptual and empirical shortcomings" (Kirkpatrick 1995: 447).

Attachment theory suggests that there exists "a behavioral system in human and other primates that has been 'designed' by natural selection to maintain proximity between infants and their care-givers" (Kirkpatrick 1995: 447–8). "Attachment" summarizes the infant monkey's or infant child's need for love, food and security, i.e. a need for an attachment to a care-giver.

Young children develop *mental* models of attachment figures. Adolescents search out models and heroes in their fields of work or sport as role models they "psychologically attach themselves to". Attachment theory also identifies a major element of explanation concerning adult relationships, e.g. husbands and wives, partners and friends.

Attachment and Religion   Greeley said "just as a story of anyone's life is the story of a relationship – so each person's religious story is the story of a relationship." Kirkpatrick continues, "From an attachment perspective the experience of a relationship with God is the experience of a deep emotional bond: . . . 'God is love'" (1995:

276

452). While God's love contains altruistic attachment and care-giving, adult romantic love contains three kinds of behavior, attachment, care-giving and sexual mating.

Recent evidence suggests that "images of God and perceptions of one's relationship with God may parallel one's orientation to other adult attachment relationships" (Kirkpatrick 1995: 454). People who feel secure rate God as more loving, less controlling and more accessible than do insecure people.

"In general, attachment theory would predict that one's relationship with God is constructed from the building blocks of one's own actual experiences in close relationships . . . just as adult love relationships are thought to be shaped by early childhood experience with attachment relationships" (Kirkpatrick 1995: 454).

Attachment to God gives emotional security. The intrinsic religious orientation positively correlates with "freedom from worry and guilt" and "personal competence and control", as mentioned earlier (Kirkpatrick 1995: 456). Other aspects of a relationship or attachment to God are less loneliness and an "oscillating relationship" through prayer which reduces stress at times of anxiety.

Dramatic Religious Conversions   "Religious conversion has been likened to 'falling in love' since the earliest days of psychology of religion" (Kirkpatrick 1995: 458). This view extends from James in 1902, through the fifty years' influence of Robert Thouless's (1923–1971) book on psychology of religion, to the present day. The attachment is central.

The Darker Side of Religious Attachments   There are unsatisfactory attachments. The picture of God in the Hebrew Old Testament, which is part of the Christian Bible, includes that of "an angry, vindictive and sometimes jealous God with whom a secure attachment relationship would seem difficult to attain" (Kirkpatrick 1995: 461). Examples of such attachments do occur in the psychoanalytic literature. However, many loving parents do punish their children in different ways, so the negative emotions which follow do not often deter attachment. What effect does the punitive element have on an attachment? "It may help to explain why conservative Christian churches, which generally give greater attention to this aspect of God, are growing while mainline and liberal denominations continue on a downward spiral" (Kirkpatrick 1995: 462).

This brief selective summary does not do justice to Kirkpatrick's case as to why he holds that attachment theory "is the kind of theory on which an adequate psychology of religion ultimately must be based" (Kirkpatrick 1995: 470). We have just noted some illuminating insights into the nature of religious experience.

We have already described Piaget's account of the development of religious thinking at the end of Chapter 4, where we were exploring his account of the development of religious thinking, the changing nature of religious knowledge at different stages of understanding, and his underlying philosophical assumptions.

Piaget and Religious Experience

As we have discovered, the idea of religion as a "schema" or "scheme" derives from Piaget's work from the 1960s onwards. A scheme is a cluster of ideas which make up a larger concept. To repeat a previous example, a farm is composed of animals, farm machinery, buildings and farm workers, and so on. A religion is a scheme consisting of beliefs, images of God, rituals, buildings, rabbis and ministers, and so on. A religious experience is a scheme which may include a view of a beautiful scene, feelings that work is boring and you need a new direction, the idea that God wants you to respond to that need you read about, and so on. Individuals use ideas or concepts from these organized structures or schemes when describing their own religious experiences.

Significance of Religious Experience at Different Ages

While Chapters 10–12 reviewed the main accounts of religious experience, Tamminen and Nurmi report some surveys of religious experience, meaning individual experiences of the Presence of God which gives a skeleton picture of the different ages at which young people have religious experiences (Tamminen and Nurmi 1995: ch. 13). A few examples are quoted to give a general picture.

In 1962, Elkind found that most American fifth grade students and 36 per cent of students in ninth grade reported experiences of the Presence of God (Tamminen and Nurmi 1995: 296–7). In Finland, Tamminen found, in 1981, in a large group aged 7–20 years, that 90 per cent of those aged 9–10 years and 55 per cent of those aged 15–16 years reported a sense of God's presence and guidance (Hyde1990: 185). Later, in 1983 and 1991, Tamminen reported that 50 per cent of students aged 13–15 years experienced the Presence of God when faced with an emergency (Tamminen and Nurmi 1995: 297).

Thomas and Cooper (1978) explored types and incidence of religious experience among people aged from 17–29 years. About a third gave a positive response to a question of the feeling of being close to a powerful spiritual force. This result was similar to Greeley's (1974), confirmed since by a later Gallup poll (Gallup and Jones 1989). In 1980 Thomas and Cooper found that 34 per cent of college, civic and religious groups reported some degree of mystical or spiritual experience (Hyde 1990: 175).

In Northern Ireland in 1982, Greer found that 33 per cent of schoolboys and 51 per cent of schoolgirls aged 12–17 years, and 38 per cent of boys aged 17–19 years and 51 per cent of girls aged 18–19 years, reported they had had "an experience of God, e.g. his presence or his help" (Tamminen and Nurmi 1995: 296).

We have already noted in Chapter 10 that the incidence of adult individuals experiencing the Presence of God in Australia, in the UK and in the United States range from 35 percent to 48 per cent (for more details see Beit-Hallahmi and Argyle 1997: ch. 5).

However, the whole point of a religious experience is that it happens at a certain point in time. It has ingredients and a structure, which will be discussed in Chapter 18. Such experiences have no "development". Those who write about "developmental theories and religious experience" are using "religious experience" to mean all experience interpreted from a religious point of view, and that is not a topic we are discussing (Tamminen and Nurmi 1995).

SOCIAL PSYCHOLOGY AND RELIGIOUS EXPERIENCE

Role Theory and Religious Experience

Clearly, religious experiences involve perceptions, concepts to describe them and the emotions experienced. Individuals also have "roles" in society. A woman may have many roles: mother, daughter, teacher, church steward, club tennis captain, and so on. Role theory has a place in social psychology, learning psychology and in many other disciplines. It has been used throughout the twentieth century, from William James onwards.

We are given or adopt a particular role in different situations: in a school, a firm, a church, in a relationship. We adopt a role and ascribe a role to God in a religious experience. A relationship involves *interaction*.

Sunden suggested that a person had to take on a "role", which involved "getting prepared for perception" and "getting prepared for action". Getting prepared involved devoting oneself to a religious tradition, with ritual support. For example, a Christian might attend a Bible class to know what the faith entailed, and attend church services, the rituals, to find out what Christian worship is all about.

Sunden felt that a period of preparation – a process of learning – would precede a religious experience. He felt that the process was similar in the natural sciences. The process involves four events:

- ✧ exciting the senses, which stimulate
- ✧ processes in the internal nervous system, which reach the brain where
- ✧ the search for a pattern begins, leading to a conscious result.
- ✧ This discovered pattern structures the perception content giving meaning (Holm 1995: 408).

Sunden pictured a religious experience as an interaction in which God and the individual each had their own role to play. A Christian, for example, learns these roles of humans, e.g. the prodigal son, and God from their reading of the Bible, in which these roles are described.

Fascinatingly, this model approaches the brilliant 1926 model of creative thinking of Wallas, which Vinacke (1952) and others showed illustrated all creative thinking in the arts, the sciences and other areas of creative thinking. Batson applied Wallas's model to religious thinking and religious experience in 1971, as described earlier in this chapter.

An individual is born into a world of social activity, which is initially outside or *external*. People give this world the status of an *objective* reality. "Through a process of socialization it comes to be *internalized* . . . " (Holm 1995: 403). In other words, the process of growing up in a family and society leads to an *internalization* of values, structures and ways of understanding and interrelating with others.

The highest level of reality is the level of "symbolic universes", which include religions. Religion provides meaningfulness and ultimate reality to people's thoughts, values, beliefs and actions. Symbolic universes give meaning to birth, life, identity and death. Mythologies – stories of our identity and significance, theology and philosophy

– are the "machineries of universe maintenance" (Holm 1995: 403), which means that these ways of knowing describe the universe to us as we understand it and our place in it.

This theory has been used to explain how religious groups like the Pentecostal movement create their own symbolic universe, as well as how some individuals create their "own mystical frame of experience" (Holm 1995: 404). People who read the astrologers, the stars or their horoscope in the newspapers may be doing the same. Is the reader's role to obey? If so, obey what?

We call the physical universe the "cosmos", while Berger calls the "symbolic universe" we each build up in our mind and inhabit a "nomos" (Berger 1973: 22–33, 62–3, 102). I say "a" nomos, because Buddhists, Christians, Hindus, humanists, Jews, Muslims and Sikhs each have a separate nomos. These "symbolic universes" are different in some ways and similar in others.

Attribution Theory and Religious Experience

Attribution theory is a convergence of six schools of thought, the first of which emerged in 1958. "Attribution theory deals with the general principles governing our selection and use of information to arrive at *causal explanation*" (444), e.g. what caused me to have an accident, fall in love, or have a religious experience? "In practice . . . people tend to make attributions quickly . . ." (Gross and McIlveen 1998: 444). This wise statement opens our review of attribution theory, where there is a perfectly legitimate tendency to enumerate all possible factors which should be *attributed* with causing a certain situation.

This approach to the study of religious experience was introduced in 1975 by Proudfoot and Shaver. It has the advantage that the attributions or cognitive descriptions, which are the interpretation attributed to the experience, are contained in reports of religious experiences. This is also the case in cognitive arousal theory, which additionally includes information brought to the situation in which the experience occurred. All this information provides the main content of the interpretation attributed to the experience. Affective experiences like religious experiences can also be affected by top-down factors. For example, known religious concepts are applied "top-down" by the subject in their account of the experience. Religious experiences are important examples for attribution theory, because often they are of an intense, impressive, awe-inspiring nature. The experience poses the question, what caused it? In searching for an answer, the individual may conclude, well, it can only have come from God: it is *attributed* to God.

Proudfoot and Shaver (1975) applied attribution theory to religious experience, as mentioned earlier. They said that "attribution theory is attractive to the student of religion . . . because it deals directly with a person's interpretation of his own experience" (Spilka and McIntosh 1995: 422).

Spilka and McIntosh (1995), in their review of attribution theory and religious experience, note the similarities in nature, aesthetic and religious experience, when discussing the nature of religious experience. Events and feelings are only "religious" if a person so interprets them, or if a person employs a "construct" or "scheme" of religious experience "top-down" onto a mysterious, awe-inspiring experience. The nature of the attribution is in all cases limited by the language available to the receiver trying to give an account of the experience.

They offer a definition: "Religious experience refers to a cognitive-emotional state in which the experiencer's understandings involve attributions or references to religious figures, roles or powers" (Spilka and McIntosh 1995: 422). "Cognitive–emotional" refers to beliefs and emotions or feelings. The "figures, roles, or powers" may refer to internal or external "actors" as the receiver indicates.

Spilka and McIntosh identify four kinds of religious experience.

1   Conforming experiences: The experiencer *attributes* the experience to either "a generalized sense of sacredness" or to "a specific awareness of the presence of a divinity" (Spilka and McIntosh 1995: 425). In the case of an encounter with Satan, sacredness becomes evil with an attendant presence "of an evil supernatural force" (425). The main *attribution* is to the truth of the experience.

2   Responsive experiences: A *responsive* experience is *attributed* to a *"mutual awareness* on the part of both the person and the supernatural. Mutuality may occur in three ways – through *salvation, positive miracles,* or *negative sanctions.* One may be saved or born again, miraculously healed, or make attributions to receive divine punishment" (Spilka and McIntosh 1995: 425). If the mutual awareness is of the devil, then the three mutual responses are *temptation, damnation* and *cursed.* In other words, a person holding such a view might say, "The devil made me do it," "Satan's reward," "Cursed by the devil," meaning good people reject the devil's offers.

3   Ecstatic experiences: An ecstatic experience is a style of experience involving intense emotion. Both the two forms of experience quoted above can be experienced in an ecstatic mode. The focus here is on the emotional intensity of the experience.

4   Revelational experiences: This category has been singled out on the criterion that a religious experience includes a divine call, such as Moses' call to free the Israelites, Jesus' call at his baptism, or St Paul's call on the Damascus Road. Many ordinary people have received the call to be a minister, priest, nurse, teacher, engineer or any other vocation.

Schachter's two-factor theory of the emotions has already been discussed. It is mentioned here just as a reminder that while the first factor was physiological arousal, the second factor was the interpretation of the arousal through attribution.

Misattribution is an important as well as an obvious idea. An individual may attribute a religious experience to a call from God to the priesthood or the medical profession. A church, nursing or medical school may not accept the strength of religious call alone to a profession, if it is properly noted that the candidate seems not to have the necessary skills, personal qualities or emotional stability to fulfil the chosen role.

Attributions are made in religious experiences after taking account of all the relevant factors the experiencer was aware of, on the basic assumption that a new unusual experience must have a cause. There are also motivational factors like basic human needs for meaning, for control and for self-esteem. Other factors of importance are the personality of the experiencer, with a history of particular influences and a character with distinctive qualities of disposition. Every individual is at the hub of a "matrix of forces" which include – across space – different social groups and personal influences, and – across time – past and present influences and future hopes and fears.

Transpersonal Theory and Religious Experience

Transpersonal theory is the name given to "the study of religious or spiritual experiences that involve assumptions not part of traditional scientific methodology".[2] "Transpersonal" literally means "across or beyond the individual person or psyche".

Susan Greenwood says, "Religious experience as addressed by transpersonal theory encompasses:

✦   the varieties named by James (1968),

✦   the kinds outlined in other chapters in this book,[3]

✦   transcendental experiences that seem to lift people outside of themselves or that engender a feeling of connectedness or cosmic unity (e.g. Underhill 1911)" (Greenwood 1995: 496), and other areas outside of our area of study.[4]

Questions are posed by Greenwood, such as, "Can scientific method be used to evaluate that which is beyond the ken of the five senses?" (Greenwood 1995: 497).

There are many perspectives on transpersonal theory outlined by Greenwood. Five will give us some of the flavour of the movement:

✦   One view suggests that transpersonal psychology should try to bring about a balance of inner and outer experience and awareness, as these are the two sides of personal wholeness (Greenwood 1995: 498, quoting Vaughan).

✦   A second view suggests that transpersonal theory should focus on integrating ancient wisdom, mysticism and rationalism with modern science (Greenwood 1995: 499, citing Sundberg and Keutzer).

✦   A third view "declares that transpersonal psychology is a synthesis of several disciplines, including religion and philosophy" (Greenwood 1995: 499, citing Washburn).

✦   A fourth view suggests that the transpersonal perspective should focus on "a deep level of subjectivity, or pure spirit, that infuses all matter and every event – the spiritual ground . . . [which] manifests itself in the world as energy" (Greenwood 1995: 500, quoting Nelson).

✦   A fifth concern notes that "one reason for creating [the movement] was . . . to study the phenomenon of religious experience independently of institutionalized religions and theological frameworks" (Greenwood 1995: 500, citing Lajoie *et al.*).

Recent work in transpersonal theory includes two important contributions from Wilber and Washburn.

Wilber in 1983 advanced a theory:

The mind and society are compound structures containing two dimensions: horizontal – surface structures, and vertical – deep structures. The tension between these dimensions provides the motive power for evolution of consciousness both for the individual and society. This evolution is hierarchical as it incorporates and transcends the previous, lower structures: it moves from prepersonal to personal to transpersonal. (The word "rational" may be substituted for "personal" to provide an added dimension to the concept.) Evolution, however, is only one half of the process; it implies matter ascending into spirit. The other half, involution, is a reverse process with spirit descending into matter. (Greenwood 1995: 500)

This is Wilber's way of describing "bottom-up" and "top-down" causation, referred to by Peacocke and Polkinghorne earlier (see Chapter 6).

In contrast to Wilber's model, which focuses on the movement to higher levels of consciousness, Washburn (1988) outlines a model which attends to the interactive activity of the ego, the individual, which becomes separate from "the ground" from which it emerged, "repressing" that memory. Then from time to time the ego reaches back beyond the "repression" to be in touch with "the ground", in pursuit of transcendence. This process is a bipolar interaction between "the ego" and "the ground" towards spiritual transcendence.

It has been said that Wilber is like a geographer "outlining the terrain of transpersonal experience", and Washburn is like the geologist "looking for the underlying volcanic activity which presumably shapes the topography . . . Washburn's theory is more psychological, whereas Wilburn's is more sociological, though each contains elements of the other" (Greenwood 1995: 511). Central to this idea is the tension of opposites, which has been taken up by other theorists, often involving ideas from Jung and a founding father of sociology – Durkheim.

Defining Religious Experience A key hypothesis of transpersonal theory is "that reconciliation of opposites is the fundamental religious experience" (Greenwood 1995: 509).

Greenwood offers a definition of religious experience, drawing on a Durkheimian–Jungian synthesis: "While religious experience has varieties of content – collective representations – which are social and historical, its essential nature has an underlying dialectical form – archetype – comprising of polar unities and the tension – the endless reconciling flow of energy between them" (1995: 509).

To the two theoretical models offered by Wilber and Washburn, which give models of the "topography" and "geology" of religious experience, Greenwood adds, from sources in the work of Jung and Durkheim we have not explored here, an interplay of "form" and "content". Jung provided the form – the archetype, while Durkheim provided the content – social religious experience.

Greenwood's picture of an archetype contains, as she says, elements in *tension*, for example:

- the rational mind   ><   the intuitive mind
- the human will   ><   God's will
- the real   ><   the imaginary or creative (cf. transitional space hypotheses)
- the profane   ><   the sacred.

The archetypal process is to redress systems which are out of balance.

In the light of these considerations, of "geography" and "geology", and "form" and "content", Greenwood offers a definition of transpersonal theory based on the previous definition of religious experience.

> Transpersonal theory is a multilevel methodology of non-duality employing simultaneous subjective and objective awareness of polar unities within an identified system. These polar unities have many forms: they can be immanent and transcendent, material and spiritual, individual and collective, psychological and sociological. Transpersonal theory is a collective representation, a current manifestation of the archetypal form of religious experience. (Greenwood 1995: 511)

In other words, a religious experience can be understood at many different levels of understanding, such as the sociological, the psychological, the personal and the religious. Greenwood has indicated some polar-unities-in-tension, like immanent–transcendent, which is a dimension of experiencing the Presence of God.

Transpersonal theory is a multidisciplinary way of examining the structure of religious experience outside ecclesiastical or theological structures. This is not to undermine proper theological insights expressed through the medium of "the story" within World Faiths. The mythological dimension of religion is not about "untrue claims", but about the literary form of "the religious story" and the kinds of insights it can convey.

PSYCHOLOGICAL APPROACHES TO RELIGIOUS CONVERSION

Conversion is the end result of a process which begins in a state of dis-ease. A problem may exercise one's mind, like Paul on the Damascus Road, or more simply, a person may feel "lost" before the conversion – with low opinion of self, without aim or hope. After the conversion they feel "found" – with a new vision, friend, community, direction, enthusiasm, help and security. There are four theoretical approaches to the study of conversion, each initially focused respectively on emotion, cognition, development and social influence.

An Approach through the Emotions:
Powerful Defence Solution to an Unconscious Conflict

This view comes from the Freudian psychoanalytic tradition, hinting that if you do not get on with your father, choose a higher wisdom – God. Freud's idea of God as being a projection of the Father into the sky does explain the *method* of forming some new concepts, but when he says, "God is nothing but a projection of father into the sky," he is confusing the *meaning* of an idea or concept with the *method* of forming or constructing it. His discovery of the *method* of forming some concepts and ideas is very helpful. It is perfectly sensible to picture God as king, friend, guide, helper, mother and father, or use other helpful word-pictures. The reality of God does not depend on the pictures, but upon a deep *intuition* of Personal Being, and the opportunity of a relationship with Personal Being held by individuals and communities which is believed to be true. The pictures are secondary and try to answer the question, "How can I put into words what I *know* is real?" Of course there is no conflict if you get on with your human father and with God. Conflicts are usually within a person, as indicated in the first approach above, i.e. you feel a failure, or cannot see the point in life, or are searching for something worthwhile to do.

An Approach through Cognition: Identity and Quest for Understanding

From Starbuck (1899) onwards "religious conversion has been rightly considered an adolescent phenomenon, or even an essential part of adolescence". There are many reasons for religious conversion, such as dissatisfaction with oneself, a loss of meaning in life, a loss of a sense of personal direction – who am I? where am I going? An individual may be glad to give up personal freedom for the sense of security gained from being a member of a community with a sense of direction, which gives support and security.

Commonly conversions are of a "personal re-commitment to a familiar religious tradition" (Beit-Hallahmi and Argyle 1997: 116). Conversion is like a love affair: it involves forming "an intense emotional attachment which may be sudden and dramatic". This leads to a rise in self-esteem and general well-being (117). In other words, after conversion you feel a lot happier with yourself.

Ordinary conversions are rare after the age of thirty, apart from some second or "born again" conversions found in some sections of the Christian Church, or "mystical conversions" when those at this stage of life find the mystical contemplative tradition spiritually helpful (Beit-Hallahmi and Argyle 1997: 116).

An Approach through Developmental Psychology: Personality Predisposition

The period before conversion is often a time of being demoralized, characterized perhaps by feelings of failure, aimlessness, despair, helplessness, anger and other downbeat feelings. Psychological readiness is therefore an important factor, so that a newfound emotional attachment will offer solutions in the form of aims in life, guidance, help, support, security and personal significance. William James describes St Augustine, Bunyan and Tolstoy as three who were demoralized before conversion (Beit-Hallahmi and Argyle 1997: 118).

An Approach through Social Psychology: Recruitment and Persuasion

"Those seeking members of a faith community try to become significant others" (Beit-Hallahmi and Argyle 1997: 123). There are three kinds of recruitment methods. The first is mind control, which means "a powerful system of persuasion based on deception, group pressure and co-ercion" (Beit-Hallahmi and Argyle 1997: 124). This frightens many people, particularly parents, but is surprisingly unsuccessful. Barker (1984) found that out of 1,000 individuals persuaded by the Unification Church to go to one of its overnight programmes in 1979, only 8 per cent joined for more than a week, and only 4 per cent remained two years later. Out of any 1,000 who had visited any Unification Church less than one would be a member five years later (Beit-Hallahmi and Argyle 1997: 124). You will soon spot this recruitment method and find a more congenial and open faith community to explore.

The second method is through those who offer genuine friendship. Invitations to visit or join a faith community are offered because those making the offer believe the vision and way of life offered is genuinely life enhancing. Peers are very successful, because such introductions grow from existing friendships across the whole age range. Ministers, priests and faith community leaders and workers are also important figures in attracting new adherents and members.

The third method is the active seeker who becomes a recruit. Many search out a religious community which meets their own particular emotional and cognitive needs. In, for example, the Christian Church, there are many different traditions and styles of worship and different patterns of group fellowship and learning.

THE EFFECTS OF RELIGIOUS EXPERIENCE [5]

Professor Ralph W. Hood Jr. in the United States began an important programme to study the psychological pattern of religious experience in 1970 and continued for over twenty years. He advanced the "theory that intrinsically orientated people would derive special experienced meanings from religion denied to extrinsically orientated contemporaries". He designed a questionnaire based on data from James's *Varieties of Religious Experience* (see Chapter 10). He later found that states of intense experience were a source of psychological strength.

In 1975 Hood constructed a test of mystical experience, based on data in Stace's analysis (see Chapter 10). This test could measure a factor of intense experience and a second factor of religious interpretation. This showed that such research on mystical experience was possible. A key result from a 1979 project was "that undergraduates reporting mystical experience showed stable personalities" (Hyde 1990: 175).

The study of religious experience from psychological perspectives can help to focus the status of the religious story, not as containing literal truth, nor as containing fictional untruths, but as archetypal signposts in story form which link the daily relationships with a relationship with Being, or God. A religious experience is, perhaps, your story, which is part of your life story. Through the insights from a deep relationship, we can improve all everyday relationships. Ideally the two-way relationship with God interacts with our two-way human relationships.

A religious experience usually has positive effects:

✧   A sense of unity: Unity within oneself is an effect after the relevant religious experience; so is the belief that "all things were working together for good". "I and what I watched were one" is an effect. "To such unity is attributed: 'beginning to understand the true meaning of the universe'; 'the power to do anything' . . . Again meaning, control and esteem enter the picture" (Spilka and McIntosh 1995: 438).

✧   Peace, joy, happiness and well-being: These are the second most important findings of the Hardy survey. These are effects. While Hardy found "a sense of security, protection and peace" to be a cluster of feelings most often experienced, the research of Spilka *et al.* (1993) linked this cluster of attributions to those clustered around a "sense of unity", quoted above, thus creating a more powerful response to the religious experience.

✧   The sense of an external power or presence: The third strongest percept from Hardy's respondents was "the sense of an external power or presence". "It comes across as a truly central theme of which James (1902) concludes, 'It is as if there were in human consciousness *a sense of reality, a feeling of objective presence, a perception* of what we may call *'something there'*" (Spilka and McIntosh 1995: 438, quoting James 1902: 58).

Spilka and McIntosh conclude their review by quoting Leuba (1929: 299): the experience we seek to understand "is one of the outstanding expressions of the creative power working in humanity. It is paralleled in the realm of reason by the development of science. Both lead, in different ways, to physical and spiritual realization."

NOTES

1   Note that the factors add to the complications of levels and methods of interpretation outlined in Chapter 11.
2   Greenwood (1995: 495). Two comprehensive texts on transpersonal psychology are Walsh and Vaughan (1993) and Tart (1992). The terms "transpersonal psychology" and "transpersonal psychotherapy" are often used interchangeably and deal with theory and practice in varying degrees.
3   This points to references to "religious experience" in R. W. Hood Jr. (1995). It also is true of the uses of "religious experience", which is the focus of this book.
4   These include: reported paranormal experiences, near-death and out-of-body experiences; religious dreams; and consciousness research on topics like meditation, healing and dying.
5   See Chapters 10 and 18.

REFERENCES

Argyle, M. (2000), *Psychology and Religion*, London and New York: Routledge.
Batson, C. D., Schoenrade, P. and Ventis, W. L. (1993), *Religion and the Individual*, New York and Oxford: Oxford University Press.
Beit-Hallahmi, B. and Argyle, M. (1997), *The Psychology of Religious Behaviour, Belief and Experience*, London and New York: Routledge.
Berger, P. L. (1973), *The Social Reality of Religion*, Harmondsworth: Penguin.

Greenwood, S. F. (1995), "Transpersonal Theory and Religious Experience", in R. W. Hood Jr. (ed.), (1995), ch. 21.

Gross, R. and McIlveen, R. (1998), *Psychology: A New Introduction*, London: Hodder & Stoughton.

Hill, P. C. (1995), "Affective Theory and Religious Experience", in R. W. Hood Jr. (ed.), (1995), ch. 15.

Holm, N. G. (1995), "Role Theory and Religious Experience" in R. W. Hood Jr. (ed.), (1995), ch. 17.

Hood Jr, R. W. (ed.), (1995), *Handbook of Religious Experience*, Birmingham, AL: Religious Education Press.

Hyde, K. (1990), *Religion in Child hood and Adolescence*, Birmingham, AL: Religious Education Press.

James, W. (1902), *The Varieties of Religious Experience*. London and New York: Longmans, Green and Co.

Kirkpatrick, L. A. (1995), "Attachment Theory and Religious Experience", in R. W. Hood Jr. (ed.), (1995), ch. 19.

Leuba, J. H. (1929), *The Psychology of Religious Mysticism*, London: Kegan, Paul, Trench & Trubner.

McCallister, B. J. (1995), "Cognitive Theory and Religious Experience", in R. W. Hood Jr. (ed.), (1995), ch. 14.

Neitz, M. J. (1995), "Feminist Theory and Religious Experience", in R. W. Hood Jr. (ed.), (1995), ch. 22.

Piaget, J. (1932), *The Moral Judgement of the Child*, London: Routledge & Kegan Paul.

Spilka, B. and McIntosh, D. N. (1995), "Attribution Theory and Religious Experience", in R. W. Hood Jr. (ed.), (1995), ch. 18.

Starbuck, E. D. (1899), *The Psychology of Religion*, New York: Scribner's.

Scott, P. and Spencer, C. (1998), *Psychology: A Contemporary Introduction*, Malden, MA, and Oxford: Blackwell.

Tamminen, K. and Nurmi, K. E. (1995), "Developmental Theory and Religious Experience", in R. W. Hood Jr. (ed.), (1995), ch. 13.

Tart, C. (ed.), (1992), *Transpersonal Psychologies*, 3rd edn, New York: HarperCollins.

Thouless, R. H. (1971), *An Introduction to the Psychology of Religion* (1923), Cambridge: Cambridge University Press.

Vinacke, V. E. (1952), *The Psychology of Thinking*, New York: McGraw-Hill.

Walsh, R. and Vaughan, F. (1993), *Paths Beyond Ego: The Transpersonal Vision*, Los Angeles: Tarcher.

Watts, F. and Williams, M. (1988), *The Psychology of Religious Knowing*, Cambridge: Cambridge University Press.

PART IV

Science and Religious Experience
Are they similar forms of knowledge?

16 Philosophy and Religious Experience

We have already discussed the non-realist views of T. R. Miles and Don Cupitt in Chapter 13, which also "belong" in this chapter. We need to distinguish a philosopher from a philosophical theologian. A philosopher is skilled in logical and analytic techniques which they apply to concepts used in religion and theology to explore what they mean. After philosophers have finished their work, they often hope that they have added a clearer understanding, but leave everything in the situation examined as it is, unchanged. A philosophical theologian uses her or his philosophical skills within "a particular religious tradition and its revelation claims" (Hebblethwaite 1987: 78). In this chapter we are largely discussing philosophical theologians.

Revelation, through Scripture, is 'top-down' talk resulting from 'bottom-up' significant religious experiences of prophets like Moses or Hosea. Modern western philosophers have been suspicious of the idea of revelation, partly because traditionally Jews, Christians and Muslims have often regarded it as a source of authoritative statements from important Teachers enshrined in Sacred Writings, sometimes called "special revelation", which is outside and "above" human reason. "General revelation", the discerning of God's presence within nature, is explored through reason. Philosophers, properly, want to make all knowledge accessible to reason. Another view of revelation holds that all knowledge is revelation, "since reality discloses itself to us as we discover and penetrate its nature" (1987: 85). This view accepts rational exploration.

But how are revelations given to human beings? "Is it just through rational reflection, mystical experience and the sense of the numinous? Or is it also through prophets and holy books, through historical events and through the developing traditions of faith?" (1987: 87). The answer is both, of course, but our concern here is only with the sense of the numinous and mystical experience.

When talking about "religious experience" we need to be clear about the meaning of the two words. "Religion" involves routes into God-talk, through Scripture, tradition, experience and reason. We have so far used "experience" practically through the whole text. We must now seek an analysis of its meaning.

"Experience" has been defined as

the state of being consciously affected by an event . . . To know something by experience is to know it by direct acquaintance . . . What is it to have experience of God? . . . To identify some experience as experience of God is to correlate my subjective states, my feelings and impressions, with a whole interpretative scheme given to me by my religion. (Hebblethwaite 1972: 265–6)

This interpretive scheme is the result of a conversation between experience and interpretation. This scheme of interpretation will have emerged as the result of an interaction between "previous interpretative schemes . . . particular events . . . and the experiences, individual and collective, of belonging to a community" (Hebblethwaite: 1972: 267–71; cf. Roberts 1991: 113–14).

From the analysis quoted above, we can see that the structure of religious experience has the same form of structure as the structure of scientific knowledge explored in Chapters 5 and 6. Religious experiences collected by Hardy and Hay, discussed earlier, have been collected from the "national community", which is largely Christian in terms of the main conceptual religious interpretive framework in society. Members of World Faith communities are likely to interpret their religious experiences from within their own traditions. However, the interpretive frameworks available have increased in number, to include New Age, eastern religions, and new religious movements. These are embraced by some in their spiritual search for the meaning of life. But boundaries can overlap, and beliefs intermingle, through dialogue.

However, the tests of the religious experiences applied by James still hold: do they give the individual delight, and do they lead to good consequential fruits? In other words do the claimed religious experiences spiritually enrich the quality of life of that individual as assessed by members of that person's community? If so, that counts as religious knowledge, just as claims of discoveries by scientists are only added to the corpus of scientific knowledge when the scientific community confers on those discoveries the status of knowledge.

Two questions need exploring: "Can I get to know God through religious experience?" and "Can I get to know God through using evidence and reason together?"

KNOWING GOD THROUGH RELIGIOUS EXPERIENCE

In Chapter 10 we mentioned Evelyn Underhill's contemplative experience. She quoted Simone Weil's experience, which was triggered by physical suffering (Weil 1973: 24). Underhill included this in her review of contemplative experience (see Chapter 10). This evoked the demand she placed on herself: "I was forcing myself to feel love," followed by her sense of God's presence, perceived through her feelings and linked to language she understood through revelation/her intuition/creative imagination. Tacit knowledge emerged into consciousness, the consciousness of God's presence, which found through verbal expression a form of knowing, accessible from language she knew, but used now with new insight and understanding. In Chapter 12 we focused on four modes of transcendental experience.

The first kind most people experience is nature mysticism. The sudden sense of a magnificent view in front and around evokes feelings of wonder and awe among other feelings. One feels lost and encompassed within the view, and such feelings may intensify so that later such an experience may modulate into a sense of mystical union with the scene, with specific details of the scene fading as one's sense of uplifting union grows.

The first of three religious modes of experience was a sense of the numinous – a feeling of awe in the Presence of God. The second kind of experience was of a conversion experience, and the third, which might grow out of the first or second, was the sense of mystical union with God.

Modern philosophical explorations of religious experience started from the recognition of many scholars that our modern idea of "religion" is a product of the eighteenth century, of the Enlightenment. When Augustine wrote of true religion, he thought of piety. A thousand years later Jonathan Edwards said that "true religion, in great part, consists in holy affections". "Only in the period of the Enlightenment did 'religion' commonly come to refer to a system of beliefs and practices, with several such systems to be found throughout the world" (quoted in Burhenn 1995: 144).

We have already described the work of the Model Makers, and noted that James shares with his forerunner, Schleiermacher, the belief "that feeling is the deeper source of religion, and that philosophic and theological formulas are secondary products" (quoted in Burhenn 1995: 147). We do not need to repeat our earlier discussion in Chapter 2 of the nature of logical positivism and its demise, or the claims of some recent thinkers, noted in Chapter 13, who hold that only language constructs knowledge. I have given reasons for rejecting that view and accepting James's view above.

Ramsey: Disclosures and Religious Knowledge

Ian Ramsey was concerned with the nature and use of religious language during the heyday of logical positivism. We are only interested in one use of religious language he explored, that where it is used to describe "disclosure situations". Everyday examples he gave included the sort of situations in which we struggle to understand something and then "the penny drops"! Ah yes! I understand, like Archimedes and his eureka experience.[1] Receiving a revelation, an insight from Sacred Writings, or having a religious experience, which may also be a revelation from God, are two other obvious religious examples. Both disclose what are new insights into the nature of God, aim in life, or the meaning of life (Ramsey 1967: 28–9), bringing knowledge in the modes of personal and/or religious discourse.

Farrer: Revelation, Religious Experience and Religious Knowledge

Austin Farrer was a philosophical theologian working, during the heyday of logical positivism, in Oxford from 1935 to 1960. For that reason we will take note of some of his contributions to this topic in his work published in the 1960s (Hebblethwaite 1983: 163–76).

Farrer holds that God works in two ways, which he called "double agency". First, God is hidden behind and within the structures of creation, which humans are free to explore. These can mediate a process of gradual indirect communication. God can be known through reflecting on nature, as in rational arguments for the existence of God, and through what human beings say and do freely: one route of God's agency. (We must return to "human freedom" in a moment with a consideration of "conscience".) This route is called "general revelation".

Farrer also notes a second way in which God acts: through a revelation or religious experience, such as Paul's Damascus road experience (Acts 9:1–18), after which he could say, "It is no longer I who live, but Christ who lives within me." This is the gift of a "special revelation", a gift of grace, which is an act of undeserved acceptance and kindness.

Note then that double agency is the act of God in creating a world within which human beings have evolved with *personal freedom* to speak and act. They later discover in their religious experience that they perceive the Presence of God, as a revelation in word or event – a gracious undeserved gift.

Hebblethwaite draws attention to elements that Farrer highlights when looking at the mode of divine action in special revelation. Farrer had just been talking of the images Jesus used in his parables. "Just such human creativity and imagination in the context of a particular tradition of faith are interpreted as the media of divine revelation . . . It is not the process of inspiration that is to be construed as revelation. Rather it is what results" (Hebblethwaite 1994b: 155). Revelation is the result of inspired communication.

Both kinds of revelation have to be "uncovered" and "accessed" through human activity, using human imagination and creativity. Against the demands of the logical positivists for verifiable proof of the existence of God, Farrer said "the hand of God is perfectly hidden" as "God makes the creature make itself". Also, "the causal joint between the Creator's action and that of the creature is absolutely inaccessible to us". However Farrer does indicate that the way of verifying Christian belief, received through revelation, is through one's experience as response. What is hidden is "the divine first cause in itself", though the effects are not hidden, as is seen in the lives of prophets, Christ and saints (Hebblethwaite 1983: 165). This is one piece of evidence for those seeking "proof".

The second piece of evidence follows: Farrer spelt out how coming "to will and do" God's will is evidence of God's action in the world, "the point where God's will touches us is in our present existence everywhere present; it is experienced by being obeyed" (quoted by Hebblethwaite 1983: 166). Now this assumes one knows God's will. Cumulative evidence comes from revelation through Scriptures, sanctity evidenced through the saints, and supremely through Christ who "gave back to God the picture of his own face and the love of his own heart" (Hebblethwaite 1983: 173f).

Farrer, however, thinks that experiential verification is only adequate when one keeps all the evidence for belief together: our own experiences, the influence of the experiences of the saints, rational arguments, the recognition of God working through Jesus, and God's creative will working in the whole world. Hebblethwaite rejects Farrer's attempt to treat the inspiration of the biblical writers and reported revelations through prophets and Christ as any different from that of later discerning believers (Hebblethwaite 1983: 175–6). I agree.

Farrer offers a "top-down" account of the verification of religious experience. We may get closer to the causal joint between the moral, personal and the religious, if we explore "conscience".

Tillich: Conscience, Religious Experience and Religious Knowledge

Response to God's will clearly involves the "conscience". Reductionist accounts of conscience by psychologists and sociologists are inadequate as they reduce moral discourse to a "lower" mode of discourse which is both impersonal and mechanistic. "Responsibility", a moral quality of conscience, is the "ability to respond". These are *moral and personal* skills. Independent modes of discourse like moral, aesthetic, personal and religious discourse cannot be so reduced without destroying their integrity as distinct "higher", or more inclusive and personal ways of knowing.

The clearest glimpse of the causal link between the personal, moral and religious modes of knowing that I know is outlined by Paul Tillich. But first we need to start with Kant's concern to clarify the relations between religion and morality, by outlining his three basic principles:

1   "morality does not need religion at all" (Kant 1960: 3) – "either in the discovery of what our duty is or the motivation for doing it" (Westphale 1997: 113).

2   "Morality leads ineluctably [i.e. inevitably] to religion, through which it extends itself to the idea of a powerful Moral Lawgiver" (Kant 1960: 5).

3   "Religion is [subjectively regarded] the recognition of all duties as divine commands" (Kant 1960: 142).

Westphal comments that Kant would say "that there can be no love of God separate from the love of fellow human beings". "But the text seems to make the stronger claim that religion is exclusively concerned with our duty towards one another, that even God is nothing but a means toward human morality" (Westphal 1997: 114). This is a "bottom-up" philosophical account of the relationship of morality and religion. Farrer's was largely a "top-down" account – God first, then his creation's response.

In Chapter 9 we discussed Paul Tillich's book, which offers a "bottom-up" account of the relationship of morality and religion (Tillich 1969). He explores the religious *dimension* of the moral imperative, which Kant called the categorical imperative, the formula for which is "to act according to those rules of conduct which can be consistently willed as a universal law" (Audi 1995: 403, col. 2), e.g. "treat others as you would like them to treat you". Secondly, Tillich explores the religious *source of the moral demands*, and lastly the religious *element* in moral motivation.

For Tillich, morality upholds the spirit of a centred person, culture points to the creativity of the spirit and religion is "the self-transcendence of the spirit towards what is ultimate and unconditional in being and meaning" (Tillich 1969: 9). Let us recall what he said. "The moral imperative is the command to become what one potentially is, a *person* within a community of persons" (11). So "a moral act is not an act in obedience to an external law, human or divine. It is the inner law of our true being . . . " (12). "An antimoral act is not the transgression of one or several . . . commands, but an act that contradicts the self-realization of the person as a person and drives towards disintegration" (12–13).

"What is the religious dimension of the moral imperative? . . . the religious dimension of the moral imperative is its unconditional character" (Tillich 1969: 14). For Tillich the Will of God is shown "in our essential being; and only because of this can we accept the moral imperative as valid." However,

> the moral imperative is unconditional only if I choose to affirm my own essential nature, and this is a condition! . . . It is the awareness of our belonging to a dimension that transcends our own finite freedom and our ability to affirm or negate ourselves . . . No religious heteronomy (= external law), subjection to external demands, is implied if we maintain the immanence of religion in the moral demand. (Tillich 1969: 16–17)

In other words, the religious demand comes from within the person to be their true self, which is expressed through the moral demand coming from within.

There is only space now to point to what I take to be a transcendental link, or causal joint, between the moral and religious *which are* personal (human) and Personal (Divine) respectively.

In Chapter 4 of his book, Tillich explores the notion of a transmoral conscience:

> A conscience may be called 'transmoral' if it judges not in obedience to a moral law, but according to its participation in a reality that transcends the sphere of moral commands. A transmoral conscience does not deny the moral realm, but is driven beyond it by the unbearable tensions of the sphere of law. (Tillich 1969: 75)

I find the simplest way to understand this from a Christian perspective is to start with St Paul's recognition that we all fail to "keep the moral law". He says, "I do not do the good I want, but the evil I do not want is what I do" (Romans 6:19 NRSV). Even trying to do the morally good evokes guilt. For example, if you want to give £20 to a charity, do you choose charity *a*, or *b* or *c*? You choose *a*, only to feel guilty when you hear a heartfelt plea for the desperate needs of *c*. This illustrates the limits of moral striving.

Jesus illustrated the way to respond to St Paul's situation in the parable of the lost or prodigal son (Luke 15:11–32), where the younger son wastes all his opportunities and money, becomes a "down and out" and loses his self-respect. He comes to his senses and goes home and receives more than *justice*, in fact a reception of unconditional *grace*, i.e. undeserved acceptance and forgiveness, which includes the restoration of his self-respect. That is how God will receive us if we approach him in a similar way, says Jesus. If we take this step, we can trust we are completely accepted as we are, and will find that we are Personally graciously accepted and are given a joyful transmoral conscience.

Then we can *recognize within us* the moral imperative within to seek to become our true selves and to love one another. This involves trying, and often failing, to accept others as they are. We are accepted as such, *not* because we achieve moral heights, because we keep failing, but because we recognize that *we are accepted as we are*, murderers, thieves, dropouts, successful or not in business, harassed mothers, struggling students, or whatever we are. *Once we accept we are accepted as we are*, we are then more able to aim to become our true selves.

The mature moral conscience transcends, for example, the moral codes of any individual secular or religious community, and is a mark of moral maturity, in, for example, a humanist. The transmoral conscience is focused less on justice and more on grace, like for example the cancellation of third world debt. For the religious person, the transmoral conscience is also the link, or causal joint, between the moral world of the person and the Personal world of the Divine. This starts as a spark within every human being and is a mature fruit of the moral imperative when it reaches up into the transmoral sphere. It is at least embryonically present in every community, and yet transcends and includes all communities in the global village we live in. The Personal transcends the Universe, and many religious people refer to this "Person" as God.

Looked at from a "bottom-up" perspective, it is like heating water to ascend as steam. If we strive to become our true selves responding to the moral imperative within, part of that involves striving "to love our neighbour". That can lead to moral maturity. But at any stage during this process of moral growth, an individual may "be grasped" by or "awakened" by a religious insight. This becomes the religious imperative within, and can lead to the development of a transmoral conscience, which includes discerning the will of God. This "being grasped" is an awareness of a causal joint between moral growth and religious insight – leading to the search for God's will. This may be discerned as a religious experience or the receipt of a revelation.

Looked at from a 'top-down' perspective, if you condense water vapour or steam you will collect water. If we seek God's will, we will try to become our true selves through responding to the moral imperative within us, leading to the development of a transmoral conscience, to become our true selves and try to love our neighbour as ourselves. Jesus put it simply: "In everything do to others as you would have them do to you; for this is the law and the prophets" (i.e. the heart of God's Will) (Matthew 7:12 NRSV).

The point of exploring conscience here is that it is one human avenue which may mediate or provide a dimension of a religious experience and/or may also inform a response to it as a prime source of personal and religious knowledge as well as of human spiritual growth. Were not the moral and religious imperatives major factors in Moses' religious experience which he interpreted as God's call?

Thatcher criticizes all accounts of morality which overemphasize individuality, "the diving within, the private access to unknown depths, the refusal to consider the possibility of self-deception, delusion, etc., when bravely actualizing one's potential".[2] Tillich's account of "conscience" seems to miss two other elements. The location of "the moral law" is not simply within the individual, but is upheld by the moral community, from whom it is initially in large part derived, and of which the individual should be a participating corporate member. The "conscience" is indivisible, but has two functions, providing a moral sense and a sense of duty. In other words, I have a judgement of reason which works out what I ought to do, and then "a magisterial dictate" – my internal moral order as to what I must do. If I obey my "sense of duty", I will have a good conscience; if I disobey, I will feel "guilty" and end up with a bad conscience.[3]

Swinburne: Religious Experience is Knowledge

Professor Richard Swinburne sets out to show that religious experience is knowledge. He defines religious experience in rather empirical terms as if God were an external object. He rejects internal objects from consideration. A religious experience seems to the subject to be an experience of God or of some other external supernatural presence (Swinburne 1979: 246). Swinburne distinguishes five kinds of religious experience.

1   An experience which seems to be of an external God, or other supernatural object, or perceiving God through a natural object like a sunset which mediates God's presence. In the latter case it is "an experience as" of the presence of God. An interpretive element is an ingredient in all experiences, as we have seen throughout our explorations.

2   An experience which many can have of "very unusual public objects", like the resurrection appearances to the disciples, or St Paul's experience on the Damascus road.

3   Religious experiences caused by private sensations, like Joseph's dream in Matthew 1:20–1, which might be auditory or visual.

4   Private religious experiences with sensations analogous to normal sensations but which normal vocabulary cannot describe and seem like a sixth sense. For example, mystics, like Ezekiel, might have had such experiences with very unusual and indescribable sensations.

5   Religious experiences which involve an awareness of God but without sensations. Mystics who experience God as "darkness" or "nothingness" seem to be avoiding "sensation talk". If anyone asks how you know you had that experience, the answer can only be "I just did".

The first four kinds of religious experience have "evidence" to support the experience, and the fifth experience provides its own evidence. This last example is exactly parallel to an answer to a question about a piece of music: how do you know that is great music? "I just do". Music is its own best evidence, there is no court of appeal outside it, for knowledge of music is contained and discussed within musical discourse.

An important step in Swinburne's argument is his appeal to what he calls the Principle of Credulity. What does he mean? He says that when a person believes something seems to be present in normal sense experience, it usually is. This is also a perfectly rational principle. Swinburne (1979: 254) argues that it is therefore inconsistent not to apply this Principle of Credulity to religious experiences. Swinburne claims that "in the absence of special considerations, all religious experiences ought to be taken by their subjects as genuine, and hence as substantial ground for belief in the existence of the apparent object – God, or Mary, or Ultimate Reality, or Poseidon" (254).

Poseidon was the Greek god of the sea, worshipped in the twelfth century BCE. That community of believers no longer exists, which underpins the point that Swinburne is tacitly making, namely, that a religious experience must have a double validation: the subject must have had it and affirm it, and the related faith community must accept it. Religions, like cultures of which they are a part, experience birth and death. This underpins the fact that a religion, with its related religious experiences, must yield credible existential meaning. If it does, it has evolutionary survival value for the faithful, as Hardy affirmed. If it does not, it will fade away.

Science, too, we have seen, accepts a new discovery or theory on the basis of double validation: the affirmation of the scientist who offers it, and the acceptance by the scientific community when they are satisfied with its truthfulness or usefulness.

To complete his case, Swinburne calls upon the Principle of Testimony. This claims that "other things being equal we think that what others tell us that they perceived, probably happened" (Swinburne 1979: 271).

These two principles, of credulity and testimony, effectively place Swinburne in the seat formerly occupied by the logical positivists. These two principles underpin "the assumption that things are [probably] as others claim to have perceived them" (1979: 272). Logical positivists demanded evidence, before their theory of knowledge collapsed. Now, as in a law court, the religious experience is presumed to be true, if affirmed by the double validation of the subject and the community, until proved false.

Swinburne does mention tests for a religious experience. "If he really did have that experience, one would expect his faith in God to be much deeper and this to make a great difference to his way of living . . . a difference to . . . behaviour in appropriate circumstances . . . prayer, worship and self-sacrifice will be more natural occupations." (1979: 273). This powerful argument must however be viewed in the larger context in which Swinburne sets it. It is not part of our brief to review the earlier part of Swinburne's book, but a comment must be made on it now.

Earlier in his book, Swinburne had set out his rational arguments for the existence of God. He explored the probability of theism, the explanatory power of theism – "Is it probable there is a God and what does that explain?" The cosmological argument involves looking at the universe and discussing its origin, while the teleological argument explores the observation that there is order in the universe: is this best explained by assuming that God designed it? Further arguments for the existence of God invoke consciousness, morality and providence, which poses the question, "Does not a good God provide for the basic needs of human beings?" Evil, history and miracles are discussed before the chapter on religious experience which we are concerned with here.

Swinburne concludes, from his survey of the arguments for the existence of God, completed prior to his discussion of religious experience, that, using a probability theorem to help him decide, the case for theism is balanced, without any definite decision possible. However, with the additional evidence of religious experience, Swinburne says that "on our total evidence theism is more probable than not" (1979: 291). Swinburne's argument, then, is based on accumulating evidence from all the arguments for the existence of God, including the decisive evidence from religious experience; he calls his method "cumulative evidentialism".

Swinburne has made an enormous contribution to the establishment of religious experience as a mode of knowledge which is a source of insight, encouragement, example, and one of the means available for finding direction and encouragement in life.

In discussing empirical work on religious experience in Chapter 10, and the Model Builders in Chapter 12, there was no occasion to mention a small school of thought which opposes all attempts to find models of religious experience which are applicable across all or most religions as advocated by the Model Builders in Chapter 12.

Katz: A Pluralist Model of Religious Mysticism

Professor Stephen Katz begins his defence of "why mystical experiences are the experiences they are" by stating his basic epistemological assumption, that is, his basic stance on the nature of knowledge. "There are NO pure (i.e. unmediated) experiences" (Katz 1978: 26). We have supported this argument from Chapter 2 onwards. Katz also properly draws attention to levels of interpretation, considerably extended here in Chapter 13.

He takes as an example "the case of a Jewish mystic" (Katz 1978: 33). Katz lists ten cultural-social elements which the Jewish mystic will have learnt as a child. There "is *very* strong evidence that pre-experiential conditioning affects the nature of the experience one actually has" (35). Katz describes and compares the experience of the Jewish mystical tradition and then compares it to the Buddhist tradition. He notes that the basis of the Buddhist account is the awareness of suffering in the world, and the goal is the extinction of suffering – *nirvana* (38). This is compared to the ultimate goal of Jewish mystical experience, *devekuth*, "which literally means 'adhesion to' or 'clinging to' God" (35).

Comparison is also made between a Hindu mystic and a Christian mystic. But here Katz admits that the "experience" is "at least partially preformed" or "prefigured" itself and then "the form in which it is reported is shaped by concepts which the mystic brings to, and which shape his experience" (1978: 26). Katz also indicates his discomfort with Stace's model of the "core mystical experience" outlined in Chapter 12. He takes, for example, the concept of "reality" and shows how it means different things in different traditions. In brief, despite his recognition of the "distinct experience" of the Hindu and Christian mystics quoted, Katz defends his "pluralistic account" of religious mysticism to account for all the evidence (66). While Katz is a social constructivist, he is not as extreme as Gimello (170–99; Watts 2002: 91).

A number of comments need to be made on this thesis. Watts points out that Katz's position is "unnecessarily extreme. He fails to consider at all what seems a likely possibility, that religious experience, though not wholly unmediated, is *relatively* unmediated" (2002: 91). Watts adds the important point that "*religious experiences may be less subject to background contextual factors, than most ordinary experiences*" (91), (my italics). This gains support when Watts suggests that Zajonc, a distinguished psychologist "has argued affect is independent of cognition". This "implies at the very least the parallel position that 'religious experience is independent of cognition' and deserves serious consideration" (96). This insight is also supported by Wynn's insight, mentioned in Chapter 7, which noted that feelings may offer "a preconceptual appreciation of the world's significance", in other words, feelings may well evoke and seek out the appropriate mode of discourse with their families of concepts.

Watts makes the further point that total emphasis on the public world of religious culture and its related institutions as constituting the structure of religious experience unduly emphasizes the public over against the private world of individual experience. He prefers "to emphasize the 'to and fro' between the private and public worlds rather than attempt to claim the primacy of either" (Watts 2002: 92). I find this emphasis on religious culture better described as "religious outlook", which includes "religious experience" as defined in Chapters 10 and 12.

Mysticism as described by Katz refers to the spiritual disciplines of different faiths developed over the centuries but not now central to the daily discipline of some members of some World Faiths in the Western World. This is different from, though of course indirectly related to, the kinds of experiences described in the archives of, for example, the Religious Experience Research Unit in Oxford set up by Hardy. Any person giving an account of an experience will most likely access the language of their own community's faith, though this was not prominent in, for example, the formation of the religious experience of Simone Weil, an agnostic Jewess, quoted in Chapter 10.

Proudfoot: A Unique Model of Religious Experience

Katz defended different faiths' unique concept of mystical experience. Professor Proudfoot studies those who "had an experience of God" and assumed "that there is a God to be experienced" (1985: 220). He ends up by defending the uniqueness of each religious experience: the description must be that of the experiencer. Proudfoot begins with a criticism of Schleiermacher, the German theologian who pioneered the modern study of religious experience.

Schleiermacher defined religious experience in 1822 as "a feeling of absolute or total dependence upon a source or power that is distinct from the world . . . which is the same thing . . . [as] being in relation with God" (Schleiermacher 1928: 17). Religious experience began with this feeling. The experience was prior to language which was subsequently employed as a "natural expression" to describe it. Proudfoot also assumes that there is no such thing as uninterpreted experience. He embraces the 1990s theory of emotion outlined in Chapter 7. He criticizes James for thinking that a religious experience is sense experience, having recognized that "this sense of the presence of an unseen reality becomes the thread that runs through the *Varieties*" (Proudfoot 1985: 162).

Proudfoot wants us to give up the idea of Schleiermacher and James that a religious experience is founded on feeling, which is prior to concepts which occupy the secondary role of describing the experience. Proudfoot, on the contrary, holds "that religious experience is constituted by concepts and beliefs" (1985: 219). Language constituted the experience. This is also a social constructivist approach. "'Constructivism' is the name given to the view that the human mind, through the imposition or projection of its own concepts on experience, 'constructs' its own world" (Hebblethwaite 1988: 50).

Proudfoot focuses on "description" and "explanation". He holds that descriptions of religious experience must be in the words used by the subject. "We have concluded that the distinguishing mark of religious experience is the subject's belief that the experience can only be accounted for in religious terms" (Proudfoot 1985: 223). He opposes "descriptive reductionism", which "is the failure to identify an . . . experience under the description by which the subject identifies it" (196). If each experience is unique, generalizations are ruled out, and the results of this approach seem to indicate that "the social-scientific study of religious experience would seem out of the question" (Burhenn 1985: 157).

Turning to "explanation", Proudfoot says that "we must explain why the subject employs those particular concepts and beliefs . . . In general what we want is a historical or cultural explanation" (1985: 223). No reasons are given for this judgement, although some kinds of explanations are explored, leaving the impression that Proudfoot thinks that present theistic interpretations are "legacies" (226) of the past.

Proudfoot suggests one possible theory to account for beliefs of people which are legacies of the past. "At a time [i.e. now] in which belief in a transcendent Creator . . . [has] been rejected by many . . . the direction of justification is reversed, and attempts are made to defend the beliefs by appeal to the affective experiences and practices" (Proudfoot 1985: 226). I do not think that this suggested theory is acceptable in the light of the evidence of religious experiences found in pre- and early Christian accounts of such experiences, amongst others, showing the centrality of feeling in response to a moment of insight. Similar moments of insight in science and in the creative arts also point to the centrality of feeling within the interacting network of tradition, conceptual frameworks, reason, volition, personal, social and cultural factors.

He assents to the view that "the concept of religion and the idea of religious experience were both shaped by the conflict between religion and the growth of scientific knowledge" (Proudfoot 1985: 232) – a widely held and acceptable view of language use now. But this does not change the fact that the kinds of experience under discussion are present in the Sacred Writings of the great World Faiths.

Proudfoot's view recognizes the influence of cultural factors on some religious experiences, but accepts that some "experiential reports . . . are sufficiently similar across different traditions to warrant use of the term *mysticism* and attention to some common characteristics" (1985: 124). This supports Watts's view quoted above that "*religious experiences may be less subject to background contextual factors, than most ordinary experiences*" (Watts 2002: 91). Proudfoot also thinks that the "labels a person adopts in order to understand what is happening to him determine what he experiences" (Proudfoot 1985: 229); they are not freely and carefully chosen to describe what is happening to him from traditional sources of understanding he finds illuminating.

The constructivist approach of Katz is of limited value to a cross-religions study, but does focus on the need to attend to the integrity of individual religious experiences, and to avoid distorting and simplifying the experiences being studied when using models. Proudfoot steers a middle course between those like James who regard mystical experiences as direct experiences of mystical union, and those like Katz who say they reflect, by their use of language, a particular religious tradition's understanding of mystical experience. The same distinction would be true for religious experiences.

Questions about Interpretation

Watts provides a clear summary of the interpretive stances of evaluating religious and mystical experiences:

> The debate about whether or not religious experience is unmediated tends to go to extremes. Some claim that it is absolutely unmediated (like James); others claim that it is as much

mediated through background cultural and cognitive processes as any other experience (like Katz). The real possibility that religious experience, especially the prepared forms that arise in the context of meditation, are less constrained by background cultural and cognitive processes than most experiences, though not absolutely free of such influences is a plausible one that deserves to have more advocates than it has. (Watts 2002: 91)

Again, in support of this view, we can repeat the insights of Zajonc, cited above, and Wynn, quoted above, which noted that feelings may offer "a preconceptual appreciation of the world's significance". In other words, feelings central to religious experience may well evoke and seek out the appropriate mode or modes of discourse with their families of concepts. This of course lends partial support to both the views of James and Schleiermacher.

Watts focuses on the need to take account of cultural changes which have affected the activity of *interpretation* since the seventeenth century, when an increasing tendency emerged to distinguish the outer world around one from the inner subjective world. Religious experience seeks to link these two worlds through the use of "double aspect" terms like "light", which can link, for example, "sunlight" and Christians talking about Jesus as the "Light of the World". We are increasingly conscious that our minds are involved in constructing our experience through patterns of interpretation that an individual 'selects' and applies to sensory phenomenal experiences. Watts's way of exploring the role of interpretation is not reductionist. He steers "away from strong social reductionism of Katz and Proudfoot, without going to the other extreme of postulating absolute pure religious experience. A likely position is that religious experience is less dependent on established constructions than most other experiences" (Watts 2002: 98–100).

Alston: Perceiving God

In *Perceiving God*, Professor William Alston argues that direct "experiential awareness of God", that is "perceiving God" through "*mystical perception*" leads to "M" *beliefs* (manifestation) which are analogous to the perception of physical objects in sense experience.[8] We perceive the world in these ways in order to gain knowledge.

We explored 'perception' in Chapter 2 above, and how we make claims like "that is a duck there" – an empirical statement, based on sense experience – our sight. Alston aims to see if the general theory of perception can be applied to claims of 'an awareness of God'. We saw in Chapter 2, that Logical Positivists attempted to limit knowledge to their own narrow view of knowledge. Alston criticizes the view that general theories of perception can only be applied to sense perception, i.e. the five senses. He labels these limitations 'epistemic imperialism', meaning that some scholars have defined knowledge in a limited way like the logical positivists.

Kant, Mackinnon, Hudson and other philosophers draw attention to the fact that there is a logical gap between a 'perception' – 'I perceive a white object', and a 'knowledge statement' – 'it is a duck', made up of concepts. The 'bridge' across the gap we call 'intuition'. We discussed 'intuition' in chapter 2, and in Chapter 4 we discovered that there were different forms of knowledge, such as empirical knowledge close to scientific knowledge, aesthetic, moral, personal and religious knowledge.

Now it is a matter of common experience that if we hear a sound, we immediately, and 'intuitively' call it either 'noise' or 'music'. That is we use language from the appropriate form of knowledge to describe/interpret it, using either empirical or aesthetic language. This of course is all through 'sense perception', but there are clearly different 'forms of intuition' available to access the appropriate language in the example given, either empirical or aesthetic language. Likewise for other experiences we may select empirical or religious language – or even feel *compelled* to use religious language after a *mystical perception*.

A further point needs to be made about the gap between perception and language in the form of concepts used to express it. We call the bridge that links the two 'intuition'. But that is a label, a name, which *means* nothing, except to describe an invisible bridge – what Farrer might call a 'causal joint'. But Farrer says it is concealed – an unknown mystery – when talking about religious experience. But is it not also a concealed mystery in the case of 'the bridge' between sense perception and empirical knowledge? Philosophers define intuition as "a certain kind of intellectual or rational insight".[9] 'Intuition' is not mentioned by psychologists, just the 'information channels in the hardware in the head' and theories which do not explain *the precise moment* or *process* of illumination, or decision making when discussing perception.[10] 'Intuition' seems not to be mentioned in the discussions on science and religion.

★　★　★

Turning to Alston's book, two terms he uses need introducing: 'mystical perception' and 'doxastic practices'.

Alston argues that the general theory of perception should not only be applied to sense perception but also to cases of claimed experiential awareness of God. Alston uses the phrase "mystical perception" to mean "the experience is . . . taken by the subject to be an awareness of God" (1993: 29). "The necessary conditions for X's being God . . . would include *being the source of existence of all other than itself, goodness, justice, moral lawgiver, having a purpose for the creation, and offering salvation to mankind*" (29). When discussing 'experiential awareness' in non-theistic religions – he describes the focus of such experiences as any *supreme reality* (2004: 136).

Alston also uses the phrase "doxastic practices", which he defined as "a way of forming beliefs" (Alston 1993: 6) when interpreting "sense perception" and then "mystical perception".

Our focus is on two questions: What kinds of "mystical perceptions" concern Alston?

How does Alston defend the epistemological or knowledge status of "mystical perceptions"?

Alston suggests that a "mystical perception" has four aspects. "(1) The awareness of God is *experiential,* as contrasted with *thinking* of God or *reasoning* about him. It seems to involve a *presentation* of God" (2004: 136). "(2) The experience is *direct*. One seems to be *immediately* aware of God . . . but there are more indirect experiences of God" (137). "(3) The experience is a *non-sensory* presentation of God. But there are also experiences of God with sensory content . . . (4) It is a *focal* experience, one in which the awareness of God is so intense as to blot everything else out. But there are also milder experiences that persist over long periods of time as a background to everyday experience" (137). Discussion is limited here to "*direct, non-sensory focal experiences*, since they give rise to the strongest claims to be genuinely aware of God" (137).

Secondly, Alston focuses on the "modes of appearance of God" who may be experienced as *being*: Good, Powerful, Loving, Compassionate, Wise and Glorious. God may be experienced as *doing* any of these activities: Speaking, Forgiving, Strengthening and/or Sympathizing (1993: 43–4).

Sense perceptions or *mystical perceptions* are expressed through "M beliefs (manifestation)" in "sense perceptual practices" – the normal way of interpreting sense experience – or, for a Christian, "Christian mystical perceptual practices (CMP)". These belief formation processes, which Alston calls *epistemic practices* (knowledge forming practices), are our ways of perceiving the world to gain knowledge. The Christian "M beliefs" develop from or are formed by the initial *mystical perception*, existentially supported by "its spiritual fruits" (1993: 35, 93, 153, 193, 254).

Alston sets out to demonstrate that if this process outlined above holds for "sense perception" it should also hold for "mystical perceptions". It is a big jump from 'perceiving a duck' to 'perceiving God', with his necessary conditions for X being God, noted above. Alston talks about our experience of humans who are good, from whom we get an idea of what it is like experiencing the goodness of God. Also he makes it clear that interpersonal perception is another epistemological practice, which gives us awareness of other persons which is a practice in a particular mode of knowing.[11] We have already discovered that personal knowledge is of its own kind, a distinct form of knowing which is not reducible. So let us place between 'perceiving a duck' and 'perceiving God' a model of 'perceiving a friend'.

The reason for the intermediary model is that sense perception of a duck involves data through our five senses, but when you meet a person, with whom you instantly have a rapport, of course the five senses are involved, but you have a feeling about the person that goes beyond sensory experience, you may catch a light in their eye missed by the camera, or feel a sense of empathy and warmth which cannot be tasted or touched, of common ground, and at the end of the first meeting you agree to meet again – the start of a friendship. You are operating in the mode of personal discourse related to personal knowledge. Look back through your own memory of verbal and visual impressions of that first meeting, and later getting to know any special friend you have. Besides those memories, what ideas, fears, hopes did you bring to bear on the occasion so that they were part of it?

The point of using this model of a friend is to indicate that Alston, when he uses the notion of 'perceiving', uses it literally for sense perception of objects, but in a legitimate but extended sense in the case of perceiving a friend, or perceiving God.

Now let's take an example of 'mystical perception' we have already mentioned: the call of Moses. What may have been perceived? A bush burning unconsumed, the setting sun behind it, the thought of the plight of the Hebrews in Egypt, the problem of who would get them out? What was God suggesting? How did God suggest it? Note the ideas relevant here, as in the case of the 'personal experience' above, are more important than the physically perceived objects which may have triggered off 'the whole religious experience'.

With these models to hand let's explore Alston's argument concerning perceiving God. Remember it is a rational logical argument. We are only going to look at some important features of it.

Alston begins his logical argument with 'sense perception'. Now the key point to spot is that Alston tells us that the argument from sense experience is circular. I'll explain 'circular' after describing it (1993: 103).

You see an object (perception), your intuition 'connects' to your 'experience', you call the object 'a tree' (the concept you chose to describe the object). That is the process or practice of making a connection between you and the world of objects, which is what Alston calls a 'sense perception practice'; it is an unconscious process that we never think about. Well we now need to think how we do it in order to go further.

How can we prove that the object is a "tree"? We assume "a tree" is an object in the real world in order to perceive an object which we then call "a tree". This is *not* a logical rational proof, but a circular argument. Rational argument is useless here! Why? We all accept 'perceptual practices', like seeing trees, people, birds and boats because they give us *pragmatic* or *practical* knowledge of the world we live in – of objects, of anticipation like driving a car, of forecasting the weather, indeed of doing any science, construction, invention or getting to know a friend.

Let's state Alston's point again, in a different way. Perceptual practices cannot be rationally justified, but we trust them because they are practical, or pragmatically rational. Practical faith can work where reason, or rational logic fails!

We have looked at an example of "personal perception practices" as a transitional model before we turn to "mystical perception practices", which I need not repeat. In that example we focused attention, not just on the things perceived but on all the associated ideas which were more important in your own "personal experience" which you remembered just now.

Let us now explore 'mystical perception practices' dealing with assumptions first. Epistemologists who discuss all kinds of experience, like sense perceptions, personal perceptions, religious perceptions, often appeal to an Eighteenth Century Scottish philosopher, Thomas Reid, who summarized the ways in which we form beliefs, which when validated become knowledge. He said we had several ultimate sources of belief:

✧ self consciousness (awareness of our own current conscious state);
✧ sense perception;
✧ memory;
✧ rational intuition (seeing with the mind's eye);
✧ reasoning of various kinds. (Alston 1983: 119; 1993:151)

Are we logically justified in trusting these sources of belief? No, but we are being practical or pragmatically rational. "All normal human beings are endowed by God or nature or both with a strong tendency to trust these sources to form firm belief . . . unhesitatingly and uncritically" (1983: 119).

We have just seen that the argument in favour of sense perception is circular; it cannot be based on reasoned argument. We trust our self consciousness, because it has been reliable in the past. We trust our memory because it is usually reliable. We trust our use of reason because it has been reliable in the past: you cannot use reasoned argument to justify the use of reason! All these arguments are circular. We justify our use of these sources of knowledge on practical grounds: they have worked in the past and are usually reliable. That is being practical or pragmatically rational.

Reid might quote a legal principle: these practices are innocent until proved guilty! Guilty? Yes, sometimes. Take an example from sense perception – I cross a field in a mist, and I say, "I see a person", but when I get closer find out that it is a bush. Another example: do you ever remember wrongly (memory)?! Be honest!

Let us return to "mystical perceptual practices". If we accept that sensory perceptual practices are practically reliable or pragmatically rational, subject to the scientific or relevant community acceptance, then it follows that we are being inconsistent if we do not also accept that 'personal perceptual practices' are also practically reliable. If we accept that 'sensory perceptual practices' and 'personal perceptual practices' are practically reliable, it follows that we are being inconsistent if we do not accept that 'mystical perceptual practices' are practically reliable or pragmatically rational, again subject to the acceptance of the relevant spiritual community.

Discussion about Alston's argument in favour of "mystical perceptual practices" draws attention to a number of points which interest us. He draws attention to the fact that the 'sense perception' and the 'mystical perception' arguments are not strictly parallel because the former only includes sensory, 'phenomenal', or 'appearance' qualities, while the latter includes 'comparative' qualities, like the goodness, guidance, or love of God.

I put in the model of 'personal perceptual practices' (following Alston's textual hint 1983: 131) between the two, earlier, so that we can see that this is also true for 'personal perception practices'. If 'personal qualities' are found in 'personal perception practices', and they are found to be practically reliable, it follows that we are being inconsistent if we do not recognize that they are equally appropriated in "mystical perceptual practices", with the qualification that the 'qualities' are being used analogically, for we can only know God 'through a glass darkly', as Alston makes very clear.

Alston notes that even if there is sometimes unpredictability in personal relations, "there is the possibility of entering into communion, fellowship, competition and so on with other persons." We need to remember that we are life long masters in the skilled use of 'sense perception practices' and 'personal perception practices', but "in Christian matters we are all in early infancy" (1983: 132).

Gale, an epistemologist who reviewed Alston's book, thinks, however, that Alston's argument, when applied to 'perceiving God practices', is not guilty of circularity, which means it has greater plausibility on this criterion than the 'sense perception practices' or the 'personal perception practices'. This is because a 'perceiving God practice' is engaged in for the purpose, or aim, of gaining "guidance in how to enter a loving communion with God and as a result realize a sanctifying spiritual and moral growth".[12]

Let's explore these two thinking processes, 'the circular argument', and the philosophically more secure 'inference from assumption' argument with other examples. Circular reasoning starts from an assumption, and ends up with a 'conclusion' which simply repeats the assumption, but looks as though it has proved the assumption to be true! As a logical argument it fails; it proves nothing; it is circular. A famous example: 'Every person should have complete freedom of speech as it is to the advantage of the State; for it is in the interests of the community that every person should be free to express all opinions without restriction.'[13] Sense perception is the same. 'I am sure I see a duck, because the object I see in the pond is a duck!'. We accept 'sense perception practice' for practical reasons: sense perceptions are usually reliable even though circular.

'Perceiving God practice' like a religious experience, however is not circular. Alston tells us it provides a map for our guidance for our interaction with God.

What is the purpose of 'sense perception practices' and 'perceiving God practices'? Let Alston answer: 'Sense perception practice' "proves itself by providing us with a 'map' of the physical and social environment, that helps us find our way around it." 'Perceiving God practice' provides an opportunity for "the transformation of the individual into what God intended him or her to be" (Alston: 1983: 131).

Hence we must look outside our experience to the saints of the spiritual life, and glean from the lives, works and thoughts of the likes of Mother Theresa "as to what it is to be more than babes in the experience of God" (132).

Where has this exploration left us? After Alston published *Perceiving God* in 1991, a discussion amongst epistemologists was published three years later to which Alston was given the opportunity to reply.

Let Alston comment first in this summary of where we are. "I can say that I do hold that mystical experience can render one . . . justified in believing . . . that God exists, but that this is not the sort of thing that is typically regarded as an argument for the existence of God."[14] Gale, a fellow epistemologist, thinks he is too modest in this judgement.[15] Gale says of Alston's book that "his discussion is a *tour de force*, exceeding in thoroughness any that I know in the extant literature."[16] At the moment it seems to be the most impressive philosophical argument of its kind in the defence of the integrity of religious experience in the second half of the Twentieth Century.

Alston is also clear that "the community will refuse to accept a particular report of a perception of God because it runs into conflict with the background belief system, whatever the individual says."[17]

The key difference between Swinburne's argument from religious experience and Alston's is this: Swinburne explored all the philosophical arguments for the existence of God which he felt gave a cumulative case, leaving 'the existence of God' undecided until the religious experience argument added the necessary extra evidence to justify belief in God.

Alston's argument on the other hand argues that no independent reasons are necessary to justify the existence of God other than 'perceiving God experiences'. Just as not all sense experience knowledge claims prove to be true, so not all 'perceiving God experience' claims prove to be true. Each are subject to the tests made of them both by the experiencer and the community to which the experiencer belongs. This rings true, simply because the religious mode of knowing, analogous to personal knowing, is a *different mode of knowing*, not reducible to other forms of knowledge.

A major criticism which Alston levels against his own case is the conflict between reports of different 'perceiving God practices' and their consequent beliefs reported within different World Faiths. I agree with his self-criticism. Is it not the case that when engaged in God-talk we have two forms of God talk, the language of worship and the language of dialogue? For example, the language of worship may be used in three contexts, by football or baseball supporters when shouting support for their own teams to whom they are devoted; by spouses who claim that they have married the best girl/chap in the world; or by Christians, Jews, Hindus, Muslims, Buddhists and Sikhs and other World Religions when gathered for worship. The language of dialogue, may be used within the Football or the Baseball Association; within a committee formed to discuss Hetero-Sexual and Same-Sex Marriages; or within an Inter Faith Relations Group.

Alston's argument shows how the Christian churches' Scriptures, theologies, beliefs and practices of worship and life styles experientially cultivate the soil to provide a seed bed for "experiences of direct awareness of God". And this helpfully empowers church adherents and members towards "life in Christ", or in Alston's words: "The final test of the Christian scheme comes from trying it out in one's life, testing the promises the scheme tells us God has made, following the way enjoined on us by the Church and seeing whether it leads to new life in the Spirit" (1993: 304). The focus is on the worship of God and Christian discipleship. Incidently, Alston is following Smart's seven dimensional model of religion in placing "experience" after the "doctrinal", "ritual" and "mythic" dimensions (Smart 1997: 27, 71, 131, 166). Alston's God-talk centres on the language of worship. This is illustrated in Alston's composite portrait of the epistemic (knowledge formation) structure of a typical case of Christian belief in a person he calls "Denise" (1993: 305f.), so the focus here is inward – looking at the authentication of faith within Christianity, a World Religion.

Is there space here in an American Denise's Christian experience and belief for God-talk focused on dialogue? Yes, if she has taken on board "her study of philosophy" and "the diversity of high religions on the contemporary scene" (305). Is there space in a British Denise's Christian experience and belief for God-talk focused on dialogue? Yes, if she is under forty years of age and has experienced good Religious Education or Religious Studies in her Secondary School and explored the similarities, differences and common ground between Christianity and one or more World Faiths in Britain, well represented in many cities and towns. How does this situation relate to figure 3.1 (p. 48), produced by Martyn Goss in Chapter 3 above? This search for the Sacred Other is true of other World Faiths, while many aspects of their enlightened moral codes are shared, and their models of saintliness often share common ground.

However it is also true that Alston's argument defends a "mystical perception" which *transcends* the limits of a 'World Faith' held by the Egyptians, as we saw in the example of the God of Abraham's call of Moses, or Jesus who included Jew and Gentile in his view of God's Kingdom, "good news" spread by Paul. For members of some Eastern Religions the "mystical perception" may be theistic or non-theistic as Smart has previously demonstrated.

Finally, a key point we must focus on is that all the accounts of religious experience reported on in this book so far, or discussed in the Twentieth Century, have tended to assume that a *causal* relationship exists between *God* or *the supreme reality* and worshippers. In other words, in theistic religions for example, God is very often regarded as the "Supreme Being" who "calls us". We will return to this point later in Chapter 19.

Wynn: Emotion – The Medium for Religious Experience

Wynn adds weight to Alston's achievement by his investigation into emotions as sources of religious understanding, which we first summarized in Chapter 7, from which we must now quote again. Wynn accepts Gaita's conclusion, after investigating a nun's behaviour in treating patients equally without condescension, that "the reve- latory force of a nun's example cannot be checked against any independent standard (e.g. the idea that human beings are created in the image of God)." But Wynn recog- nizes that it can be argued that "it is the language of God as parent in particular that enables the nun's affective response and associated behaviour. That language may not be an independent check on the affective response, but the response is in some way informed by it".[18]

In other words, feelings and their attendant emotions seem to have an intuitive link with the appropriate mode of discourse, here *religious*, which has an effect on behav- iour, though the links cannot be checked by independent verification or analysis of language use.

We have earlier explored the important arguments of Swinburne, Alston and Wynn. We now need to turn to the research work of Dr Caroline Franks Davis, which adds to our understanding.

Franks Davis: The Evidential Force of Religious Experience

In exploring arguments from religious experience, Franks Davis examines argu- ments from analogy. We have already begun to explore the argument that all experience is "experiencing-as" as the sixth meaning of "religious experience" in Chapter 12, and we now need to add to that argument here. A major contributor to this discussion, not mentioned before, is Professor John Hick. He talks of "experi- encing-as", and here I simplify, in two senses.

The first sense relates to looking at perceptual images that have an equal either/or appearance. We looked at two examples in Chapter 2. They were Neckar's cube and Rubin's profile or vase images. Franks Davis refers to the similar duck/rabbit image Franks Davis 1989: 84–6; Hick 1989: 140). Any of these images can be seen first in one way and then in another: they are *equal* either/or images.

Hick uses "experiencing-as" in a second sense where he says that, "all intentional experience is 'experiencing as'. It arises from the interpreting and misinterpreting of 'information' . . . impacting us from an external source . . . We describe as religious experiences those in the formation of which distinctively religious concepts are employed" (Hick 1989: 153). In other words, what Hick is saying is that when you describe your religious experience you select from your memory religious concepts that best interpret and describe it. Now this sense of talking about 'experiencing-as' involves one or more layers of interpretation like those we described in Chapter 13. In this use there is no *equal* take it or leave it, either/or option. The experiencer is offering you only one option, the one they described, supported by the three princi- ples of credulity, testimony and inference – all rational principles.

Franks Davis cannot dismiss Hick's second argument on the grounds that "experiencing-as" "is usually reserved for permanently ambiguous figures such as the duck-rabbit" (Franks Davis 1989: 86). That is, she cannot dismiss his "second kind of argument" on the grounds that it is really "the first kind of argument", and therefore weakens the case for "religious experience". It is perfectly legitimate to engage in the "second use" kind of argument when clearly outlined. Hick is clear.

Earlier in this chapter we explored the sense of "personal encounter" argument outlined by Ian Ramsey and the argument from analogy with sense experience outlined by Alston. We now turn to the arguments using the analogy with aesthetic and moral experience, and an argument using the concept of basic beliefs.

However, before turning to these arguments it is worth outlining the criteria which the mystics used to test their mystical experiences.

Criteria for Testing Religious and Mystical Experiences

Franks Davis suggests that mystics use four criteria by which they test purported mystical experiences (1989: 71):

1   internal and external consistency;
2   moral and spiritual fruits;
3   consistency with orthodox doctrine;
4   evaluation of the subject's general psychological and mental condition.

1   Tests of internal consistency include "the coherence of the concepts used" in the receiver's descriptions of his or her religious experience, "the internal consistency of that description, the way the experience fits into the pattern of other experiences, and consistency with non-religious background knowledge" (Franks Davis 1989: 71).
2   The "fruits" criterion was mentioned by James and discussed in Chapter 10. He faithfully echoed the experience of many mystics.
3   The "consistency with doctrine" criterion we have noted previously in a looser form. We noted when discussing the nature of scientific knowledge that a scientific "knowledge" claim can only be affirmed finally when the discoveries of pioneer scientists are not only supported by the evidence of experimental results where relevant, but also by the acclaim of the relevant scientific community. So also with religious experience, which needs to accord with understandable patterns of experience and fruits deemed desirable by the religious community. Nowadays the "fruits" criterion will be more important in society than a narrow doctrinal account of the religious experience, because the "fruits" criterion is more universal than doctrinal tests. We have previously covered some of this ground in Chapter 10.
4   The last criterion, concerned with psychological well-being, has tended to offer considerably more support to the intrinsic religious orientation than to the extrinsic religious orientation. This is because the latter has been shown to be immature and under the external authority of others. The intrinsic religious orientation shows a mature digestion of a personally understood faith

stance focused on serving God through seeking to meet the needs of others, which is life enhancing for all. This was discussed in Chapter 3.

The Argument for Religious Experience from Basic Beliefs

We have previously discovered that a sufficient number of examples is adequate support for accepting perceptual beliefs, memory beliefs, and beliefs about other people's thoughts. Likewise, a sufficient number of examples is adequate support for the "properly basic belief" that "God exists". Plantinga is the reformed epistemologist who advanced this argument.

Franks Davis holds that this is an argument for the defence of religious experience (1989: 87). She cites Plantinga's comment in support, which is that "God has implanted in us a natural tendency to see his hand in the world around us" (88). Plantinga also illustrates the natural human "disposition to believe in God with some examples of religious experiences".

Plantinga says:

> there is in us a disposition to believe propositions of the sort this flower was created by God or this vast and intricate universe was created by God when we contemplate the flower or behold the starry heavens or think about the vast reaches of the universe . . . Upon reading the Bible, one may be impressed with a deep sense that God is speaking to him. Upon having done what I know is cheap, or wrong or wicked, I may feel guilty in God's sight, and form the belief God disapproves of what I have done. Upon confession and repentance I may feel forgiven, forming the belief God forgives me for what I have done. A person in grave danger may turn to God, asking for his protection and help, and of course he or she then has the belief that God is indeed able to hear and help if he sees fit. When life is sweet and satisfying, a spontaneous sense of gratitude may well up within the soul, someone in this position may thank and praise the Lord for his goodness, and will of course have the accompanying belief that indeed the Lord is to be thanked and praised. (Franks Davis 1989: 89, citing Plantinga 1983: 80)

The quotation from Plantinga above needs reading carefully. In it he cites seven specific beliefs derived from the religious experiences mentioned. These he calls "basic beliefs". He admits that the belief "God exists" is entailed, assumed or inferred from those basic beliefs. "God exists" is assumed by those basic beliefs, without which they make no sense.

These arguments are not meant to persuade an atheist, but rather to reassure a "natural theist" that his or her beliefs are in a weak sense rational, i.e. they are not irrational.

Franks Davis, like Alston, recognizes that no philosopher has managed to provide inductive confidence for our reasoning processes, experiences, memories, and the claims of other persons, because, as we saw in Alston's discussion earlier, such attempted proofs get involved in circular arguments (Franks Davis 1989: 100).

We need to return to our discussion of Swinburne's principles of credulity and testimony and inference, which are all rational and have a particular strength we can now discuss. These principles effectively overrule the non-realist assumptions and arguments advanced by thinkers like T. R. Miles, discussed in Chapter 13, and the arguments of constructivist thinkers like Proudfoot, which were mentioned earlier in this chapter. That means that non-realists and constructivists are sceptical and refuse to accept the priority of experience before its articulation. The priority of experience is assumed in science as viewed by critical realists, as are the principles outlined by Swinburne mentioned at the beginning of this paragraph, and supported by the reformed epistemologists, Franks Davis and others. The priority of experience is also accepted in the law courts, though of course it is subject to critical scrutiny.

Put another way, the non-realists *refuse* to accept at face value the reports of those who say they have had a religious experience. They think they know better, as do all who employ the "hermeneutics of suspicion" we discussed in Chapter 13: the non-realists say things like, "What the experiencers say they have experienced are not really 'experiences of God' but feelings which they picture by using God–talk." If non-realists said that, they would be reducing 'the claimed report from an experiencer making a critical realist use of language' to saying that 'the language use was really the emotive use of language'. This is pure reductionism in a positivist mould, unless of course the non-realists and constructivists are *only describing their own language use*! However, we need creative thinkers like non-realists to explore different understandings for individuals and communities to evaluate.

Now, while these three principles of Swinburne should be applied to "religious experiences" with caution, they can be shown to work in the same way that they work in scientific discovery and in artistic creativity. They are tested for "results" and are "accepted" by the relevant scientific or artistic community. Likewise, religious experiences are accepted after inspection of the effects or results of them on the lives of the experiencers and those around them, and they are found congenial to the shared beliefs of the relevant faith community. Put bluntly, religious experiences can only be shown to be real if they produce real effects.

After her researches were completed into religious experiences from Hindu, Buddhist, Muslim, medieval Christian, and contemporary sources of Orthodox, Catholic and Protestant Christianity, Franks Davis could say that she accepted four types of irreducible religious experiences:

1 bipolar numinous experiences;
2 experiences of a loving (etc.) relationship with a personal "other";
3 extrovertive mystical experiences of unity of many kinds;
4 introvertive mystical experiences of unity (1989: 177).

I wish to conflate the third and fourth categories from this list, on the grounds that this book is about those kinds of religious or mystical experience that seem to happen spontaneously. That means removing from these categories those mystical experiences of either kind which are prepared for or facilitated by similar meditation techniques (Franks Davis 1989: 186).

In sum, Franks Davis says that there are three broad categories of religious experience: the numinous experience, the mystical experience and the sense of the presence of a personal "other" experience, and these kinds of experience provide good grounds for accepting "belief in God" (Franks Davis 1989: 186). In her survey of many different numinous and mystical experiences, Franks Davis shows that they can provide good evidence on their own for the following claims (190):

1   The world of physical bodies and narrow consciousness is not the whole of reality.

2   The "everyday ego", the everyday me, has a "deeper self" which depends on ultimate reality.

3   Whatever *is* the ultimate reality is holy, of supreme value and underpins everything else.

4   This holy power can be experienced as an awesome, loving, pardoning, guiding (etc.) presence with whom individuals can have a personal relationship.

5   An experience of awe before the numinous presence of God can slip into mystical union as the sense of self dissolves.

6   Some kinds of union with God provide the "greatest good" one can attain, giving one final freedom, or salvation, from egocentricism, as one discovers in that union one's "true self" or "true home" (Franks Davis 1989: 191, abbreviated).

These six claims from the three kinds of religious experiences, the numinous, the mystical and the "sense of a Presence" experiences, provide strong evidence for theistic belief (Franks Davis 1989: 191).

The key claims are that human beings have a "true self" beyond their "everyday self", intimately related to the divine nature which is the ultimate ground of being.

We have seen that there are complex interactions between "raw" experience and different levels of interpretation in Chapter 13.

Dr Janet Martin Soskice records that in her "study of models and metaphors, religious experiences are noted as an important way of grounding reference to God" (Franks Davis 1989: 240, citing Soskice 1985: 137–40, 150–3). Soskice is a critical realist, defending theological realism, and makes very plain that "models and metaphorical theory terms may, in both the scientific and religious cases, be reality depicting without pretending to be directly descriptive, and by doing so to support the Christian's right to make metaphysical claims" (Soskice 1985: 145).

Soskice clarifies her notion of "experience":

> It is a portmanteau term to cover two sorts, the first being the dramatic or pointed religious experiences of the kind which might prompt one to say, "whatever appeared to me on the mountain was God", or "whatever caused me to change my life was God". The second are the diffuse experiences which form the subject of subsequent metaphysical speculation . . . (Soskice 1985: 150)

In other words, Soskice identifies two kinds of experiences, the first being the individual religious experiences which individuals can identify in time and situation. All experience and experiences which are together interpreted from a theistic standpoint are the second kind of religious experience. This latter kind I described in Chapter 10 as the seventh use of the term, and describe as "a religious outlook".

Ward (2004, Ch. 11) points out that within the global village in which we live there will continue to be dialogue between the World Faiths. He suggests that a converging account of the qualities of God will emerge through the many organizations committed to dialogue. One of the extraordinary characteristics of saints across many faiths is the way they penetrate the symbols of their own faith towards the One Within and lead lives exemplifying characteristics shared by many faiths.

The spiritual community of the global village will affirm those "perceived God practices" and "consequent beliefs", which are life affirming and life enhancing for individuals in a community. These "consequent beliefs" will be considered by those exploring a world theology. How else can one make sense of that visionary Carpenter's words when he shared one of his most important pieces of teaching with a despised, and female, member of another faith? "Believe me, the time is coming when you won't worship the Father either on this mountain or in Jerusalem . . . God is Spirit, and those who worship God must be led by the Spirit to worship him according to the truth" (John 4:21, 24). Sometimes the spokespersons or "prophets" are pop singers, sometimes journalists, sometimes politicians, sometimes clerics: the Spirit is not limited in her outreach. Mahatma Gandhi, Martin Luther King, Nelson Mandela, Mother Theresa and Sir Bob Geldof shared their visions in the twentieth century. What "perceiving God practices" will lure and inspire the visionaries of the twenty-first century?

NOTES

1   Archimedes, a Greek mathematician (*c.* 287–212 BCE), was given a problem by King Heiron of Syracuse. He was asked to discover whether the king's crown, which was meant to be made of solid gold, had in fact been fraudulently made of cheaper silver and then gold plated. He was not allowed to damage the crown by cutting off a tiny piece of metal and analysing it. He was thinking about this problem when he went to the bath house. There he observed that the amount of water which flowed out of the bath was equal to the amount displaced by his body when it was immersed. This gave him a method by which he could test the density of the metal used in the crown. He knew that gold is nearly twice as dense as silver, so he could lower the crown into a vessel exactly full of water, measure the volume of water that flowed out, and then repeat the experiment, using a lump of pure gold of the same weight. If the crown really was gold, the two volumes would be the same, but if it was gold-plated silver, it would have nearly twice the volume of the gold lump, and would therefore displace nearly twice as much water. Vitruvius describes Archimedes' behaviour at his discovery: "In his joy, he leapt out of his tub, and rushing naked towards his home, he cried out with a loud voice that he had found what he had sought. For as he ran he repeatedly shouted in Greek, '*heureka, heureka!*' ['I've found it, I've found it!']." Source: Dr J. R. Waldram.

2   Personal communication from Professor Adrian Thatcher.

3   I am also grateful to Professor Adrian Thatcher for these two insights. Cf. Reardon (1966: 274–8).

4   I can find no reference in the works of I. Barbour, P. Davies, A. Peacocke, J. Polkinghorne, M. Hesse (Newton-Smith 1986).

5   Alston (1983: 131; 1991: 154). What applies to "sense perception" applies to "personal perception" according to Wittgenstein's principle here (cf. 1991: 175, 185). For a brief account cf. Peterson *et al.* (1991: 17–20).

6   Gale (1994: 143). Alston (1991: 250): "A manifestation perception practice has the analogous function of providing a 'map' of the 'divine environment' providing guidance for our interaction with God."

7   M. Wynn, personal communication, 22 June 2004.
8   W. P. Alston (1993) *Perceiving God*; W. P. Alston (1983), "Christian Experience and Christian Belief"; W. P. Alston (2004), "Does Religious Experience Justify Religious Belief?", pp. 134–63.
9   B. Russell (1995), 'Intuition', in Audi (1995), p. 382.
10 R. Gross and R. McIlveen (1998), *Psychology: A New Introduction*; P. Scott and C. Spencer (1998), *Psychology: A Contemporary Introduction*.
11 W. P. Alston (1983), "Christian Experience and Christian Belief", p. 131; W. P. Alston (1993) *Perceiving God*, p. 154. What applies to 'sense perception' applies to 'personal perception' according to Wittgenstein's principle here. Cf. pp. 175, 185;
12 *Religious Studies* (1994), 30, pp. 135–81, p. 143. W. P. Alston (1993) *Perceiving God*, p. 250. "A manifestation perception practice has the analagous function of providing a 'map' of the 'divine environment'.providing guidance for our interaction with God."
13 R. Audi (ed.), (1995), *The Cambridge Dictionary of Philosophy*, p. 124 (adapted).
14 *Religious Studies* (1994), 30, pp. 135–81, p. 175.
15 *Religious Studies* (1994), 30, pp. 135–81, p. 137.
16 *Religious Studies* (1994), 30, pp. 135–81, p. 138.
17 *Religious Studies* (1994), 30, pp. 135–81, p. 176.
18 M. Wynn, personal communication, 22 June 2004.

REFERENCES

Alston, W. P. (1983), "Christian Experience and Christian Belief", in A. Plantinga and N. Wolterstorff (1983), *Faith and Rationality*, Notre Dame, IN: University of Notre Dame Press.
Alston, W. P. (1993), *Perceiving God: The Epistemology of Religious Experience* (1991), Ithaca, NY: Cornell University Press.
Alston, W. P. (2004), "Does Religious Experience Justify Religious Belief?" in M. L. Peterson and R. J Vanarragon (eds), (2004), *Contemporary Debates in Philosophy of Religion*, Oxford: Blackwell Publishing; *Religious Studies*, (1994) vol. 30, Cambridge: Cambridge University Press.
Audi, R. (ed.), (1995), *The Cambridge Dictionary of Philosophy*, Cambridge: Cambridge University Press.
Burhenn, H. (1995), "Philosophy and Religious Experience", in R. W. Hood Jr (ed.), (1995), ch. 7
Franks Davis, C. F. (1989), *The Evidential Force of Religious Experience*, Oxford: Clarendon Press.
Gale, R. (1994), "The Overall Argument of Alston 'Perceiving God', *Religious Studies* Vol. 30: 2, pp. 135–49; 150–81, Cambridge: Cambridge University Press.
Gross, R. and McIlveen, R. (1998), *Psychology: A New Introduction*, London: Hodder & Stoughton.
Hebblethwaite, B. L. (1972), "The Appeal to Experience in Christology", in S. Sykes and J. P. Clayton (eds), (1972), *Christ, Faith and History*, Cambridge: Cambridge University Press.
Hebblethwaite, B. L. (1983), "The Experiential Verification of Religious Belief in the Theology of Austin Farrer", in J. C. Eaton and A. Loades (eds), (1983), *For God and Clarity*, Allison Park, PA: Pickwick Publications.
Hebblethwaite, B. L. (1987), *The Problems of Theology* (1980), Cambridge: Cambridge University Press.
Hebblethwaite, B. L. (1994a), *Modern Theology*, vol. 10, Oxford: Blackwell.
Hebblethwaite, B. L. (1994b), "The Communication of Divine Revelation", in A. G. Padgett (ed.), (1994), *Reason and the Christian Religion*, Oxford: Clarendon Press.

Hick, J. (1989), *An Interpretation of Religion*, Basingstoke: Macmillan.

Kant, E. (1960), *Religion Within the Limits of Reason Alone* (1793), New York: Harper & Brothers.

Katz, S. T. (1978), "Language, Epistemology and Mysticism", in S. T. Katz (ed.), (1978), *Mysticism and Philosophical Analysis*, London: Sheldon Press.

Newton-Smith, W. H. (1986), *The Rationality of Science*, London: Routledge & Kegan Paul.

Peterson, M., Hasker, W., Reichenbach, B. and Basinger, D. (1991), *Reason and Religious Belief*, Oxford: Oxford University Press.

Plantinga, A. (1983), "Reason and Belief in God", in A. Plantinga and N. Wolterstorff (eds), (1983), *Faith and Rationality*, Notre Dame, IN: University of Notre Dame Press.

Proudfoot, W. (1985), *Religious Experience*, Berkeley, CA: University of California Press.

Ramsey, I. T. (1967), *Religious Language*, London: SCM Press.

Reardon, B. M. G. (1966), *Religious Thought in the Nineteenth Century*, Cambridge: Cambridge University Press.

Roberts, J. D. (1991), *A Philosophical Introduction to Theology*, London: SCM Press; Philadelphia, PA: Trinity Press International.

Russell, B. (1995), "Intuition", in R. Audi (ed.), (1995).

Schleiermacher, F. (1928), *The Christian Faith*, Edinburgh: T. & T. Clark.

Scott, P. and Spencer, C. (1998), *Psychology: A Contemporary Introduction*, Oxford: Blackwell.

Soskice, J. M. (1985), *Metaphor and Religious Language*, Oxford: Clarendon Press.

Swinburne, R. (1979), *The Existence of God*, Oxford: Clarendon Press.

Tillich, P. (1969), *Morality and Beyond* (1964), London: Fontana.

Watts, F. (2002), *Theology and Psychology*, Aldershot: Ashgate.

Weil, S. (1973), *Waiting for God*, New York: Harper & Row.

Westphal, M. (1997), "The Emergence of Modern Philosophy of Religion", in P. L. Quinn and C. Taliaferro (eds), (1999), *A Companion to Philosophy of Religion*, Oxford: Blackwell.

17 | Gathering Threads

The previous chapters are relevant threads to our understanding of both the nature and function of religious experience and whether religious experience is knowledge or not. What else can we draw on to highlight aspects of religious experience not yet considered, before outlining an account of the nature and structure of religious experience in the next chapter?

We need to turn to the sociology and phenomenology of religious experience, and the papers of the fourth conference on "Scientific Perspectives on Divine Action". The conference was concerned with how scientists understand the idea of God working in the world, and was held in June 1998, being co-sponsored by the Vatican Observatory and the Centre for Theology and Natural Sciences. We only need to gather some of the insights from these three sources, and then retrieve some salient points we have explored before for further reflection.

THE SOCIOLOGICAL CONTEXT OF RELIGIOUS EXPERIENCE

Sociology looks at individuals and their experiences in the light of their societies. Professor Margaret Poloma starts her review of the sociological context of religious experience by quoting Berger's reminder that society defines us, but is in turn defined by us. Berger draws attention to the fact that modern society is pluralist, so "all religious traditions tend to be undermined". The traditions therefore have three options:

1   deduction – assume the authority of the faith and deduce insights from the unchanging tradition;
2   reduction – try to reinterpret the tradition in secular terms;
3   induction – begin from retrieved experience from the tradition and contemporary human experience, and explore signals of transcendence in modern society. Berger does this, as we recorded in Chapter 9 (Poloma 1995: 164).

Besides Berger's account of five signals of transcendence, just mentioned above, we have also reviewed the two sociological accounts of religious experience of Glock and Stark and of Greeley in Chapter 10.

Poloma notes that Spickard, in his criteria of the sociology of religious experience, identifies five approaches to the topic: Jamesian, labelling, constructivism, learning and sharing time.

1 *Jamesian approach* – Spickard recognizes that the Jamesian approach is the most commonly used social scientific perspective on religious experience, but he is critical of it because it does not attend to the social origins of religious experience. He praises Poloma for her work on the Assemblies of God for recognizing the effect of religious experiences on a religious institution.

2 *Labelling Experiences* – Spickard criticizes Labelling Experience also because "although our vivid experiences are physiologically 'real', they are framed by labels that we attached to our experience to explain them" (Poloma 1995: 173).

3 *Constructivism* – While the Jamesian and Labelling approaches distinguish beliefs from religious experience, Spickard argues that the constructivists try to show how experiences and ideas or beliefs interact, using Proudfoot's (1985) work as an example. Constructivists recognize that we cannot accept religious experiences as just given. Spickard argues that we must ask about origins, which may well include beliefs in their formation (Poloma 1995: 173).

4 *Social Learning* – If the Jamesian approach regarded religious experiences as a unique kind of irreducible experience, and the Labelling approach is of a unique kind of experience,[1] this approach holds that you can learn to have a religious experience and produce a similar state of mind.

5 *Living in Shared Time* – Spickard, rather than accepting James's argument that religious experiences are private events, suggests that such experiences are primarily social. But this is an argument after the event, not a description of it, which is what James offers. It is true that when describing an initial experience the individual who had the religious experience uses "shared language". However, sharing is a consequence of the initial experience which readers and communities can access through their inclusive accounts of the nature of their religious traditions. This notion is outside the concerns of this study.

A Case for the Social Nature of the Jamesian Approach

Poloma strongly defends the approach of James on sociological grounds, in three crucial points.

1 James was trying to refute the reductionist bias in some of his contemporaries in the scientific community. He was also refuting the charge that religion was an anachronistic survival from the prescientific age (James 1928: 498–9).

2 While James noted that the private realities of an individual's religion may lead to egocentrism, that is better than a scientific view which took "no account of anything private at all", a telling anticipation of behaviourism, now a discarded methodology (Argyle 2000: 7; see below).

3   While "James may have overstated his case for 'private' religion, he did recognize that religious experiences had social consequences" (Poloma 1995: 175). Religious experiences produce "fruits for life". Conversion "should produce *Saintliness*, 'the collective name for the ripe fruits of religion' (James 1961: 220). Saintliness, in turn, increased charity or 'tenderness for fellow creatures'. Whether religious experience is public or private, people who have them live in a social world which they help to create and modify. Thus, at least indirectly, they would appear to have social consequences" (Poloma 1995: 175).

Consistent with this argument, Poloma, in her own work, always seeks to make clear the institutional and interpersonal social consequences of religious experiences. She is clear that identifying the social consequences is as important as identifying the social facts which help to promote them. For example, in her own work Poloma has found that religious experience is a most important predictor of well-being. She also found that those who prayed were more likely to forgive than those who did not, a finding confirmed in a follow-up study (Poloma 1995: 176).

Bridging Psychological and Sociological Approaches to Religious Experience: Poloma's Sociological Model

Poloma seeks to build links between small-scale sociology, the affect of the social context on how an individual expresses their religious experience, and large-scale sociology, the exploration of a religious culture from a sociological perspective. That is, she seeks to link the "agents or actors" and their experiences – which are defined by the actors and become "social facts" – to the "social structures". Different schools of sociology focus on either "social facts" or "social structures". Poloma recognizes that "religious experience and the social context in which it occurs are inseparable . . . Religious experience is [at least in part] dependent upon the community, particularly on the language of the community" (Poloma 1995: 177–8).

Sociologists who are concerned with "social structure" ask the question, Are religious experiences a normal mode of experience amongst the range or kinds of experience found in society? The answer is yes, as witnessed by the statistical evidence given earlier of the proportions of populations who report having had religious experiences. Other modes of experience include all forms of experiencing and knowing, like an artist experiencing or seeing a landscape as an impressionist, or a concert-goer enjoying sound experienced as music, or experiencing a baseball or football match.

Sociologists who are concerned with "social facts" focus on different groups who have religious experiences, whether such groups are classified on demographic criteria like sex, age and race, or on type of religious group. For example, research has shown that more women than men have reported having religious experiences, and more conservative Protestants have reported having religious experiences than Catholics or broad church Protestants (Poloma 1995: 178).

While the social context is an enabling context for religious experience, it is also the case that "meaningful religious experiences help to support the edifice of society."[2]

Our understanding of our religious experience, and our religious outlook on life, are likely to fall into the category of an extrinsic, intrinsic or quest mode of religious orientation. Those of us in a faith community need "quest" thinkers in the faith community who outline different ways of understanding, or expressing in words, a Faith Tradition.

For example, the active Christian community has included distinguished thinkers like Professor Braithwaite (1966), who largely accepted logical positivism as his philosophical framework and suggested that the Bible largely contained stories in order to feed and support our moral life. However, he worshipped in the Chapel of King's College, Cambridge, a little-known fact (Ward 2002: 202). Likewise, the non-cognitivist, T. R. Miles, whose views were discussed in Chapter 13, is a member of a Quaker community. Quest thinkers (as defined by Batson *et al.* 1993: 166) are found as members of other World Faiths, like the late Rabbi Hugo Gryn and the late Mr Kishan Singh Panesar (a Sikh member of the Ramgarhia Board Gurdwara in Leicester). There are living quest thinkers in all World Faiths. All quest thinkers seek to bridge the gulf between the positivist, secular or materialist ethos of society and reflective faith within and between Faith Communities. Berger's inductive approach seems to offer one valuable route for dialogue and convergence between World Faiths as well as dialogue and convergence between the denominations within World Faiths.

PHENOMENOLOGICAL PSYCHOLOGY AND RELIGIOUS EXPERIENCE

At the beginning of the twentieth century religious experience was a central subject in the psychology of religion. The best tools of enquiry available then were the written accounts used by James, and introspection – explored through questionnaires and written accounts. These methods were replaced when psychoanalysis gained influence, whose new conviction was that "virtually all significant psychological processes are unconscious and thus well out of reach of introspective awareness" (Wulff 1995: 183).

The next shift came with behaviourism, which would allow discussion only if based on observable, measurable data or information. Both "the depth psychologists and behaviorists agreed that the study of experience . . . is radically insufficient if not entirely futile in the conduct of psychological investigation" (Wulff 1995: 183). However, "there are no behaviourists anymore; they are extinct . . . Behaviourism was abandoned during the 'cognitive revolution' in psychology which recognized not only consciousness but also its contents – plans, rules, values, theories and explanations, and whole worlds of experience, such as morals, mathematics, and the study of science" (Argyle 2000: 7).

"Phenomenology", as we discovered in Chapter 9, is the study "from outside" or "from inside" of phenomena, or things, beliefs and/or persons as they are. In such study you initially suspend your own biases, prejudices and judgements and seek to describe and understand impartially what you observe, "standing in the shoes" of those you are describing. "Phenomenological psychology seeks to reclaim for experience the pre-eminent position it possessed at the end of the nineteenth century, but without returning to a naïve introspectionism" (Wulff 1995: 184). Phenomenology refers to "essentially descriptive approaches of experience" (184).

There are two main approaches of phenomenological psychology. In the first, the investigator draws on his own experience. The work of Otto is a classical exposition of this method, which we explored in Chapter 12. Otto analyses the highly complex feeling state that has a dual structure: the feeling and the object of the feeling. The feeling is pictured in terms of its object, the *mysterium,* which the rational mind calls the "wholly Other".

> On the one hand, it is the *mysterium tremendum,* an awesome and unapproachable object characterized by awefulness, majesty and energy. On the other hand, it is the *mysterium fascinans,* an alluring and fascinating object that is schematized by such notions as perfect love, mercy and salvation . . . It is an inborn potential best evoked by the personal example of others.[3]

In the second approach the investigator draws on the experience of others. The second approach is well illustrated by James's work, which we reviewed in Chapter 10. James "once confessed to having enough of a 'mystical germ' to respond sympathetically to the testimony of others . . . [but] that his constitution so limited his capacity to enjoy mystical states, that he was forced to 'speak of them only at second hand'" (Wulff 1995: 189, citing James 1902; 1995: 301). However, to speak with authority James turned to "cases where the religious spirit is unmistakable and extreme" (James 1928: 39) described by those who are "specialists" (1928: 486).

It has been said "that James was the first to undertake a strictly descriptive phenomenology of religious experience without the biasing effect of some theory or doctrine" (Wulff 1995: 190). Wulff celebrates the return of this method of exploring religious experience as of central importance to its future study.

FEMINIST THEORY REVISITED

The work of Neitz evokes further reflection (see ch. 15; Neitz 1995: 532). Carole Rayburn makes an opening contribution (1995: ch. 20). She raises the question of an embodied God, noting immediately that God has no body. However there are frequent references to parts of the body used poetically as analogies for different divine moods, attitudes or actions. God has ears to listen, nostrils to express anger, a mouth to pronounce punishment (477, citing 2 Samuel 22:7–16; Psalm 18:7–9). One reflects that God expresses many other messages, such as mercy, love or a divine call.

The topic of God and sexuality invites reflections on Genesis 1:27, and I use the scholarly translation of the New Revised Standard Version.

> So God created humankind in his image
> in the image of God he created them,
> male and female he created them.

I accept that this translation is an agreed removal of the original male Hebrew language used to refer to humankind, which was in its turn a product of a male-dominated society. The translators make it clear that the mandate they received said that "the churches have become sensitive to the danger of linguistic sexism arising from the inherent bias of the English language towards the masculine gender . . . in references to men and women, masculine-oriented language should be eliminated as far as this can be done" (NRSV: xiv).

But this is a feeble mandate, as the translators still describe God as a male three times in the passage above. The word "his" could so easily be replaced by "divine's", and "he" by "divine", "him" by "divine", but better to have answers from feminist theologians.[4] This provides another miniature example of knowledge formation. It would seem that the ecclesiastical powers decided that inclusive language would only be acceptable by the community at an intermediate stage, pending full acceptance of full inclusive language by the ecclesiastical community later, but I hope soon.

If God incorporates, metaphorically, all life-enhancing human characteristics and created "humankind in his image", we need to see where these life-enhancing characteristics have been distributed. One can turn to a psychotherapist like Dr John Gray, who has made a considerable study of male and female characteristics, tested on more than 25,000 participants in his relationship seminars. He has outlined ways in which there are male and female personality differences which the other sex needs to know about in order to improve their own marital and interpersonal relationships (Gray 1993: 4).

He suggests from his findings that men mistakenly expect women to *think,* communicate and react the way that they do, and that women mistakenly expect men to *feel,* communicate and respond the way women do. Major differences are that men mistakenly offer solutions to women and invalidate their feelings, when women only want to be listened to and their *feelings* to be understood, nothing more. They do not want solutions. Women mistakenly offer unsolicited advice and direction to men which makes them feel devalued and incompetent. Men when stressed withdraw into their "psychological cave". Women need to talk about their worries without being given a solution.

Gray maintains that men are motivated when they feel needed, women when they feel cherished. Gray outlines the different languages men and women use; the different ways they need and express intimacy. Men and women tend to offer each other the kind of love they themselves need, not the kind the opposite sex needs. Men need "a kind of loving that is trusting, accepting and appreciative". Women need "a kind of loving that is caring, understanding and respectful" (Gray 1993: 12).

In brief Gray has clarified that men's sense of self is based on power, competency, efficiency and achievement. They do things to prove themselves: their sense of self is defined by their ability to achieve results. Women value love, communication, beauty, relationships, in which, as women, they support, help and nurture each other. They experience fulfilment most through sharing and relating. In brief men fantasize about powerful cars and powerful achievement. Women seek fulfilment through sharing and relating. Their concerns are living in harmony, sharing in the community and developing loving relationships. Men are goal-oriented; women are relation-oriented.

If a study of differences between the sexes uncovers these differences, and a study of male and female clergy shows that the former have more female characteristic tendencies than the average male, and that female clergy have more male characteristics than the average female (Francis 1991: 1133–40), two insights may emerge. The first is that God shared some of the characteristics of Godself between males and females. The second insight can have two sides, one of which seems to be that some clergy have already begun the task of seeking to incorporate the characteristics of the other sex empathically. This points towards a goal of personal maturity which is the incorporation empathically of the human life-enhancing characteristics of the other sex, and the developing of both sets of qualities which enable and facilitate personal and communal flourishing.

This takes us back to the male and female goals suggested by Neitz, which are respectively "autonomy" and "connectedness with others", which gain considerable support from Gray's analysis.

"Autonomy" however, needs further discussion. Three factors have converged to weaken the traditional sense of community that used to be found in stable communities. If the eighteenth century was the age in which liberalism began to flourish, it also nurtured individualism, which seems to have been fostered at the expense of notions of community.

Autonomy reached a high point in educational philosophy when Dearden articulated the educational aims for many, stating that

> a valid aim of education should be personal autonomy based on reason, of which there are two aspects, one negative and one positive. The negative aspect is concerned with independence of authorities who prescribe what I believe, and who direct what I am to do. The positive aspects include testing the truth of things for myself a) by experience and b) by the critical estimate of the testimony of others. It also includes deliberately forming intentions and choosing what I shall do according to a scale of values which I myself appreciate. Both understanding and choice, or thought and action are to be independent of authority and based instead on reason. This is the ideal. (Dearden 1968: 46)

Within this aim there is tension between two concerns, one positive and one negative. The positive concern is that the fostering of an intrinsic orientation to life is the ideal: ideas, values and aspirations have to be "made one's own" to have the ring of personal integrity about them. These need to be socially regulated and accepted. The negative concern is that this could be taken as a manifesto for rampant individualism: indeed, it is a totally male statement, goal oriented, which Gray would recognize. Personal autonomy is not unbridled individualism. "What governs the autonomous person is not feeling but reason . . . To be morally responsibly I have to be basically autonomous, self governed . . . [and] give reasonable explanations and justifications of my actions . . . and engage in reasonable co-operation with my fellows."[5] John White's review of four critics of this liberal position highlights a change of emphasis within the liberal school away from personal choice, hinting that autonomy cannot be separated from family values, community cohesion and democratic choice (White 2003).

While "autonomy" is a bulwark against "socialization" and "indoctrination", the "four critics" support a female agenda concerning the development of relationships, teamwork and community, for feminists properly want to reconstruct a form of "autonomy" "that develops through relationships" (White 2003).

Secondly, the male focus on the car is an important symbol in another sense: the development of a mobile society. Is the weekend more focused on where the car can go than on how to develop community activities?

We have briefly noted the development of individualism since the eighteenth century, and how the educational espousal in the 1960s of "personal autonomy based on freedom" seemed to emphasize personal choice more than a sense of community.

If the community had too much control of individual freedom in the medieval period because of transport limitations, career limitations and social limitations, has the pendulum begun to swing back from the 1960s–1980s focus on individual freedom towards a sharing of this with a concern for family and community? Should not the feminist voice be heard less as a voice from one sex than as a holistic call to better relationships in home and community?

From an individual's perspective, a religious experience is about a relationship with God. Sociologically, it may be a means for the survival of *Homo sapiens* as Hardy affirmed. It should also be a catalyst for improving relationships, as some research programmes have indicated. Perhaps Neitz has highlighted a mature goal for individuals and society.

PSYCHOTHERAPY REVISITED

Three issues need further reflection and highlighting: the difficulty of interpreting an intense emotional experience, the conflict between bottom-up and top-down accounts of religious experience, and the important idea of "transitional space", which we discovered from object-relations theory in Chapter 14.

Interpreting Intense Emotional Experience

Since the disappearance of logical positivism there has been an increasing recognition that emotional knowledge is as important as rational or intellectual knowledge, and that they are in fact integrated into personal knowledge. In Chapter 15 we learnt from Watts and Williams, whose research work was in clinical psychology, that there are two ways of knowing. If you gain insights, whether from psychotherapy or religious knowledge, to which you can bring only a little experience, these insights are really just intellectual knowledge. Call it *half-knowledge* if you like. You only know it through your reason. You simply possess ideas, which you understand intellectually, but for which you "feel" very little.

But what if you have an intense emotional experience which arouses you to serious thought – what is this experience of? What do I do about it? Such an experience involves a serious emotional and intellectual struggle to understand it in the full sense. That is emotional knowledge + rational knowledge = personal knowledge = *full knowledge.*

This is the kind of knowledge one gains through psychotherapy in order to understand oneself fully – emotionally and intellectually – so that you then passionately want to do something about whatever the "problem" was. It is the same with religious experience say Watts and Williams. This insight adds greatly to our understanding of the importance and power of a religious experience to the individual who receives it. James was clear: the tests of the religious experience are the enabling power or delight it gives you and the fruit of it in your life.

There is a vast difference between these two ways of knowing. Rational or intellectual knowledge not based on emotion or experience is "half knowledge", although it is still important. Knowledge which is based on or related to emotions or emotional experience is "full knowledge".

Bottom-up and Top-down Interpretations of Emotional Experience

The American defence of the integrity of personal knowledge was underpinned by exploring in some detail in Chapter 3 the views on human nature and its irreducibility by one of two best-selling authors. This was followed by a theoretical analysis of the process of psychotherapy in the same chapter.

In Chapter 6 we were guided by Drusilla Scott and Richard Gelwick through some of Polanyi's ideas, of which a central insight was into the nature of the scientific process. Russell and others looked at science from the end backwards and came up with the *objective impersonal* white-coated model of observation, hypothesis, experiment, verification. The data proved the hypothesis. As you look back on the results, they appear inevitable.

Wallas, Polanyi, Vinacke and others looked at science from in front – into the foggy problem ahead, driven by an intuition. As you look forward to the results, they look unachievable. It is a bottom-up process. The act of discovery is an act of the creative imagination exploring reality – the real world. The start is a *subjective personal* problem, which is followed by hard work and failure, and ignoring the problem for a while as it simmers in the unconscious; then, illumination, verification and celebration!

Back to insights from clinical psychology. Watts and Williams are looking, from the 'bottom-up), at the problem of understanding an intense emotion experience and the process of slowly discovering what language best describes and interprets it: it is a creative process – a bottom-up process. If you look at the solution of the constructivist, they have a designer-ready top-down solution, which is screwed down onto the data to make sure it fits, rather like the objective view of science. Of course it fits if the view is from the end backwards to the problem. But that is not how it appeared to the mystic struggling to put their experience into words – bottom-up – or to the man who receives a religious experience out of the blue, trying to put it into words. Of course the words are chosen from the available vocabulary, but you are not predetermined to choose humanistic or religious vocabulary.

To argue that the experience is "constructed", as the constructivists do, is to confuse "nominalism" = "What I call it is what it is", with critical realism, which seeks to select the vocabulary, i.e. the concepts, which best describe and interpret the data or the experience being studied.[6]

Transitional Space

We were told, when discussing object relations theory in Chapter 14, that the infant has an object like a comfort blanket or a teddy bear, which they use to form an emotional bridge between the safe world with mother and the outside world without her. Pruyser used this idea to suggest that we all need a means of bridging the gap between the world centred on ourselves and the real world which one needs to understand. If the first world is the autistic world and the third world is the realistic world, the middle world was called the illusionist world, from "illusion" meaning "play".

Pruyser suggested that this middle world provided "transitional space" in which all creative thinking took place: scientists formulating theories, composers writing music, Shakespeare writing his plays, a patient seeing a psychotherapist to understand powerful disturbing emotions, or a person seeking meaning in a powerful emotional experience which only religious language described adequately.

This world of transitional space is where the meaning of what is real in the world is discovered through scientific theory, where meaning – discovered in life – is expressed, for example, in sound through a composer writing an opera, a pop song, a blues song or a sea song. This transitional space is filled with the literary form of myth – stories with profound religious meanings, models or theories in science, experiences which only make sense when regarded as a relationship with the Other some call God.

These pictures of science, of the literary world, of the musical world, of the ultimate meaning of the world are all reality depicting, i.e. they picture reality to make sense of it. This world of transitional space is the world we know and understand for it makes sense of all we find around us. Without it, all we find around us makes no sense. The world of meaning which inhabits transitional space is provisional, approximate and revisable, but is engaged with, and seeking to picture and understand, what is real.

All forms of knowledge, or modes of knowing, have links between what is perceived and how these perceptions are described. These links are intuitions of different kinds, each appropriate to its form of knowledge. The labels marked "intuitions" contain unfathomable mysteries, whether in registering the beauty of a scene, the love of another person, the meaning of chaos theory or the insights of a religious experience.

Intuitions are the "interpretation joints" which trigger off reason to give a full account of the perception in the chosen mode of discourse or form of knowledge. I think that these modes of discourse are "maps" of forms of knowing, and in the case of religious discourse or the religious form of knowledge, I believe these maps, symbols, stories, experiences and rituals are all signposts towards what is, and Who is real.

RELIGION AND NEUROSCIENCE

Neuroscience is a relatively new discipline. It is concerned to link particular mental activities with appropriate parts of the brain. A cluster of accounts of similar models illustrate the tentative kinds of exploration taking place.

First, however, it is important to start with how a human being is viewed. The predominant view of human nature up to the end of the twentieth century was formulated by Descartes in the eighteenth century. Simply, he said that a person consists of a mind and a body, a dualist view. This view is now being replaced by a cluster of related views of a person.

Views about the relationship between mind and brain have often gone to extremes. One extreme, associated with Descartes, and known as dualism, sees mind and body as radically different substances. At the other extreme there is physicalism, that reduces mind to brain in the sense of seeing mind as entirely explicable in terms of brain, and sometimes not quite as "real" as brain. An example is Frances Crick's view that we are each just "a bundle of neurones". An increasing number of people now see the need for a middle position, which is sometimes known as "non reductive physicalism". However the challenge is to find a good, detailed way of fleshing out that middle position. It is a technical problem in philosophy that has not been fully solved. However there are various ideas currently under discussion that help with this.[7]

The view called "non-reductive physicalism" (Watts 1999: 329) means "seeing the human being as a multilevel psychosomatic unity, who is both a logical organism and a responsible self" (Murphy 1999: 147). Professor Nancey Murphy outlines an account of how some mental processes are linked to the environment, some are stimulated by the environment, and other are independent of it (1999: 157).

So, first, there is a *structural coupling* between an organism and its environment (Clayton 1999: 195). On the one hand, "The organism cannot be decoupled from its environment without dying" (195). On the other hand, those links constitute what it means *to be* a fish, a monkey or a human. Experiences *linked* to the environment, like watching cows eating grass, can be expressed in empirical discourse. Some mental processes are *stimulated* by the environment evoking, for example, Darwin's reflections on his observations of natural life, leading to the theory of evolution, expressed in scientific (biological) discourse. Other mental processes in, for example, aesthetic, moral or religious discourse, are *interdependent* with the environment. So, living in the real world, a human sees a boy taking someone else's apple (empirical discourse) and calls it stealing (moral discourse).

Murphy argues that "*causal* reduction of the mental to the neurobiological *fails*" (1999: 147). She uses the philosophical term *supervenience* to explain. Her technical neurological examples are not easy to understand. This one is: "the *supervenience* of moral on nonmoral properties involves what we may call 'conceptual supervenience'". (151). But note we have already explored this concept without knowledge of it! Moral discourse *supervenes* empirical discourse, personal discourse supervenes moral discourse, religious discourse supervenes all modes of discourse. In each case the "higher" mode of discourse "includes" the lower forms of discourse.

We all know that when you see a boy take an apple without paying for it, and say he is dishonest, moral discourse does not change empirical discourse, but *supervenes*. Likewise, this illustrates another concept used in this neurobiological argument, the use of *downward causation*. If a boy picks up an apple in the orchard to help you collect them, you are grateful. If a boy takes an apple off a fruit stall without paying for it, your moral principle (honesty) exercises *downward causation* on the situation – you may ask him if he *ought* to pay for it. We will discuss *downward causation* again later.

A person is able to effect downward causation by supervening or imposing a higher level of knowing on a lower level of knowing. For example a religious call can influence the choice of a career; a moral principle can change business practice; a musical composer imposes aesthetic taste on a sequence of notes (music superimposed onto noises or sound frequencies). In each example, a human *purpose* (not a mechanical cause), is superimposed on physical actions through a mental process.

One more principle is highlighted by Murphy: *the emergentist principle.* "Biological evolution . . . encounters laws, operating as selective systems, which are not described by the laws of physics and inorganic chemistry, and which will not be described by the future substitutes for the present approximations of physics and inorganic chemistry" (Murphy 1999: 156–7). This is a refutation of *reductionism.*

We can return to an example of *emergentism* we have already described: Piaget's account of the development of stages of thinking in Chapter 4. The child emerges from the sensorimotor stage to the preconceptual stage, and then continues on through the concrete operational stage to the stage of formal operational thinking. But in each *emergent* stage *downward causation* is involved. A simple example: a very young child has to *learn* the concept of the conservation of matter (it doesn't disappear when out of sight): *thought is internalized action,* Piaget has taught us. A baby drops or pushes a rattle out of reach. You cover it with a handkerchief. It no longer exists for the child. The child turns to something else in sight. Keep doing it! One day the child will pull the handkerchief away, uncovering the rattle, and give you a triumphant smile! *Emergentism* and *downward causation* are frequently interconnected. The rattle does not exist until the child *causes* the handkerchief to be removed!

So how does "a brain with a structure like ours . . . and responds to electrochemical stimulation and viruses . . . produce religious ideas and religious experiences as ours does"? (Clayton 1999: 182)

Professor Philip Clayton lists what neuroscientists know, and in putting the pieces together recognizes that physical sciences like neuroscience "tend to push one in the direction of *physicalism,* the view that all things that exist are physical" (1999: 184). He recognizes that this is a *methodological procedure,* but also an *ontological* and therefore a *metaphysical* statement. In other words, to say that "all things are only physical" is a metaphysical statement. That means it is the foundation assumption of all of that person's thought.

Clayton rejects the two poles of the argument about the nature of a human being – *dualism* and *physicalism* – which he calls *metaphysical dualism* and *strong reductionism* (1999: 191). Clayton then distinguishes between *intentions* or purposes, which are the desires of organic species, especially of human beings, and *causes,* which are about *physical* relationships, "the bread and butter of the physical sciences" (191).

Clayton then turns to information biology and virtual reality. A key insight he draws from these fields of study is this:

> The causal line seems to move "up" from the physical inputs and the environment to the mental level, then along the line of mental causation – the influence of one thought on another – then "down" again to the influence of other physical actions, to make new records and synaptic connections (the place where nerve cells join) within the brain to produce new verbal behaviors and so forth. (Clayton 1999: 196)

He continues:

> this view is monist, not dualist: there is only one physical system, and no energy is introduced into that system by some spiritual substance external to it. At the same time, it seems,

subsequent states of the entire system cannot be specified without reference to the causal influence exercised by the higher-level phenomena. (Clayton 1999: 196)

Clayton moves towards a theory of the Person, taking account of both the physical inputs into the brain – physical processes – but also "the emergent level of thought" (1999: 197). He finds no difficulty in describing a *person* in terms we all accept, as "one who is able to enter into human social interaction". "Personhood is therefore a level of analysis that has no complete translation into the state of the body or the brain" (197). He now explores *emergentist supervenience* and *emergentist monism*, noting that this begins in chemistry and works up through the evolution of sentient organisms (213).

Thus, "nonreductive physicalism" is not a reductionist view like that of Dawkins. It is a broad view of human nature with complementary levels of description, as we saw in Harry Williams's discussion of his friend "Fred" in Chapter 6, to which is now added a level of description from neurology – i.e. a brain-level view.

These new views of human beings are holistic or unitary views of human nature focused on physical nature, even though explored at complementary levels of understanding, which also indicates the kinds of upward and downward causation that can operate in human life. They replace Descartes' dualistic view. A human being is a psychosomatic whole who can employ downward causation, but can be understood in different modes of knowledge, which in their complementary way take account of both causes of physical effects and reasons for human action. This new cluster of similar views are individually called by different names, like 'nonreductive physicalism' or 'emergent monism' (Barbour 1999: 278).

These views would be acceptable to theologians like Karl Rahner, whom Fraser Watts quotes as saying, "We are transcendent creatures, but this transcendence arises out of the natural world that is God's creation; there is no opposition between our transcendence and our naturalness" (Watts 1999: 330). These are different levels of description in different modes of discourse which are complementary, e.g. theological discourse (God talk) refers to transcendence (Ellis 1999: 449–74), while empirical and scientific discourses refer to our natural world, though each mode of discourse includes those lower in the hierarchy of the forms of knowledge (see figure 6.1 in Chapter 6).

Watts emphasizes that "complementary descriptions are paralleled by complementary explanations" (1999: 330) and continues, "the basic solution to this problem that I favour is 'perspectivism': regarding theological and naturalistic perspectives as different in character, but complementary to one another" (1999: 331). We have already noted in Chapters 4 and 6 that different modes of discourse address the different kinds of understanding which form the different levels of understanding in Watts's 'perspectivism', a term I endorse.

Reductionism and Religious Experience

We now have three forms of reductionism focusing on religious experience, i.e. the attempt to say that, for example, "religious experience is *nothing but* a reflection of human needs", Freud's view we explored earlier. We have also explored the most vigorous current critique, that from social constructivists. A new reductionist critique from neuropsychology attempts to explain religious experience in terms of the parts of the brain which "hosts" the activity and brain function.

Each form of reductionism involves a *genetic fallacy,* i.e. it confuses contributory factors with the *nature* of a human activity or experience and also the different modes of discourse to which each is related (Watts 2002: 78).

Religious Experience, Temporal Lobes and Epilepsy

There is a suggestion of a relationship between religion, the temporal lobes of the brain and epilepsy. The press has talked about the "God spot" in the brain. However, Watts, who formerly did research work in clinical psychology, points out that the suggested link between religion and epilepsy is uncertain for five reasons. First, one has to distinguish between religious experiences which take place *during* a seizure (an epileptic fit), and religious experiences which take place when the person is normal. Secondly, religious experiences which take place during a seizure are unlike those normally recorded, as the former are "associated with an element of anxiety", and the latter have "a positive emotional tone, even when people have been severely stressed before the experience". Thirdly, patients who have a religious experience during a seizure later describe what they have experienced as "no more real than, say, a dream", whereas people who have powerful religious experiences are generally convinced of the reality of what they are experiencing. Fourthly, those who experience the Presence of God do not report the weird quality of a person who has had a seizure. Fifthly, "other studies that have controlled for the brain damage, psychiatric illness, etc., also have not found a link between religious experiences and TLE (Temporal Lobe Epilepsy)".

Watts concludes: "At the present time (1998), there seem to be no compelling scientific reasons for linking the neural basis of religion and TLE" (Watts 1999: 334). However, as Watts makes clear, this specific argument does not rule out the fact that there will in the future be a theory which in time does offer an account of which parts of the brain are involved in religious experiences. It is likely that this will be accepted by most of the community, just as the theory of evolution is so accepted.

NEUROPSYCHOLOGY AND RELIGIOUS EXPERIENCE

Professor Eugene d'Aquili and his associates have proposed the first tentative model, that the frontal lobes of the brain have links with religious experience. It has yet to be tested and accepted or rejected by the community of neuropsychologists. The frontal lobes of the brain also process other self-conscious and high-level human activities, like recognizing the mental powers of another person, "processing emotional experience, and the capacity for empathy and moral insight" (Watts 2002: 82). He suggests that there are seven operating regions of the brain, which he calls "cognitive operators". We are interested in two of them, a Causal Operator and a Holistic (integrating) Operator.

"The Causal Operator" in the brain is concerned to trace the origin of any "strip of reality", e.g. What causes the sun to rise? What causes water to boil? What caused the Universe to come into existence? What causes you to fall in love? What causes you to have a religious experience?

> In essence the causal operator performs its functions on any given strip of reality in the same way that a mathematical operator functions. It organizes that strip of reality into what is subjectively perceived as causal sequences back to the initial terminus of that strip. In view of the apparently universal trait . . . of positing causes for any given strip of reality, we postulate that if the initial terminus is not given by sense data, the causal operator generates *automatically* [my italics] an initial terminus . . . Science refuses to postulate an initial terminus or first cause of any strip of reality. (d'Aquili and Newberg 1998: 77)

In contrast however, d'Aquili postulates that "under more usual (non-scientific) conditions the causal operator generates the initial terminus or first cause of any strip of reality" (d'Aquili 1998: 77). So from a mental construct in a person's memory, like 'God', d'Aquili says, "We are proposing that gods, powers, spirits, or what we have come to call personalized power sources . . . are automatically generated by the causal operator." Thus, "when the strip of reality to be analysed is the totality of the universe, then the initial terminus or first cause which is automatically produced by the causal operator is Aristotle's *First Mover Unmoved*" (78). This d'Aguili calls the control aspect of religion, allowing humans to have a sense of control over the environment through relating to a personal power source, God. How does d'Aquili accept that science does *not* have a Causal Operator, but religion has an *automatic* Causal Operator?

The second centre of operations d'Aquili calls the Holistic Operator, (d'Aquili and Newberg: 1998: 84) which processes two aspects of religion. The first is ceremonial ritual and individual meditation, which both create a sense of wholeness over the "multiplicity of baseline reality" (84). Secondly, this Operator also makes possible a perception of wholeness, which "depending on its intensity, can be experienced as beauty, romantic love, numinosity or religious awe" (86). D'Aquili outlines five important kinds of experience here: aesthetic experience, personal experience in the form of romantic love, extrovert mystical experience and introvert mystical experience (cf. Stace in Chapter 12), and "the state of religious exaltation which Bucke has called Cosmic Consciousness" (d'Aquili and Newberg 1998: 87; Bucke 1969). The last three states monitored by the Holistic Operator, d'Aguili suggests, are not about "control", the role of the Causal Operator, but self-transformation through altered states of consciousness in the form of surrender to an absolute reality. The last of the three forms mentioned is described as mystical union with Absolute Unitary Being (AUB), which is also described in the literature of the world's great religions concerning religious experience and mysticism.

Watts comments on this whole theory. He says that d'Aquili introduces the terms Causal Operator and Holistic Operator, which are not generally accepted terms in neuroscience, not least because "there is no generally accepted comprehensive theory of human cognitive architecture and its neural substrate". However, he sympathizes with d'Aquili that in grappling with high-level functions, he needs to make up his own comprehensive theory (Watts 2002: 84). However, he is not convinced that one Causal Operator can account for all types of causes, least of all the more complex causal problems. Watts is more impressed with the Holistic Operator in relation to ritual cere-mony, e.g. religious services, and personal meditation. Watts notes that a more serious limitation of d'Aquili's theory "is that mystical experience may not be central to reli-gious beliefs and practices". Secondly, "the ascription of causality to God, and the accompanying sense of control over the world" are features of primitive religion rather than, for example, contemporary Christianity. However, Watts recognizes that d'Aquili and his colleagues have evolved over twenty years a plausible theory of the neural basis of mystical experience, the best theory at present available, though it is early days to suggest that we have "the definitive scientific account" (Watts 2002: 85).

My own comment focuses on the need to distinguish between the scientific modes of discourse and knowledge used appropriately in working out neuropsychological accounts of the brain base of, on the one hand, aesthetic experience, romantic love, mystical extrinsic and intrinsic and cosmic union experiences leading to personal transformation, and, on the other hand, the fact that *persons* can make very clear distinctions between different forms of discourse. Scientific accounts, when reduc-tionist, reduce the boundaries of other human forms of knowledge by dismissing them. Human understanding includes as central the appropriate personal forms of knowl-edge, like aesthetic, personal, moral and religious knowledge, which lie at the heart of individual and community life, for human beings are also spiritual beings.

PHYSICAL NETWORKS AND RELIGIOUS EXPERIENCE

Peacocke opens his paper on "How Does God Communicate with Humanity?" with the story of Elijah waiting for a message from God, which did not come through a great wind, earthquake or the fire of a volcano or burning bush, but through "a sound of sheer silence" (1999: 215). Peacocke acknowledges and endorses the move to replace Descartes' dualistic view of human beings – as mind and soul – with a holistic or unitary view of human nature.[8]

To support his unitary view of human nature, Peacocke quotes Philip Clayton, who outlines the mind–brain–body relation to personhood:

> We have thoughts, wishes and desires that together constitute our character. We express these mental states through our bodies, which are simultaneously our organs of perception and our means of affecting other things and persons in the world . . . Embodiment . . . [is] the precondition for perception and action, moral agency, community and freedom – all aspects that philosophers take as indispensable to human personhood and theologians have viewed as part of the *imago dei* (image of God). (Peacocke 1999: 231)

In other words, if I want to buy a scarf (thought), I have to walk to the shop (using senses to find the way and the body to get there) and then ask for a scarf, please (mouth + idea of scarf + language to express idea of wanting it). I will not try to steal it, but pay for it (moral agency). I will smile and make conversation about the weather (community relations). Then I decide I have time to go elsewhere (freedom). All the people I meet, whether I like them or not, I recognize as made in "the image of God". In this example, a person's action involves different levels of operation, e.g. physical, biochemical and psychical, and complementary modes of knowledge to achieve a holistic understanding, e.g. scientific, personal, moral and philosophical under-standing, which in this case is religious. The philosophical response of an agnostic could well be that of an ethical humanist.

Many accept that God created the world in the beginning and continues to create it through processes open to scientific investigation. Peacocke, and modern scientists generally, accept that relationships in the world are caused through bottom-up, top-down or whole-part causation. Fewer think now that one cause can be identified. Increasingly, causation is thought of in terms of networks of complicated interactive systems whose causal routes can be traced, with difficulty. Peacocke assumes that God works in and through these systems: there is no direct divine interventions which disrupt them. The term for this view is panentheism (God-in-all-things-ism).

But God, though ineffable – too great for description in words – and ultimately unknowable in terms of what Godself is like, "is at least personal, and personal language attributed to God is less misleading than saying nothing!" (Peacocke 1999: 237).

How can God communicate with human beings? Interpersonal relationships occur as the result of the interaction of mental and physical networks in harmony. "This suggests that religious experience that is *mediated* through sensory experience is intel-ligible in the same terms as that of the interpersonal experience of human beings" (Peacocke 1999: 243).

"What about those forms of religious experience which are *unmediated* through sense experience?" (Peacocke 1999: 243).

I note that we already have different ways of communicating. Before we had speech we had tasting, encouraging or discouraging smells, eye contact, making pleasant or unpleasant noises, gentle or aggressive touch. Then we developed speech, an evolved mental-physical sense. When you are at a party, you may suddenly feel that someone behind your back is looking at you. You turn round quickly and she or he is! Is that the sixth sense we have mentioned?

If you are aware of a sense of the numinous, or a sense of the Presence of God, is not this a kind of sense perception? If our means of communicating evolve in more complex ways, might this not be an evolved form of communication? Is this the sev-enth kind? If you are in a quiet mode of "waiting upon God", moving into an unmediated unitary experience of mystical union, "all of this can be mediated through patterns in the constituent world including brain patterns. Experience of God indeed often seems to be ineffable, incapable of description in terms of any other known experiences or by means of any accessible metaphors or analogies" (Peacocke 1999: 244).

Comments

We have been looking at neuroscientists in dialogue with theologians, and some of the participants are both. They have been exploring the middle way between dual-ism and physicalism. Their contribution has been enriched through the introduction of the concepts of *nonreductive physicalism, supervenience, downward causation* (top-down causation, sometimes by sentient purpose, e.g. of humans), *the emergentist principle* (= bottom-up causation/evolution), *perspectivism* and *emergent monism*.

Intentionality (= human purpose) has always been present. We need to attend to this helpful movement.

The really amazing mystery is not whether there are or are not religious and mystical experiences. The evidence we have explored and collected makes the "yes" verdict irrefutable. The real mystery is how to choose the appropriate "intuition", the "bridge" or "connector" between any perception or experience and its appropriate mode of discourse. Here are two examples I have already mentioned.

When I was in a classroom with twelve-year-old pupils, and we were talking about Genesis 3:8, "God walking in the garden in the cool of the day" I asked my "serious" question, "Was God wearing evening dress or jeans and a tee shirt?" Why did they all roar with laughter at my question? When I asked 16–17 year-old students to sort out the groups of words mentioned in Chapter 2, table 2.1, why did they have no difficulty at all in sorting out the principles of classification?

How do people's minds/brains know which "intuitive connector" to select to connect the right perception to the appropriate mode of discourse, whether the experience is empirical, scientific, aesthetic, moral, mythological, personal or religious? The twelve-year-olds were able to search for words to indicate that my question was in the wrong mode of discourse (empirical) when we were discussing a story (mythological discourse) which tries to picture aspects of the relationship between God and humankind. They knew it was not a historical account of a meeting (empirical discourse), as I was assuming with my question.

A second point needs to be emphasized, one that was expressed by some of those involved in the discussion of neuroscience and the Person. The physicalists are right in *one* respect: if you are discussing, for example, music, morals, relationships or religious experience, "*there is only one physical system, and no energy is introduced into that system by some spiritual substance external to it*" (Clayton 1999: 182).

Is it the case then that all our most important experiences, as well as all our thinking, take place in transitional space, and that our subsequent observable behaviour is a result of them?

Earlier we asked a question which has not yet been answered: So how does 'a brain with a structure like ours . . . which responds to electrochemical stimulation and viruses . . . produce religious ideas and religious experiences as ours does?' We will try to explore this in the final chapters.

NOTES

1   The Labelling approach oddly calls this "an odd brain state".
2   Poloma (1995: 179). For an account of the social sources of personal religion, its nature and personal and social consequences, consult Batson *et al.* (1993).
3   Wulff (1995 186); for a fuller account see Wulff (1991: 528–32, 573–4).
4   "Godself" is a clumsy word, though better than he, him and his.
5   Bailey (1984: 237); "fellows", like "guys", in contemporary *use* include both sexes.
6   Kaufman (1993) is not "constructivist" in this sense. He offers a contemporary reconceptualization of Christian theology.
7   Personal communication from Dr Fraser Watts.
8   Earlier we called this view "non-reductive physicalism", but Peacocke calls it "emergent monism". Both terms mean a unitary view of human nature.

REFERENCES

Argyle, M. (2000), *Psychology and Religion*, London and New York: Routledge.

Bailey, C. (1984), *Beyond the Present and the Particular*, London and Boston: Routledge & Kegan Paul.

Barbour, I. (1999), "Neuroscience, Artificial Intelligence and Human Life: Theological and Philosophical Reflections", in Russell *et al.* (1999).

Batson, C. D., Schoenrade, P. and Ventis, W. L. (1993), *Religion and the Individual*, Oxford and New York: Oxford University Press.

Braithwaite, R. (1955), "An Empiricist's View of the Nature of Religious Belief ", in I. T. Ramsey (ed.), (1966), *Christian Ethics and Contemporary Philosophy*, London: SCM Press.

Bucke, R. M. (1969), *Cosmic Consciousness: A Study in the Evolution of the Human Mind* (1901), New York: Dutton.

Clayton, P. (1999), "Neuroscience, the Person and God", in R. J. Russell *et al.* (eds), (1999).

D'Aquili, E. G. and Newberg, A. B. (1998), "The Neuropsychology of Religion" (1996), in R. Watts (ed.), (1998), *Science Meets Faith*, London: SPCK.

Dearden, R. F. (1968), *The Philosophy of Primary Education*, London: Routledge & Kegan Paul.

Ellis, G. F. R. (1999), "Intimations of Transcendence: Relations of the Mind and God", in Russell *et al.* (eds), (1999).

Francis, L. J. (1991), "The Personality Characteristics of Anglican Ordinands: Feminine Men and Masculine Women?", *Personality and Individual Differences* 12, 1133–40.

Gray, J. (1993), *Men Are from Mars and Women Are from Venus*, London: Thorsons.

Hood Jr, R. W. (ed.), (1995), *Handbook of Religious Experience*, Birmingham, AL: Religious Education Press.

James, W. (1928), *The Varieties of Religious Experience* (1902), New York and London: Longmans Green & Co.

James, W. (1968), *The Varieties of Religious Experience* (1902), London/Glasgow: Fontana.

James, W. (1985), *The Varieties of Religious Experience* (1902), Cambridge MA: Harvard University Press.

Kaufman, G. D. (1993), *The Face of Mystery*, Cambridge, MA: Harvard University Press.

Murphy, N. (1999), "Supervenience and the Downward Efficacy of the Mental: A Nonreductive Physicalist Account of Human Nature", in R. J. Russell *et al.* (eds), (1999).

Neitz, M. J. (1995), "Feminist Theory and Religious Experience", in R. W. Hood Jr. (ed.), (1995), ch. 22.

Peacocke, A. (1999), "The Sound of Sheer Silence: How Does God Communicate with Humanity?", in R. J. Russell *et al.* (eds), (1999).

Poloma, M. M. (1995), "The Sociological Context of Religious Experience" in R. W. Hood Jr. (ed.), (1995), ch. 8.

Proudfoot, W. (1985), *Religious Experience*, Berkeley, Los Angeles, London: University of California Press.

Rayburn, C. A. (1995), "The Body in Religious Experience", in R.W. Hood Jr. (ed.), (1995), ch. 20.

Russell, R. J., Murphy, N., Meyering, T. C. and Arbib, M. A. (eds), (1999), *Neuroscience and the Person*, Vatican City State: Vatican Observatory Publications and Berkeley, CA: Center for Theology and Natural Sciences; University of Notre Dame Press.

Ward, K. (2002), *God: A Guide for the Perplexed*, Oxford: One World.

Watts, F. (1999), "Cognitive Neuroscience and Religious Consciousness", in R. J. Russell *et al.* (eds), (1999).

Watts, F. (2002), *Theology and Psychology*, Aldershot: Ashgate.

White, J. (2003), "Five Critical Stances towards Liberal Philosophy of Education in Britain", *Journal of Philosophy of Education* 37 (1), pp. 147–184, Blackwell Publishing.

Wulff, D. M. (1991), *Psychology of Religion: Classic and Contemporary Views*, New York: Wiley.

Wulff, D. M. (1995), "Phenomenological Psychology and Religious Experience", in R. W. Hood Jr. (ed.), (1995), ch. 9.

18 The Wallas Models of Religious Experience in Context

BATSON'S MODEL: RELIGIOUS EXPERIENCE AND PERSONAL TRANSFORMATION

In Chapter 15 we recorded the fact that one way of studying some kinds of religious experience can be based on a model of creative thinking, which Wallas outlined in 1926. The four stages in his model are:

✧ *preparation* – the unsuccessful attempt to solve a problem by using old ways of thinking;
✧ *incubation* – give up the attempt to solve the problem, so it sinks into the unconscious;
✧ *illumination* – the sudden emergence of a new way of looking and a new structure of thinking which reorganizes the elements in the 'problem', leading to a new solution;
✧ *verification* – test the value of the new solution: does it work?

In 1971 Batson applied Wallas' 1926 model successfully to religious experience and religious conversion (Batson *et al.* 1993, ch. 4).

Commenting on the four stages of Wallas' creative sequence, Professor Daniel Batson describes how the process may have a physiological base in the brain (Batson *et al.* 1993, 99–100). He concludes, "During the preparation stage of the creative process, the individual attempts to deal with the problem in terms of the logical, linear thought of the dominant left hemisphere. During the incubation stage, active processing by the left hemisphere is relaxed, permitting the less dominant, perceptually orientated right hemisphere to go to work reorganising the cognitive structures. This reorganization leads to the new insight or illumination. Verification involves a return to left hemisphere dominance; the individual logically tests the functional value of the new insight" (101). Batson notes that at the time of writing "the application of research on hemisphere specificity in creativity is, at present, highly speculative" (101). Batson recognizes that this four-stage sequence in the creative process can be found in many religious experiences, and resolves a state of unease and its solution in many religious experiences identified by William James, himself, and others.

Batson suggested that the stages of the sequence can be interpreted as:

✧ *existential crisis:* e.g. "What is the meaning of my life? How do I deal with the fact that I am going to die?" (1993: 103).

✧ *self surrender:* "Trying and failing to regain existential meaning within one's existing reality, one is driven to a point of despair and hopelessness" (104).

✧ *new vision:* "Into this 'dark night of the soul' may blaze the light of new vision" (104).

✧ *new life:* "If the new vision . . . is to be effective . . . the individual must live the vision. A new life must follow" (105).

Since 1971, Batson has devoted over twenty years to the development of our understanding of religious experience outlined in his publications. Summarising this four-stage model he says, *"religious experience involves cognitive restructuring in an attempt to deal with one or more existential questions"* (Batson *et al.* 1993: 106). He shows how the creative analogy of his model has been validated by convincing evidence that new religious vision often follows an existential crisis. From the evidence he collected, he has shown that this model successfully provides a psychological model for "what happens psychologically during dramatic, reality-transforming religious experiences" (115).

This model is well established, for it has evidential support, widespread acceptance, and is well documented in his publications listed in his books.

MILES' MODEL: RELIGIOUS EXPERIENCE – PERSONAL ENRICHMENT OR PERSONAL TRANSFORMATION

Miles (1983) began his small research project concerning religious experience with sixth form pupils (aged 17–19 years), in the classroom, in 1977, with his preformed conceptual framework based on Vinacke's use of Wallas' model (Vinacke 1952: 243–4). Miles focused on the independent elements coordinated in a religious experience, which he described as "a collage", but he unknowingly followed in the steps of Batson who had successfully focused on the holistic concepts of his analogical use of Wallas' model outlined above. Miles' conceptual framework of analysis was independently formed in 1975–6, and for that reason, many of the results of his research unconsciously formed *independent support* for Batson's work, which Miles read on microfilm when writing up his results in 1982.

These two models are complementary, not contradictory, as they explore different aspects of Wallas' model, already mentioned, as indicated by the fact that Miles' results support those of Batson. As Miles' work has not previously been published, a few aspects of it will be briefly outlined here, to illustrate Miles' focus on the analysis of distinctive constituent elements in religious experience and in spiritual psychic functions (Miles 1983, ch. 11).

Kinds of Religious Experience

We have seen in Chapters 1, 10, 11 and 12 that transcendental experiences are generally of two kinds. The first kind consists of those which are described as experiences of the transcendent, the divine (James 1968; Otto 1959; Proudfoot 1985; Hick 1989; Smart 1997). The second kind consists of those described as experiences of self-transcendence. These are often in the form of naturalistic mystical union with nature, humanistic personal union in love with a partner, self-transcendence experienced as true friendship within a humanistic outlook, or perhaps even as humour (Laski 1961; Maslow 1970; Berger 1971). Of course these three forms of personal relationships can be interpreted as gifts from God and held within a religious outlook on life.

For the moment, we will confine our interest to two forms of religious experience, both bipolar, i.e. an experience, usually alone, of an individual in the presence of the Other, or a conversion experience usually associated with a religious community. This may occur in a gathering in a place of worship, or afterwards possibly alone following a period of deep, though sometimes intermittent, reflection. The third form of experience of interest to us in this study is religious mysticism, i.e. personal union with the Divine Presence. Sometimes such mystical experiences in some Eastern religions may be described as experiences of Non-Self, the Void, Nirvana or as union with the Universal Self (Happold 1975).

We noted earlier how James and others have found that sometimes the bipolar relationship can develop into mystical union.

The Roots of Religious Experience

It is quite clear from the literature that a whole range of events, emotions, incidents or associations may mark the genesis of such an experience. Some have argued for one root or cause, perhaps "total dependence", a sense of the numinous, or "a surprise by joy".[1] Others have argued for "guilt", a "religious instinct", "inborn temperament" or "reason" amongst other suggestions for a single root or cause.

The other school of thought is that there are many roots or causes, which may include those mentioned and others. Sometimes qualities of mind are emphasized, like the reason, emotion or will. Others emphasize personal needs – security, recognition, new experience, responsive relationships, perhaps personal aspiration to embody one or more of the higher values, or to engage in the search for the meaning of life. No attempt is made here to summarize all possible roots (Clark 1958: ch. 4; Thouless 1971: ch. 3), rather the central point needs to be made that there are many and various kinds, and different clusters may be important to different persons.

Triggers of Religious Experience

Various stimuli which evoked religious experiences are listed in Greeley's 1975 work from a survey of 1,467 people in the United States. They range from 49 per cent who cited music down to 15 per cent of respondents who cited looking at a painting. Then, in order of descending importance on this scale, were prayer, the beauties of nature, such as sunsets, moments of quiet reflection, attending services, listening to sermons, watching little children, reading the Bible, being alone in church, reading a poem or novel, childbirth, sexual activity and creative activity.[2]

Hardy in the UK, reviewing the analysis of the first 3,000 reports of religious experiences,[3] noted that the fifteen main triggers of these experiences were depression (18 per cent), down to solitude (1.5 per cent). In between, in descending order, were prayer and meditation, natural beauty, sharing in religious worship, literature-drama-film, illness, music, crisis in personal relationships, the death of others, sacred places, visual art, creative work, relaxation, the prospect of death and solitude. Other triggers mentioned included happiness, childbirth and sexual relations.[4]

The Elements of Religious Experience

A wide range of objects, incidents or associations may "trigger off" a religious experience. These triggers are received by the individual through "perception" (discussed in Chapter 2). Our senses are the organs of perception: they give us information of the external world and are the means through which the trigger-perceptions evoke the response of the subject. Sometimes these triggers came from within – guilt, joy, curiosity, cognitive dissonance – i.e. something you are thinking about "doesn't fit" into your scheme of things. The elements listed below may "appear" in any order or all together: every person is different.

The first element in such an experience is "emotion". Some argue that a particular emotion, like "wonder" or "awe", is central (Wilson 1971). Our present concern is simply to register the fact that emotion is a central element – emotions of all kinds.

The second element is "sensation". Sensations are different from emotions. Emotions usually have an object or focus of attention but no bodily location, whereas "sensation" refers to location. People can describe where they feel pain, or what sensations they have after being betrayed or falling in love. Emotions like love, awe, joy, fear, anger, regret, guilt and hatred are focused on an object or cause of concern.

The third element is "mood". Moods can also be distinguished from "emotions". Moods or "states of mind" have causes, as do emotions, but they do not have an identical object or even a network of unidentifiable causes. Moods are the states of mind in which we find ourselves after something has happened: they are not sources of energy. Emotions evoked, like awe, anger, love, guilt, can become sources of energy which frequently motivate a person to follow a course of action which may be long term or short term.

The fourth element is the cognitive element: it refers to the concepts and modes of discourse used to describe and interpret the experience. In describing these experiences, we need to remember that we often recognize that our descriptions or discussions are often at the edge or limits of language, so that symbol, analogy, metaphor, paradox and myth are among the oblique forms of linguistic usage which may be employed. Philosophers help to clarify these issues for us by exploring how these language uses convey meaning.

The fifth element, which overlaps in different ways with two elements already mentioned, is "attitude". An attitude might be, for example, "hostile", "friendly", "cool", "distant", "familiar" or "respectful". Attitudes are linked to emotions, but emotions are more primitive and less public. One may be angry without adopting a hostile attitude, or feel afraid and yet strike an attitude of bravery or indifference. Attitudes are shown, while emotions need not be. Basically, an attitude is a learned disposition to respond in a consistently favourable or unfavourable manner with respect to a given object or cause of concern.

Attitude is often thought to have three main components: the cognitive component which has to do with beliefs about an object or cause of concern; the affective component which has to do with feelings; and the "action-tendency" component which refers to a person's "intention" or "readiness to behave in a particular way" (Halloran 1967; Fishbein and Ajzen 1975). Feelings are sometimes equated with emotions, though strictly they can be distinguished. Emotions are primitive, as has been said, while feelings relate to emotions tempered by thought.

These five elements are characteristically found in transcendental extrovert or mystical experience. The cognitive element involved in the interpretation may operate at any of the eleven "levels" of interpretation previously noted, and use all the skills discussed in the process of interpretation of an experience, called hermeneutics, described earlier. The cognitive element is linked to the other elements, and the interaction is facilitated by the sixth element – the creative imagination.

The creative imagination is frequently mentioned in the literature (Vinacke 1952: 243–4; Barnes 1960; Happold 1975: 118; Bowker 1978; Kaufman 1993: ix). Wallas (1926) outlined a model to describe the process of creative thinking (Vinacke 1952: 243–4). This was first applied to religious experience by Batson in 1971 and developed in 1982 and 1993 (Batson 1971; Batson and Ventis 1982; Batson *et al.* 1993: 98–9; Loder 1981: ch. 2). Here are the four stages again:

1   Preparation: this is a time when the individual with his or her own skills and predispositions reflects on a problem and its complexities.
2   Incubation: after the conscious mind has done its work in the first phase, the unconscious takes over and makes the extraordinary connections which constitute new discovery.
3   Illumination: the solution is intuited, grasped or disclosed at the moment of inspiration.
4   Verification: the stage of testing against the evidence or through deductive thought.

This theory has widespread support from scientists and artists as a model for understanding much of their own creative work.[5] It also illuminates an understanding of religious experience. It is not reductionist. On the one hand, it neither prescribes nor precludes the operation of the Holy Spirit. It is open to anyone to interpret a situation as one in which the Holy Spirit is working through the systems network of causal interconnections, with reasons for the argument. The use of this model makes total sense in the light of Berger's statements, as a sociologist, that "religion is . . . [a] human enterprise" (Berger 1973: 34).

The Static Structure of Religious and Mystical Experience

Models suggested to account for the structure of religious experience range from single unit structures like Schleiermacher's (1988) "total dependence" model, or Otto's (1959) "numinous experience" model, to those who "treat the term 'religious sentiment' as a collective name for many sentiments which religious objects may arouse" (James 1968: 47). What is required is one model which is flexible. I suggest that "collage"[6] is an appropriate model for religious experience. The collage can have any combination of elements, which will vary according to the nature of the particular experience.

The appropriateness of the model "collage" needs more comment because it is not only appropriate for a religious experience, but also for every kind of experience. As used in relation to experience, collage can be defined as a centre of meaning often described by a concept which co-ordinates perceptions, sensations, emotions, previous beliefs, expectations, intentions, evaluations, interpretations, moods, attitudes and the creative imagination. An examination of a range of mystical, transcendental and religious experiences lend support to this model, as does an analysis of other forms of experience. The particular clustering or association of elements which make up the collage, the static structure, is related to the function of the experience and its dynamic structure.

The Function of Religious Experience

In the earlier review of the principal works concerning religious experience in Chapters 10, 11 and 12, a number of functions were noted which frequently overlap, but perhaps two main categories emerged. They were the function of "personal enrichment" and the function of "personal transformation".

"Personal enrichment" seems to cover James's model of "the religion of the healthy minded", Leuba's "will to live", Huxley's "aesthetic enrichments", Wordsworth's, Zaehner's and Paffard's nature mysticism, and Maslow's "incorporation of B (being) values". "Personal enrichment" includes aesthetic delight as well as moral and spiritual encouragement. Hardy, after the analysis of the first 3,000 accounts of religious experiences received by his Religious Experience Research Unit in Oxford, found that 24 per cent of them reported steady spiritual growth within themselves as a result of their experience(s) (Hardy 1979: 27, 8(i)(b) + (e)).

"Personal transformation" probably includes the models of Starbuck, James's "sick soul", Otto, Underhill, Zaehner's monistic and theistic categories, the models of Laski, Stace, Smart, Kee, Glock and Stark, and Hardy.[7] Hardy reported that 18 per cent, or one in five, recorded a sudden change to a new sense of awareness, a conversion experience, an awareness of a "moment of truth", which changed the quality direction of their lives.[8] The transformation is from a state of personal dis-ease to one of inner harmony or personal integration through the transformation of the unsatisfactory relationship or state of affairs.

The development of a religious experience which took place within the individual was distinguished into the two categories of "gradual growth" and "conversion" in

Hardy's study, as we noted above. Sometimes a person's religious experience developed as the result of a relationship to another person who influenced them considerably. These kinds of experiences were not separated into the two categories just mentioned.

Interestingly, less than one in a hundred reported the admiration of an ideal human figure or hero leading to a religious experience, but 11 per cent, just over one in ten, developed spiritually as a result of an encounter with another person who greatly influenced them, while 12 per cent grew spiritually as a result of the influence of literature or the arts, though Hardy thinks that the real influence of literature is larger than this (Hardy 1979: 27, 8(ii) and following examples).

Another central function of religious experience, as Starbuck and James indicated, is that it shortens the period of storm and stress, both in puberty and in later life, and makes easy what is necessary, namely, the movement which transfers the centre of activity from self-interest to interest in the whole community of which the self is but a part. Starbuck's three precepts will bear repetition: "in childhood, conform; in youth, be thyself; in maturity, lose thyself" (1899: 415). This function is one aspect of "personal enrichment" for some or "personal transformation" for others, depending on the pattern of personal maturation.

The Dynamic Structure of Religious and Mystical Experience

Transcendental experiences can lead to "personal enrichment", which is largely a continuous healthy process of growth in the direction of maturity, with personal dis-ease creating no major retardation. "Personal transformation" will probably have a developmental structure, and from the various discussions of mystical and religious experience and of creative thinking, it seemed possible to support a five-stage theory developed from the four-stage Wallas model (Batson *et al.* 1993: 98–9; Loder 1981: ch. 2; see p. 337 above):

✧ *Stage 1   The Problem*: The individual has a problem involving some aspect of personal life which needs attention and may seem insoluble, such as feelings of inadequacy, failure, meaninglessness, boredom, guilt, suffering or one or more broken relationships.

✧ *Stage 2   Preparation and Action*: The individual thinks through and sets out to implement the culturally appropriate remedial measures for the problem. These may include engaging in some statement or act or repentance, act of atonement or compensation, a period of probationary reformed behaviour, or it may involve forgetting the problem or avoiding it by concentrating on work or leisure. However, the course of action or inaction fails to resolve the problem.

✧ *Stage 3   Frustration and incubation*: Failure leads to frustration and a temporary shelving of the problem which is then handled by the unconscious through which a series of associations, often extraordinary, are made. The collage is gradually assembled, apparently without conscious thought, until the last crucial pieces are about to be slipped into place.

✧ *Stage 4   Illumination*: The individual now consciously realizes that all previous conscious efforts to resolve the dis-ease or problem have failed, but that, as the unconscious finally allows the last pieces to be placed in the collage, the individual suddenly becomes aware that the problem is solved. This involves either a conscious recognition of either the moment or process of personal transformation. Perhaps a change of perspective on life follows if the experience is a "conversion", "revelation" or "disclosure" with attendant changes in attitude. It may well be accompanied by the generation of, or release of, new energy, which finds expression in a new way of life.

✧ *Stage 5* *Verification*: It is through living the new way of life (e.g. marriage means a change from a single state to a married state; a "spiritual marriage", "partnership", or "friendship" with the Other, however named, equally requires a dialogue of sharing and learning), which brings delight, personal integration or harmony, together with the development of "fruits of the Spirit", which belong to every World Faith (Galatians 5:16–26). By living in this new way, with many a slip and recovery, the individual tests their "revelation", "religious experience", "insight" or "disclosure" leading to the new way of living. Starbuck was talking sense when he said in 1899 "in childhood, conform; in youth, be thyself; in maturity lose thyself". We will explore the evidence that this way of living, focused on the needs of others, is life enhancing later in the chapter.

This model of spiritual or personal transformation is shared with scientific and artistic creative thinking: it describes the processes of artistic creation, e.g. in writing a new opera or pop song, and in discoveries in science, e.g. outlining the theory of relativity. Who first said "Eureka"?![9] This model should have the practical value of adequately linking the static structure of the experience, the collage, to functions and a dynamic structure of the experience. The relationship between the collage and personal enrichment is a simple dynamic of the accumulation of pieces and their interrelationship, using the processes of assimilation and accommodation. The relationship between the collage and personal transformation is one of replacement, or dismantling and reconstruction: the new perspective on life replacing or transforming the old.

PHENOMENOLOGICAL PSYCHOLOGY AND RELIGIOUS EXPERIENCE

This subject was explored briefly in the last chapter. Professor David Wulff, here, focuses on the problems of classification when he writes that "Contemporary psychologists use the term phenomenology with widely varying degrees of precision . . . for example, in many current psychological textbooks the personality psychologies of Rogers, Maslow, and Kelly are denominated phenomenological theories" (Wulff 1995: 184).

I quoted two of the most popular UK Psychology textbooks to classify the different kinds of psychology in Chapter 13 (Gross & McIlveen 1998; Scott & Spencer 1998). They have no sections on phenomenology. Spencer includes a chapter on "Personality – The Individual and Society" (p. 548f.), exactly illustrating a use and examples that Wulff quotes above (Wulff 1995: 184). Wulff (1991) is one of the most inclusive and rigorous academic scientific authorities on the psychology of religion, so we need to listen to him when he says that "Phenomenological psychology seeks to reclaim for experience the pre-eminent position it possessed at the end of the last century . . . " (Wulff 1995: 184)

But what is 'phenomenology'? Strictly, the word means "the study of that which appears" (184). But Wulff goes on to tell us that this requires that a researcher/observer describes what she sees and hears, that is, the activities and the interpretations of the practitioners, but "bracketing out" her own. But the literature is rather uneven, depending on what self-imposed rules a researcher/observer exercises. Wulff outlines the history of rigorous or 'classical phenomenology' based on the "bracketing out" of the observer's assumptions. So in order to understand another's experience, "I must relive it or imaginatively reconstruct it, by drawing on similar experiences of my own". This also means the avoidance of "pejorative comparisons among religious traditions" (Wulff 1995: 185). But it also means a "psychological description" based on "a systematic introspection" (186). Wulff suggests that Otto, quoted earlier (ch. 13, p.214) is an example of one who chose analogies and ideograms from everyday life to illustrate the *mysterium tremendum et fascinans*, the tremendous and fascinating mystery, 'the numinous' at the heart of a religious experience (186). This may be just psychological, but Otto's interests were wider as we will see in the next section.

Phenomenological Psychology of Religion

My brief account of what Wulff has just described above shows that some of the early phenomenologists were also interested in the psychology of religions, which became intermingled. This he discerns in Otto's work. Wulff reports, that "all psychologies of religion limit themselves to what is 'psychical'" (187).

Wulff distinguishes a clear difference between *phenomenology of religion* and *the phenomenological psychology of religion*. The former discipline is usually adopted by those interested in the *history of religions*, and so, in order to get inside the spirit of other religious traditions, they adopt a historical and cross cultural perspective. This helps them avoid "distorting or judging them in term's of the investigator's own faith" (187).

Smart (1968; 1997) seems to successfully inhabit the position of a phenomenologist, and he seems to include Otto in this category (1997: 11). Perhaps that is because Otto's model of *mysterium tremendum et fascinans*, is an 'impartial' concept of the 'numinous' (Otto 1959 (1917): chs. 2–4) which is central to many world faiths, and in his later books perhaps he more closely embraces the position of a phenomenologist in his books on Christianity and Indian Religion (e.g Otto 1932).

Wulff's criterion of impartiality as the final test for this distinction is very helpful, namely that "if the primary identification of the investigator (i.e. his religious bias) . . . shines through at various points, in the kinds of comparisons made . . . and even the conclusions that are drawn" (1995: 187), it falls into the category of 'a subjective psychology of religion', and not in the two authentic categories Wulff is describing.

Wulff identifies three methods used in the phenomenological psychological of religion:

1. The Direct Approach: Using one's firsthand mystical or religious experiences which includes one's own and also one's patients' spoken experiences (188).
2. The Indirect or Vicarious Approach: Turning to the experiences of others as the 'source of self-observations', which was the approach of William James and Marghanita Laski who collected their religious and mystical experiences from both friends and published accounts. Wulff reports, that "James was the first to undertake a strictly descriptive phenomenology of religious experience without the biasing effects of some theory or doctrine" (190).
3. An Empirical Phenomenological Analysis of Spiritual Experience is based on gathering written descriptions or transcribed interviews and processed to discern the "essential general structure" of the experience being explored (Wulff 1995, p. 190f.).

SCHOOLS OF THOUGHT CONCERNING
PSYCHOLOGY OF RELIGION

"Ralph Hood (1998), a major figure in American psychology of religion, suggests that there are six psychological schools of thought regarding religion" (Nielsen, 2000: 6).

1. **The Psychoanalytical Schools.** The strengths and weakness of the work of Freud, the founder of this movement, have been explored in Chapter 14. Freud regarded religion as "the black tide . . . of occultism" (258 above). He attempted to reduce religious belief to "its material constituent parts" (257 above), which was flawed in different ways. Nielsen, however, reminds us that "contemporary psychoanalytic interpretations are not necessarily hostile to religious faith."
2. **The Analytic Schools.** The strengths and weaknesses of the work of Jung, the founder of this movement were explored in Chapter 14. Nielsen observes that "most psychologists, however, consider such descriptions to be undemonstrated by scientific research, and therefore it plays a limited role in psychology".
3. **Object Relations Schools** developed from psychoanalysis focusing on maternal influences in the development of inter-personal relationships, giving us the concept of 'transitional space' – a possible element in religious experience (see chs. 14 & 17).

"Each of these three schools rely on clinical case studies and other descriptive methods based on small samples which run counter to the prevailing practice of psychology in America" (Nielsen, 2000: 6).

4. **Transpersonal Schools.** Greenwood suggests that transpersonal research is the latest movement within psychology, its fourth force, succeeding the first three forces, behaviourism, psychoanalysis, and humanist psychology (Greenwood in Hood 1995: 495–6). Its orientation is holistic, focusing on the integration of mind, body and spirit. Transpersonal theory and Religious Experience was discussed in Chapter 15 in the Social Psychology section.

5. **Phenomenological Schools.** Wulff, in the last chapter, and earlier in this chapter, has encouraged us to distinguish between a Phenomenology of Religion and a Phenomenological Psychology of Religion. We have recognized that Smart's and possibly Otto's work is in the first category, while James and Laski are in the second category, with others, who collected descriptions of particular religious and mystical experiences and focused reflection on them or assessed responses before and after particular religious experiences seeking their shared underlying structures and responses.

6. **Measurement Schools** form the main approach in American psychology of religion. The construction of scales to measure religious belief and practice has been an ongoing concern as demonstrated by Hill and Hood's compilation of scales (1999). They are very important, and represent the tremendous effort that has been devoted to the measurement of religious phenomena, though little has been previously said in this book on this topic (e.g. p. 278; p. 286), as it is outside the constraints of the title. Measurement scales have successfully assessed 'types of religiosity' designed by different researchers, including Hood (1995) and Hill & Hood (1999). Batson and colleagues designed many scales using their intrinsic – extrinsic – quest orientations for this purpose (Batson *et al.* 1993: chs. 6–11). However, measurement has also been used widely and successfully in studies concerning, for example, religious development, attitude change or cognitive theory, e.g. Goldman, Batson, Miles, Zachry.

TRANSCENDENTAL EXPERIENCE, LANGUAGE AND STUDENT UNDERSTANDING

The Hardy key question asks, "Have you been aware of or influenced by a presence or power, whether you call it God or not, which is different from your everyday self?" (Hay 1990: 56). Hay and Morisy found that when the Hardy key question was placed in the National Opinion Poll in 1976, 36 per cent of the population replied that they had had such an experience. In that analysis they found that if students had remained in education until the age of nineteen, the respondents who gave a positive answer rose to 44 per cent. If education continued beyond the age of twenty, then the proportion who said "yes" increased to 56 per cent.

When Hay and Heald placed the Hardy key question in a Gallup Poll in 1987, the proportion who said "yes" rose to 48 per cent, compared to the figure of 36 per cent in 1976. When Hay engaged in a personal interview with respondents in Nottingham, the proportion of positive responses rose to 58 per cent (Hay 1990: 79).

As the larger proportion of the sixth form population who were asked the Hardy key question by Miles intended to enter higher education (i.e. full-time education after twenty years of age), his result of a 59 per cent positive response (Miles 1983: 193) accords with the findings of Hay.

In the exploration of the use of religious language earlier, we explored some of the non-realist uses of languages, suggesting that religious language is used just to express human feelings like pleasure or dis-ease, harmony or disharmony. Non-realist thinkers say that religious language does not refer to "God who is" – there is no

God, they say. Such a view is rejected by the vast majority of serious religious thinkers of the second half of the twentieth century who are critical realists, including all those quoted in this work except the named non-realists. Janet Martin Soskice well represents critical realists when she says that religious language is "reality depicting" (Soskice 1985: 145).

Two important philosophical theologians already quoted are Paul Tillich and Keith Ward. Tillich takes the view that revelation should be regarded as an "inner experience". "There are no revealed doctrines, but there are relevatory events" (Tillich 1951, i, 130 quoted by Ward 1994, 229). These inner experiences of 'essential mystery' are beyond subject–object relations, releasing 'the power of being which gives the courage to be', an ecstatic experience, a miraculous sign-event, disclosing the 'ground of being', which is a 'personal transformative experience' (cf. Ward 1994: 230).

Ward goes further, and takes the view that

> revelation may be characterized as a direct intention of God . . . to communicate truths beyond normal human cognitive capacity. Its form is influenced by the cultures and histories in which it occurs. It is properly received only by commitment, trust and hope. It is a blend of human imagination and reflection and Divine (or enlightened) co-operative persuasion, which leads, in the great scriptural traditions, to a canonical model of one supreme value and goal. In the Christian tradition, revelation takes the form of a historical self-disclosure of the Divine. The source of revelation lies in an authoritative empowerment, experience and inspiration, to which Scripture witnesses. It thus involves elements of reflective thought, experiential awareness, and cognitive, moral and mental empowerment. (Ward 1994: 343)

From the contributions in the last paragraph, we can see that the concepts "revelation" and "religious experience" can actually apply to the same event. When they do, they are just different viewpoints: "revelation" is "looking down", and "religious experience" is "looking up",[10] but they can both be talking about the one event. This parallels "top-down" and "bottom-up" causation noted in other forms of knowledge. When one talks about Scripture as "revelation", then that refers to digested and edited material from the experiences of those who received or discovered the original insights.[11]

Now we must face the central philosophical problem that all humans should face, and are frequently totally unaware of. As we have been discovering in our exploration so far, it is a facile and inaccurate assumption in society that "science and real knowledge are based on facts", but "religion is just opinion based on faith".

By now, I hope we have together explored the view that all knowledge is based on perception and interpretation: there is no such thing as an uninterpreted event or fact. Perceptions are interpreted by concepts applied to them which may then be taken to be facts.

THE STRUCTURES OF FORMS OF KNOWLEDGE

Now, there is a logical gap between statements about sense experience and those about physical objects, for "no statement about sense experience will serve as the necessary or sufficient condition for a statement about physical objects".[12] There is a logical gap – no link – between: what you sense – *gap* – and what you say you experience through your senses (using concepts). This gap is "jumped" or "crossed" by "pure rational faith [which] has an absolute claim on us" (Ameriks 1995: 399). This is true

✧ of your sense experience of a moving object in the water – *gap* – and your judgement that it is a fish, a mackerel;

✧ of your sense experience of sound – *gap* – and the statement expressing your judgement that you are listening to a great piece of music;

✧ of your sense experience of a picture – *gap* – and the statement expressing your judgement on a great painting;

✧ of your sense experience of TV pictures – *gap* – and the statement expressing your moral judgement about war crimes in Kosovo;

✧ of your sense experience of a person – *gap* – and the statement expressing your judgement that you are in love with him or her;

✧ and finally of awe-filled sense experience – *gap* – and your judgement that this was a profound religious experience.

In each case there is a logical gap between the sense experience we have of what occupies our attention, – *gap* – and the mode of discourse, such as empirical, aesthetic, moral, personal or religious, from which we select concepts to give an account of the matter of our present concern.

Now at this point it is necessary to discuss this logical gap between our perceptions, or sense experience, and statements about our experience. Baelz, invoking the support of Strawson, argues that

> we apprehend the existence of the external world by an act of *intellectual intuition* [my italics; cf. Ameriks 1995: 400] . . . we cannot help believing that there is an external world . . . it is a basic assumption underlying the whole of our thinking and acting. It is built into the structure of our language: thing-words and person-words are logical primitives, in terms of which other particulars must be designed. The human community would not be what in fact it is if men did not believe in the existence of things and persons. (Baelz 1968: 91)

In Chapter 7, when we explored *emotion*, we learnt from Wynn that *emotions* run ahead of our *intellectual* explorations. Are we emotionally attracted to a subject area before we explore it, an attraction which can grow into a love for our subject? It seems like it.

We must now examine the meaning of *intellectual intuition*. Bruce Russell holds that *intellectual intuition* refers to the activity of the mind without any emotional component (Russell 1995: 382). The contemporary *use* of the concept *intellect* is as Russell indicates. However, Professor Denys Turner explains that there is "an older conception of *intellect*, according to which faith can be genuinely present only within a mind compelled by its immanent energies to engage with the mysterious 'givenness' of creation" (Turner 2004: xv). The view of *intellect* without an emotional component emerged in the fourteenth century, when it was contrasted with "the knowledge of love", the emotional seed of faith (77). In other words, it is a "matter of faith that reason can know God" (xvi). Saint Thomas Aquinas knew this in the thirteenth century, as Turner points out. This indicates that Aquinas is using faith to refer to his emotional insight as trust, which stimulates his intuitional search, which is articulated through *reason* (see Chapter 7). Wynn reminded us that William James, at the beginning of the twentieth century, stressed that "feeling is the deeper source of religion" before he highlighted the link between *emotion, intuition* and *intellect* for the twenty-first century (see Chapter 7).

Emotion inclines our *intuition* to suggest to us which mode of discourse we should use to express our understanding of the matter in hand, e.g. religious, aesthetic, moral, personal, scientific or empirical. We employ rational faith, in Kant's sense, without any emotional content, to jump the *gap* between our *sense experience* in the vehicle of our *emotionally guided intuition* to the mode of discourse, or form of knowledge we employ to express our *understanding*.

Happold talks of "mystical *intuition*" (my italics) as that which closes the logical gap when talking about religious and mystical experience (Happold 1975: 55). Similarly, the statement that "there is a God . . . would have to imply that there are certain experiences available at least to certain people" (Hudson 1974: 70). This seems to have been true of most of the great religious leaders of the world, as well as for over half of the English-speaking population each side of the Atlantic who continue their full-time education past the age of twenty years. This seems to be sufficient evidence for us to support the inductive hypothesis that religious language obliquely refers to Transcendent Reality, which is Being itself, just as Darwin collected together his data to support his inductive hypothesis, which we call the theory of evolution in which I also believe.

Others say this logical gap is crossed by the creative imagination (Happold 1975: 118; Bowker 1978; Hooker 1982: 50–6), or by the unconscious (Jung 1973). It hardly matters whether we call the gap jumper "intuition" or "the creative imagination", though "the unconscious" sounds more like a "location" or "faculty" in which the activity of "gap jumping" takes place. It is through crossing the logical gap that everyone learns that "I am real", and most people learn that "God is real" (Buber 1958). The "gap jumping", please note again, is an act of faith.

A further point needs to be made. If you are asked to tell a family how to use the Underground in London, I am sure you will give them good instructions, but I bet you *do not* tell them to "mind the gap" between the train and the platform when getting on or off the train. Passengers can hear this as a recorded announcement many times on a single journey. Similarly, I did not even know that there was a logical gap between my sense perceptions and the concepts I use to describe and interpret them until the philosophers told me! We have learnt how to "cross the gap" automatically, intuitively: we are not even conscious that it is there! Then why am I making such heavy weather of this small, apparently trivial point? The "connector" or "bridge" across the gap is intuition, as we have said before.

It is not a trivial matter for philosophically materialistic scientists, or any other logical positivist sympathizer, to repeat the inaccurate assumption in society that "science and real knowledge are based on facts", but "religion is just opinion based on faith". Now we see that all forms of knowledge are based on *intuition*. Each mode of knowledge is built on related, adequate and relevant evidence. "Proof" is strictly *only* a relevant concept in logic, mathematics or in examples of deductive thinking. The concept of "proof" is alien to most forms of knowledge, which are underpinned by inductive hypotheses, unless one needs to engage in a piece of deductive thinking.

All areas of understanding and forms of knowledge are underpinned by faith: you only jump over a gap, through intuition, by *faith*.

All this involves a paradigm shift[13] in how we think of modes of discourse and forms of knowledge. If pupils and students are given access to these insights, the religious education teacher will never again be faced with the *sincere* question, "Can you *prove* it, Miss?" or "Can you *prove* it, Sir?" That was my question in my teens, and now I see that sufficient evidence and reasons are adequate for all forms of knowledge apart from logic and mathematics, where "proof" can be achieved.

Questions of this kind indicate the constraints created by logical positivism in the last three quarters of the twentieth century to an understanding of the nature of knowledge, and particularly to the arts, moral knowledge, personal knowledge and religious knowledge.

Reflections on the Structure of the Forms of Knowledge

The skeleton model of the structures of the forms of knowledge was set out in such a way as to show the similar pattern of the structures. A comment is needed to show the quite different ways in which they may operate. When, in outlining the "Structures of Forms of Knowledge", I used terms like "faith", "gap" and "intuition", each term takes on quite different nuances appropriate to its own form of knowledge, and each form of knowledge is essential to make up the complete "rainbow". Let us take science and religious experience as at the extremes of the range. The procedure in science was outlined in some detail at the beginning of Chapter 6, indicating its rigorous procedure in establishing scientific knowledge. That contrasts with the formation and structure of a religious experience outlined in this chapter.

Let me try to portray these differences through an imaginary camera lens. In science the data gathered "out front" by the lens and processed behind it is likely to converge onto the film into an agreed photograph – a robust conclusion after the processing of the data agreed by "the international community of scientists". In religious experience, the data gathered by the lens (of experiences with similar foci), is likely to form a "strip of pictures", that would be processed differently by, for example, different World Faiths, or, in personal terms, it would be a video clip record of a life-changing experience, which has in some cases changed the course of history. But these differences in formation and testing do not invalidate the integrity of the different forms of knowledge: there are greater degrees of freedom of interpretation as one moves up the hierarchy of forms of knowledge.

IS RELIGIOUS EXPERIENCE KNOWLEDGE?

Let us remind ourselves of the three-stage structure of the "activity of knowing" an object we can see. First, we see something, i.e. we make a perception, e.g. it is a white object. *Third*, we apply a concept to the perception, it is a duck. But what makes the link between the perception, stage 1, and the allocation of a concept, stage 3? And I am not talking about the electrical "wiring" in the brain, which we will be able to have a map of one day, though that is involved, and we could not make the connection without it. I am talking about the process of transformation of the perception which takes place between the act of perceiving and the allocation of a concept to it, the *logical gap between percept and concept,* which we call intuition. Making the transformation through intuition is stage 2. The three stages below are separated by dashes.

Plantinga, one of the moderate foundationalists, has given us some practical insights into what can count as knowledge. There are three kinds of basic beliefs which can practically establish knowledge:

1 perceptual beliefs: I see a white object – intuition – it is a duck;
2 memory belief: I had lunch yesterday – intuition – it was enjoyable;
3 beliefs about other people's thoughts: "Great news" – intuition – he's happy;

These are foundation beliefs, and you do not need any supporting arguments, just enough supporting examples to give you adequate evidence that your procedure works – the procedures of "knowing" just outlined are practical, and reliable.

Alston builds on this. He seeks to defend the integrity of religious experience as a form of knowledge, but stops short of making that claim. His philosopher followers feel his claim has been established and is free standing, i.e. his case does establish that religious experience is knowledge. Let us summarize the arguments we have previously examined. Alston argues that the general theory of perception can be applied to religious experience. I shall include my intermediate step in the argument.

1 A sense perception: you see a white object – intuition – it is a duck. In this example, the only activity is searching for a concept to apply to the white object. It involved empirical information sorting only – an intellectual exercise only.
2 A person perception: she meets a really nice fellow – intuition – she would like to marry him. In this example, much more is involved. Information, yes, but also feelings of delight, dreams, worries about him, practical problems of work or where to live. This perceptual intuition involves information, feelings, will – it is a fully multidimensional personal complex of intuitions, or one intuition with a network of dimensions attached.

Now in the case of both the duck and the fellow, one can get it wrong. It may not have been a duck, and she did not marry that chap, in fact she married the fifth fellow she fancied. She was attracted to each man because of different good qualities. However, the two of them had different interests, and so on. But that does not invalidate the procedure, because it is the only procedure each of us has anyway. However, we have tests available to apply to our perception of others: how do we get on? how does our family react? how do our friends react?

On balance, perception works both in terms of empirical perceptions, i.e. of things you see, but also of personal perceptions, apprehensions about people you meet and their qualities. Both sets of perceptions, together with the perception that you both feel that you "click", you intuitively turn into concepts. You then test out both whether your perceptions are accurate, and not just idealized projections, and whether you used the right concepts to interpret those perceptions, and whether you do in fact both "click".

When we get to perceiving God experiences, these include some of the characteristics of the perceiving person experiences, namely, in God's case, the extending of "perception" to include qualities like goodness, guidance or the love of God. This usually happens within the social context of the religious ethos in society in general, or in a religious community in particular. Now that extension is perfectly acceptable because it holds for the personal perception experience: the girl or man may have human qualities of goodness, guidance and love, and support of family and community around them.

As all three cases are acceptable on pragmatic or practical grounds, i.e. because they work on the whole, then they are all acceptable. Except that being a doubting Thomas, I have dragged you through a long argument to establish the nature of knowledge in a range of forms of knowledge. Namely, scientific knowledge is only established, not just on experimental evidence, but when the scientific community give it a majority nod of approval and have accepted "a new way of looking", as has been the case with such different theories as evolution, relativity, quantum theory and chaos theory. The same thing happens to moral principles, e.g. compare St Paul and Wilberforce on slavery – they are accepted when the community give a majority nod of approval. Now the same is true of religious experience and mystical experience. Let us list the results of religious experiences which will persuade a community of the integrity of a religious experience:

- ✧ James: does it give you delight? does it bear fruit in your life?
- ✧ St Paul's fruits of the spirit: 1 Corinthians 13 – Love is kind and patient, never jealous, boastful, proud or rude. Love is not selfish or quick tempered. It does not keep a record of wrongs others do. Love rejoices in the truth, not in evil. Love is always supportive, loyal, hopeful and trusting. Love never fails! Galatians 5:22–23 – be loving, happy, patient, kind, good, faithful, gentle and self-controlled.
- ✧ Berger: in a pluralist society, induction begins from retrieved experience from the tradition and contemporary religious experience.
- ✧ Poloma's defence of James on sociological grounds: Religious experience should produce "fruits for life"; conversion should produce saintliness, "the collective name for the ripe fruits of religion". Saintliness increases charity, or "tenderness for fellow creatures". People who have religious experiences live in a social world which they help to create and modify; thus religious experiences have social consequences.

I make a partisan value judgement when I say I hope the religious community will say 'yes' to religious experiences which break barriers, develop the intrinsic religious orientation, or the quest religious orientation, and are life enhancing, but that leads into the next section.

But before we leave this question, we need to ask if religious experience is knowledge. A key reason why religious experience can count as knowledge is this. The intuitive leap of faith involved in sense perception is exactly parallel with the intuitive leap of faith in person perception and in God perception. The fact that enough examples of religious experience – i.e. 36–50 per cent of the named populations – give sufficient support to affirm that religious experience is a form of knowledge which, when tested and affirmed by the community, is a good foundation on which to base one's life.

Are Transcendental Experiences Life Enhancing?

. . . Realistic News . . .

Sir Alister Hardy reports on the analysis of the first 3,000 religious experiences studied, that about 5 per cent of them reported experiences of negative destructive forces (Hardy 1979: 28, 9(ii)(m)). This darker side of human experience leads to "feelings of remorse and guilt, of fear and horror" (51). He says, "It seems likely that the proportion of people who have such experience may be much greater than our figures would suggest, for our appeal was for records of religious or spiritual experience rather than those of an evil nature" (78). Hardy noted that Glock and Stark (1965), in their Californian survey, found that an average of 34 per cent of Catholic and Protestant respondents were "sure that they had experienced a feeling of being tempted by the Devil" – i.e. they knew the almost irresistible power of "the evil pull inside them" when tempted.

These reports of an evil experience received by Hardy have been analysed (Jakobsen 1999). Evil can be recognized by individuals and groups in the intent of another person, and sometimes "in the surrounding spirit world". In some traditional societies, a person called a shaman mediated between the good and evil spirit worlds to maintain a balance. In modern society there is "if not a denial then definitely an understatement of the evil of negative aspects of human spirituality" (Jakobsen 1999: iv).

In her study Jakobsen examined 4,000 of the 6,000 reports of religious experiences now in the Hardy archive. She does not analyse the nature of evil, just categorizes the views she found. Some evil experiences happened to religious people, or in religious places. The main group of experiences of evil occurred in the room where the individual was sleeping, though other places were mentioned for their waking state.

Evil was described in many ways, such as vile, rotting, black, crackling, as Satan or the Devil, a figure with a tail or horns, or as having a grinning evil face. Reactions to evil included ice-cold shivering, tingling scalp, sweating with terror, paralyzed lips, vomiting from fear, shaking, being unable to speak or move. Immediate protection from evil was sought through reading the Bible, making the sign of the cross, or praying – like saying the Lord's Prayer.

Each individual faced by evil forces or temptations was aware that the temptation faced involved a recognition that this was a battle for their soul, and that "evil actions" would "take control" through "a power without". This process some described psychologically as a collapse of personality, and others as an attack or possession by evil spirits or the Devil.

Jakobsen relates her analysis to the work of Scott Peck (see Chapter 3), and quotes a character in one of Susan Howatch's novels: "Evil exists. Those who forget that fact or ignore it or reject it are at best taking a big risk and at worst conniving at their own destruction. All creation has its dark side".[14]

. . . Better News . . .

Sir Alister Hardy reports on the analysis of the first 3,000 religious experiences studied that about a fifth of them reported that they received a sense of purpose or new meaning to life as a result of their religious experience(s). Improved attitudes towards others were a consequence of their religious experience for 8 per cent of the respondents, while 4 per cent were influenced to *change* their beliefs (Hardy 1979: 29, 98–103). Another project found that there had been a persistent and positive change in the development of more positive attitudes towards other people (Beit-Hallahmi and Argyle 1997: 89–90).

I must quote one example from Hardy:

> In November of 1968 my eyes chanced to light upon "Honest to God" on the library shelves . . . In due course I reached the passage, or one of them, which Robinson quotes from Tillich's "The Shaking of the Foundations" about God being depth and so on, which ends up "He who knows about depth knows about God." The whole passage immediately impressed me as throwing, for me, a new light upon the situation and as being perhaps possibly of application to myself . . . As however the afternoon and evening progressed, I became increasingly suffused, as it were, with the joyous realization that I knew that God existed, that I could as I suppose in retrospect I had wanted to be able to do, say "I believe in God". From that evening I lived in a state of what I can only describe as exaltation for the next few weeks, overwhelmingly aware of the feeling of closeness to God, to a reality that was constantly present in the midst of other activities . . . I am also aware of a desire to pass on the good news – culminating in teaching a few periods of scripture in the school where I already teach. Prayer has become a reality, not just something you were told you ought to do, and many sentences from the Prayer Book have taken on a quite new meaning – the one that springs to mind is "Whose service is perfect freedom". This and others mean something real instead of being words which you understand but don't convey a known truth. (2387, F, 51) (Hardy 1979: 101)

I quote this experience, because when Robinson's brilliant book was published in 1963 it was heralded by a whole-page article in the *Observer* Newspaper one Sunday. I was a religious education teacher then. I bought the book immediately: it was liberating, after which teaching became a joyous challenge, with mountains of intellectual problems ahead. Why? Because it demonstrated, first, that it was not an empirical fact that God was up in the sky, and, secondly, the inadequacy of spatial concepts to "portray God", e.g. "God is up there". Thirdly, I gained greater insight from "psychological images", e.g. "God is in the depth of your being".

Some challenges were earmarked by searching pupils and students. When one problem is resolved another appears, as in traversing the Lake District! I am struggling with current problems still!

In the analysis of the Hardy material, some religious experiences were classified as "cognitive and affective elements". We must look at these experiences, but a preliminary comment is necessary first. Let us take the experience of Moses and the Burning Bush as an example to analyse the emotional and cognitive elements involved (Exodus 1–3). Let us briefly summarize a reconstruction of the situation. Moses, a Hebrew, is brought up in the Egyptian court. He sees an Egyptian overseer beating a fellow Hebrew and kills him, believing that no one has seen him do it. The next day he sees two Hebrews fighting and remonstrates with them. One of them asks if he is going to be killed as he killed the Egyptian yesterday. Realizing that if this man knew what had happened the knowledge was more than likely to get back to the king, Moses fled for his life. In exile he pondered on what could be done about his enslaved fellow Hebrews in Egypt. Then one day he saw a "bush was on fire, but it was not burning up". "This is strange!" he must have said to himself. "I'll go over and see why the bush is not burning up" (*Into the Light*, The Bible, C.E.V: 56). It looked alight, but was not consumed. Perhaps he saw a bush at sunset with the sun behind it. He approached it with feelings of curiosity, awe, and wonder. His unconscious digestion of his problem surfaced: God was calling him to share in solving the problem – eureka! Except that Moses did not say eureka: he did not want to do it. List his feelings and insights after his religious experience, once you have read the account of it! My point is that in the Hardy-reported experiences, most of the feelings and insights are the result of the experience, and therefore I include a brief note of them here.

As a result of their reported religious experiences, they received a whole range of new life-enhancing feelings: 25 per cent, or one in four, had a sense of security, protection and peace; 21 per cent, or one in five, had a sense of joy, happiness and well-being; 7 per cent reported feelings of awe, reverence and wonder; 7 per cent had a sense of harmony, order and unity; 6 per cent had feelings of love and affection; 5 per cent felt exaltation, excitement and ecstasy; while 4 per cent received feelings of forgiveness, restoration and renewal.

Other feelings reported slightly less frequently include a sense of integration, wholeness and fulfilment; hope and optimism; a sense of release from the fear of death; yearning, desire and nostalgia; a sense of timelessness; a sense of being at a loss for words. The smallest reported category, of 1 per cent, felt a sense of indifference and detachment as a result of their experience, though it was of sufficient importance to them to write it down and send it to Sir Alister.

The experiences had considerable effect on the volition or new intentions of the recipients, as well as on their new understandings and sense of direction. Thus, 20 per cent were sure they experienced a sense of presence (non-human); 19 per cent had a sense of certainty, clarity and enlightenment after their experiences; while 16 per cent mentioned a sense of guidance, vocation and inspiration; 7 per cent felt they had gained new strength in themselves. Some gained a larger picture: 14 per cent felt a sense of prayer answered in events; and 11 per cent discerned a sense of purpose behind events (Hardy 1979: 26–7, 51–67).

Happiness is a major effect of having a religious experience. This is often expressed in different ways, e.g. feeling bathed in light, being at peace or restored, being happy or elated, being uplifted or awestruck. Sometimes a religious experience occurred in prayer, or led to prayer, leading to strong feelings of happiness and existential well-being (Beit-Hallahmi and Argyle 1997: 89).

Wuthnow found that those who had had peak experiences, religious or natural-istic, were less interested in having a highly paid job, job security or a beautiful house and valued more working with people in need or working for social change. They had less interest in social status, fame or having lots of friends than the non-peakers (Glock and Wuthnow 1979).

Four other studies have shown that there is a considerable increase in commitment to the religious life, a closer sense of communion with God, and more involvement with activities of the worshipping community.

In one study more than half of those who had had a religious experience reported an increase in their levels of self-esteem (Beit-Hallahmi and Argyle 1997: 90).

You can judge for yourself how life-enhancing these experiences were. I shall simply add the two quotations below.

The conclusion of Beit-Hallahmi and Argyle is worth quoting:

> Religious Experiences convey to those who have them, that they have been in contact with a very powerful being or force, "whether they call this God or not", that there is a unity in the whole of creation; they feel united and have love towards other people; they feel more integrated, perhaps "forgiven"; they are happier; they have had experience of timelessness, perhaps eternity; and they believe that they have been in contact with some kind of reality. (Beit-Hallahmi and Argyle 1997: 96)

Finally, is "religious experience" a biological endowment, which is of survival value to *Homo sapiens*? The last word must be left to two distinguished biologists, who have been the two great advocates of "religious experience" in the second half of the twentieth century, Sir Alister Hardy and Dr David Hay: "What then of 'religious awareness'? Has that evolved because it fits us for the world? Hardy certainly believed it to be the case" (Hay 1990). "For Hardy, the feeling of being in touch with a reality beyond the self is as much something biologically real, as for example, the experience of being in love" (26).

Hardy and Hay have helped to modernize James's good ship, collecting materials to repair the damage inflicted by logical positivism, behaviourism, psychoanalysis and scientism. Alston is the chief architect in redesigning the structure of religious experience, with Swinburne's help. The model builders have put in new decks. Together they, with others, have restored "religious experience" as a form of knowledge. Together they have steered this rebuilt ship through the treacherous waters between the cave of Scylla and the whirlpool of Charybdis[15] – Barthian theology and the selfish gene – safely to the beginning of the twenty first century.[16] *The Faithful, for the last three-quarters of a century, have **known** more than they could **say**.*

NOTES

1 Schleiermacher, Otto and C. S. Lewis respectively.
2 Greeley (1975). No respondent mentioned drugs.
3 Hardy's respondents defined "religious experience" in their own terms (see Chapter 10).

4   Hardy (1979: 28, 81–98). Drugs were also mentioned: 1 per cent had used anaesthetic drugs, and 0.6 per cent had used psychedelic drugs.

5   Harding (1940); Vinacke (1952: 243–4); Ghiselin (1952); Vernon (1970); and Hill (1979). See also the place of the principle of association in creativity, psychoanalytic theory and cognitive theory (Bolton 1972: 183–4).

6   See Smart (1973: 149). Here Smart uses the term "collage" as a model for a religious system. I am borrowing it for the use indicated above.

7   As well as Streng (1978); Horne (1978); and Lyttkens (1979: 15, 211–12), not reviewed here because of shortage of space, Horne shows that mysticism, creative insight, psycho-therapy and ritual are all rational activities and use insights about life for the purpose of personal transformation.

8   Hardy (1979: 27, 8(i)(c)) and following examples.

9   Cf. the exultant exclamation "I have found it!" of Archimedes in Chapter 16.

10 The spatial images are just that, pictures which portray "God's viewpoint" and "a person's viewpoint".

11 Note that "received" is the top-down, or God-giving perspective, and "discovered" is the bottom-up, or human perspective. A dialogue with God involves both processes operating simultaneously. If one word or the other is used alone, it simply indicates the perspective from which the writer is talking.

12 Hudson (1974: 68); cf. " . . . in a full treatment of the relation of perception to . . . wholes . . . the interweaving of the elements of Kant's threefold synthesis of apprehension in intu-ition, of reproduction through imagination, and of recognition by concept should receive extended treatment" (Mackinnon 1978: 133).

13 A paradigm shift occurs when scientists reject one theory, hypothesis or viewpoint for another. Here this would involve rejecting the hypotheses that "science and real knowledge are based on facts", but "religion is just opinion based on faith". These would be replaced by the model or hypotheses that "all forms of knowledge, other than Mathematics and Logic, are based on *intuition*". "Each mode of knowledge is built on necessary assump-tions, and related, adequate reasons and sufficient relevant evidence."

14 Howatch (1997) *A Question of Integrity*: 458; cf. her brilliant descriptions of ecclesiastical temptations in her six Starbridge novels, e.g. *Glittering Images*.

15 Greek mythology: a monster inhabited a cave traditionally situated in the Straits of Messina and devoured sailors if they sailed too close to it (Homer, *Odyssey* 12.85–6).

16 Hay (1990: 26); Hardy (1978: 198–9). Barth, a theologian, dogmatically argued: "Knowledge of God . . . consists . . . in the knowledge of the God who deals with man in this Revelation in Jesus Christ" (Hardy 1978: 199); "On a Barthian view, the only appro-priate religious language is language which bears witness to God's self-revelation in Jesus Christ" (Hebblethwaite 1987: 108). Barth argued against "natural theology" because it treats God as a "subject" or "object" to be investigated by "independent" humans. I support Hardy in defending natural theology (cf. Tillich 1968: 535–9; Morgan 1998: ch. 8).

REFERENCES

Ameriks, K. (1995), 'Kant, Immanuel', in R. Audi (ed.), (1995), *The Cambridge Dictionary of Philosophy*, Cambridge: Cambridge University Press.

Baelz, P. R. (1968), *Christian Theology and Metaphysics*, Peterborough: Epworth Press.

Barnes, K. C. (1960), *The Creative Imagination*, London: Allen & Unwin.

Batson, C. D. (1971), "Creativity and Religious Development", PhD thesis, University of Princeton.

Batson, C. D. and Ventis, W. L. (1982), *The Religious Experience*, Oxford: Oxford University Press.

Batson, C. D., Schoenrade, P., & Ventis, W. L. (1993) *Religion and the Individual: A Social Psychological Perspective*, New York and Oxford: Oxford University Press.

Beit-Hallahmi, B. and Argyle, M. (1997), *The Psychology of Religious Behaviour, Belief and Experience*, London and New York: Routledge.

Berger, P. L. (1971), *A Rumour of Angels* (1969), London: Pelican.

Berger, P. L. (1973), *The Social Reality of Religion* (1969), Harmondsworth: Penguin.

Bolton, N. (1972), *The Psychology of Thinking*, London: Constable.

Bowker, J. (1978), *The Religious Imagination and the Sense of God*, Oxford: Oxford University Press.

Buber, M. (1958), *I and Thou*, Edinburgh: T. & T. Clark.

Clark, W. H. (1958), *The Psychology of Religion*, New York: Macmillan.

Fishbein, M. and Ajzen, I. (1975), *Beliefs, Attitude, Intentions and Behavior*, Reading, MA: Addison-Wesley.

Ghiselin, B. (1952), *The Creative Process*, Berkeley and Los Angeles, CA: California University Press.

Glock, C. Y. and Stark R. (1965), *Religion and Society in Tension*, Chicago: Rand McNally.

Glock, C. Y. and Wuthnow, R. (1979), "Departures from Conventional Religion: the Nominally Religious, the Non-religious, and the Alternatively Religious", in R. Wuthnow (ed.), (1979), *The Religious Dimension*, New York: Academic Press.

Goldman, R. (1964), *Religious Thinking from Childhood to Adolescence*. London: Routledge & Kegan Paul.

Greeley, A. M. (1975), *The Sociology of the Paranormal*, London: Sage.

Gross, R, & McIlveen, R. (1998) *Psychology: A New Introduction*, London: Hodder and Stoughton.

Halloran, J. D. (1967), *Attitude Formation and Change*, Leicester: Leicester University Press.

Happold, F. C. (1975), *Mysticism* (1963), Harmondsworth: Pelican.

Harding, R. (1940), *The Anatomy of Inspiration*, London: Cassell.

Hardy, A. (1978), *The Divine Flame* (Gifford Lectures of 1965 London and Glasgow: Collins, 1966), Oxford: Religious Experience Research Unit, Westminster College. Available from Alister Hardy Research Centre, Department of Theology and Religious Studies, University of Wales, Lampeter, Ceredigion, Wales SA48 7ED.

Hardy, A. (1979), *The Spiritual Nature of Man*, Oxford: Clarendon Press.

Hay, D. (1990), *Religious Experience Today*, London: Mowbrays.

Hebblethwaite, B. L. (1987), "Religious Language and Religious Pluralism", *Anvil* 4 (2), 101–11.

Hick, J. (1989), *An Interpretation of Religion*, Basingstoke: Macmillan.

Hill, C. C. (1979), *Problem Solving*, London: Pinter.

Hill, P. C. & Hood Jr., R. W. (eds.), (1999) *Measures of Religiosity*, Birmingham, AL: Religious Education Press.

Hood Jr., R. W. (ed.), (1995), *Handbook of Religious Experience*. Birmingham AL: Religious Education Press.

Hood Jr., R. W. (1999), *American Psychology of Religion and the Journal for the Scientific Study of Religion*. Paper presented at the annual convention of the Society for the Scientific Study of Religion, Boston.

Hooker, M. (1982), "Myth, Imagination and History", *Epworth Review* 9 (1), 50–6.

Horne, J. R. (1978), *Beyond Mysticism*, Ontario: Laurien University Press.

Howatch, S. (1997), *A Question of Integrity*, London: Little, Brown and Co.

Hudson, W. D. (1974), *A Philosophical Approach to Religion*, London: Macmillan.

Jakobsen, M. J. (1999), *Negative Spiritual Experiences: Encounters with Evil*, Oxford: Religious Experience Research Unit, Westminster College, Oxford.

James, W. (1968), *The Varieties of Religious Experience* (1902), London and Glasgow: Fontana.

Jung, C. (1973), *Psychology and Religion* (1938), New Haven: Yale University Press.

Kaufman, G. D. (1993), *The Face of Mystery*, Cambridge, MA: Harvard University Press.

Laski, M. (1961), *Ecstasy*, London: Cresset Press.

Loder, J. E. (1981), *The Transforming Moment*, San Francisco, CA: Harper & Row.

Lyttkens, H. (1979), "Religious Experience and Transcendence", *Religious Studies* 15, 211–12.

Mackinnon, D. (1978), "Some Epistemological Reflections on Mystical Experience", in S. T. Katz (ed.), (1978), *Mysticism and Philosophical Analysis*, London: Sheldon Press.

Maslow, A. H. (1970), *Religions, Values and Peak Experiences* (1964), New York: Viking Press.

Miles, G. B. (1981), 'Transcendental and Religious Experiences of Sixth Form Pupils: An Analytic Model'. Paper presented at 'Implicit Religion' Conference, Denton Hall, Yorks. May 1981; available from Canon Professor Edward Bailey, The Old School, Church Lane, Yarnton, Oxford OX5 1PY <www.implicitreligion.org>.

Miles, G. B, (1983), 'A Critical and Experimental Study of Adolescents' Attitudes to and Understanding of Transcendental Experience', unpublished Ph.D., School of Education, University of Leeds, U.K.

Miles, G. B. (1993), 'Transcendental and Religious Experiences of Sixth Form Pupils: An Analytic Model.' Paper presented at RIMSCUE Conference September 1993, printed in SPES: Newsletter No. 1 of RIMSCUE Centre, October 1994, Faculty of Education, University of Plymouth, Douglas Avenue, Exmouth, Devon EX8 2AT.

Morgan, R. (1998), "The Bible and Christian Theology", in J. Barton (ed.), (1998), *The Cambridge Companion to Biblical Interpretation*, Cambridge: Cambridge University Press.

Nielsen, M. E. (2000), "Psychology of Religion in the USA" (pp. 1–12).Internet: 'Back to the Psychology of Religion Home Page'.

Otto, R. (1932), *Mysticism East and West: A Comparative Analysis of the Nature of Mysticism*, New York: Macmillan.

Otto, R. (1959), *The Idea of the Holy* (1917), Harmondsworth: Pelican.

Proudfoot, W. (1985), *Religious Experience*, Berkeley, CA: University of California Press.

Russell, B. (1995), "Intuition", in R. Audi (ed.), (1995).

Schleiermacher, F. D. E. (1988), *On Religion: Speeches to Its Cultured Despisers* (1821), trans. R. Crouter, Cambridge: Cambridge University Press.

Scott, P. & Spencer, C. (1998), *Psychology: A Contemporary Introduction*, Oxford: Blackwell.

Smart, N. (1973), *The Science of Religion and the Sociology of Knowledge*, New Haven: Princeton University Press

Smart, N. (1997), *Dimensions of the Sacred* (1996), London: Fontana.

Soskice, J. M. (1985), *Metaphor and Religious Language*, Oxford: Clarendon Press.

Starbuck, E. D. (1899), *The Psychology of Religion*, New York: Scribner's.

Streng, F. J. (1978), "Language and Mystical Awareness", in S. T. Katz (ed.), (1978), *Mysticism and Philosophical Analysis*, London: Sheldon Press.

Thouless, R. H. (1971), *An Introduction to the Psychology of Religion* (1923), Cambridge: Cambridge University Press.

Tillich, P. (1951), *Systematic Theology*, vol. 1, Chicago: Chicago University Press.

Tillich, P. (1968), *A History of Christian Thought*, New York: Simon & Schuster.

Turner, D. (2004), *Faith, Reason and the Existence of God*, Cambridge: Cambridge University Press.

Vernon, P. E. (ed.), (1970), *Creativity: Selected Readings*, Harmondsworth: Penguin.

Vinacke, V. E. (1952), *The Psychology of Thinking*, New York: McGraw-Hill.

Wallas, G. (1926), *The Art of Thought*, New York: Harcourt.

Ward, K. (1994), *Religion and Revelation*, Oxford: Clarendon Press.

Wilson, J. (1961), *Philosophy and Religion*, Oxford: Oxford University Press.

Wilson, J. (1971), *Education in Religion and the Emotions*, London: Heinemann.

Wulff, D. M., (1991), *Psychology of Religion: Classic and Contemporary Views*, New York: Wiley & Sons.

Wulff, D. M. (1995), 'Phenomenological Psychology and Religious Experience' in R. W. Hood Jr. (ed.), (1995), ch. 9.

Zachry, W. (1990), "Correlation of abstract religious thought and formal operations in high school and college students", *Review of Religious Research* 31: 405–412; quoted in R. W. Hood Jr. (ed.), (1995).

19 | Science and Religious Experience

Are they similar forms of knowledge?

PHILOSOPHICAL ASSUMPTIONS AND PSYCHOLOGICAL STRUCTURES

The account of religious experience outlined in Chapter 18 is underpinned by the phenomenological and social science studies reported earlier, all of which are based on Descartes' dualist view of persons as made of two substances, mind (or soul) and body.

The emergence of middle views between dualism and physicalism, briefly described in Chapter 17, does not affect the *personal* views of religious experiences, as described in Chapter 18. That is the way they were experienced, as described by those who had the experiences, underpinned by dualist assumptions. Middle views accept those personal descriptions, but not their dualist assumptions, and so suggest a change of metaphysical foundation principles or assumptions.

All three philosophical views – dualist, middle and physicalist – agree that all experiences are based on the extremely complex electrical network of routes and connections between and within brain cells. The middle views reject the idea of a separate mind or soul advocated by the dualists, but agree with the physicalists that "there is only one physical system and no energy is introduced into the system by some spiritual substance external to it", as Clayton said in Chapter 17. I now accept the middle views.

If this is so, we need to reflect on the status of the *intuitions,* which link sense experience to the different modes of discourse and their forms of knowledge. It is clearly the case that the different intuitions, which link sense experience to empirical, scientific, aesthetic, moral and personal forms of knowledge, are self-evidently universally approved, because every *human* being alive *uses* them.

Therefore, we all have almost overpowering evidence that these intuitions are trustworthy and lead towards forms of truth, personal maturity and wholeness. The aesthete may perhaps have an unconditional call to be an artist or a poet (cf. Wordsworth's call in Chapter 1), and we all love beautiful forms. Most people feel the power of an absolute moral imperative – to be honest, to search for justice, or to be helpful – whether we meet its requirements or feel guilty when we fail. We have explored the call to *personal* communion in Chapter 6, when we discussed Fred. We have evidence that the different intuitions, which lead us to empirical, scientific, aesthetic, moral and personal knowledge, are trustworthy. It follows that if the majority of the world's population is aware of the sacred, the religious intuition is likely to be trustworthy as Alston had demonstrated (ch. 16), as are the other intuitions mentioned. No mode of discourse needs external special substances to be introduced into a person's physical system. The scientific, aesthetic, moral, personal and spiritual voices shout from within – all are Ultimate voices. From where?

Given the overwhelming evidence of support for the religious intuition, this provides support for the claim that "the soul" is the spiritual centre of a person, in the context of the middle views, but we cannot pursue this topic here (Ward 1992: ch. 7; 2000: 30–1).

I must return to the defence of the religious intuition, for it needs qualification. The last chapter ended with the conclusion "*The Faithful . . . have **known** more than they could **say**.*" Having defended religious experience as a form of knowledge when the religious experience of a person is accepted by the appropriate "faith community" as knowledge, in that sense that statement is true. But it is true in other ways, as we will explore later, and in another sense it is an excessive claim, which must be balanced by the fact that "*the Faithful have always **said** more than they could **know***".

Look at the range of World Faiths – atheistic, theistic and naturalistic. Buddhism says that "there is nothing lying behind the showing up of Ultimacy; the point is to journey on an ultimate path towards enlightened acceptance of the world in its wondrous actuality". The theist holds that "God in various conceptions is what shows up in this most amazing way and God is even more terrifying and wonderful than ultimacy experiences suggest". Evolutionary naturalism, or physicalism, affirms that "nature, needing no creator being and subject to no ultimate purpose, is richer and more wondrous than we can imagine" (Wildman and Brothers 1999: 412). Is there any "common ground" to remove barriers between these positions?

If all that seems a bit abstract, let us take an example from Christianity which evokes reflection, on the topic of "saying more than we can know". I quote a Professor of Divinity at Harvard University: "The significance of the category of Christ was expressed in classical Christianity by claims about his deity . . . Christ can properly be understood only as somehow identical with God . . . " (Kaufman 1993: 83). He notes some of the problems this causes: "from the beginning Jews, and later on also Moslems, held that Christians . . . were guilty of idolatry; that is in their talk about Christ they seriously confused and compromised the most fundamental of the monotheistic categories, *God*. (I think they were substantially correct on that point . . .)" (84). Could this be a move towards "common ground"?

There is no space here to discuss this subject in any depth, only to make one key point. Some forms of religion in all World Faiths create barriers between different faith groups within and between communities and nations and have been identified as "licensed insanities" (Bowker 1987). These are always forms of the "authoritarian" or "extrinsic" religious orientations we discussed in Chapters 3, 8 and 14. Responsible members of World Faiths, conscious that we all now live in a global village, with near-instant communication with people anywhere in the world, are pursuing the task of dialogue, exploring common ground and differences in an atmosphere of goodwill.[1]

WILDMAN AND BROTHERS: A SIGN-TRANSFORMATION MODEL
OF RELIGIOUS EXPERIENCE

In our previous discussion of models of religious experience, these models were largely causal, indicating a chain or network of factors and events. Wildman and Brothers propose "to model ultimacy experiences[2] by tracing sign-transformation rather than causes and effects" (Wildman and Brothers 1999: 398). "When one sign stands for another, we speak loosely of sign transformation" (399). Talk of causation suggests causal chains, but Wildman and Brothers rather think of a surging river than a single thread.

Let us take an example. Look back at the call of Moses in Chapter 18. Is this not an example of the transformation of a sign – a continuously lit-up bush linked to his pressing questions about how to help his people? Does not the moment of illumination occur when he receives his call from God, the God within? Did Moses take the bush as a metaphor for God's illumination? The *Concise Oxford Dictionary* says that metaphor is the "application of name or descriptive term or phrase to an object or action to which it is imaginatively but not literally applicable". How does this work? First a person strains inward to grasp an aspect of personal experience of which one had hardly been aware. Next comes a stage of "unfolding" when "new" feelings are put into words. Third, there is then a phase when relevant associations emerge into conscious awareness. Lastly, the experience is expressed in ways in which the individual can better understand both the experience and her or himself. Perhaps this is one way in which religious ideas are put into words (Watts and Williams 1988: 148).

May this experience have happened in "transitional space" – "a transitional sphere . . . that mediates between the spheres of wish and of external reality"?

> God is to be found, not wholly in the world of inner fantasy nor wholly in the world of external reality, but in the transitional world that is . . . "outside, inside, at the border" . . . The world in which God is experienced is a world of "play" and "trust" in which man can come to understand both himself and external reality. (Watts and Williams 1988: 35–6)

After all, one's behaviour could only be observed before and after the experience, so where did it happen? Did you discover or decide that you wanted to be a train driver, a nurse, or that you were in love? Where? And I am not talking about the geographical place! You can simply answer "in your mind". But that makes you a dualist. What can you say if you are sympathetic to the position of the middle views? "Transitional space"? Does God inhabit the "transitional sphere" (see Watts and Williams 1988: 34–7)? Perhaps "transitional space" is a valuable metaphor which facilitates any person in handling the full range of forms of knowledge and their complementary levels of understanding "a person", who is an undivided whole.

BOTTOM-UP AND TOP-DOWN CAUSATION

We have already discussed how intellectual development, as described by Piaget, emerges through stages (in Chapter 4) and that in each stage there is an interaction between bottom-up and top-down causation (in Chapter 17).

The same processes can be observed in the development of moral thinking and religious understanding. Take Piaget's theory of moral development, for example (Piaget 1932). Briefly, it goes through four stages, the egocentric stage (ages 3–5), in which rituals, like washing hands, and rules for playing games, which are introduced to the child, but the child will change the rules to suit its game! The heteronomous stage (= the rule of another) is that stage when rules are imposed on the child by adults, e.g. one older small child quotes a rule to another small child: "You can't have a sweet before lunch." "Why not?" "Because Mummy says so." Then emerge two autonomous (= own rule) stages (internalized rules – now my own). If the older child was at one of these stages, she would have answered the younger child, "Because it will spoil your appetite for lunch." The stage of equality emerges when the child accepts that *each* child gets one *or* two pieces of chocolate – the same for each. So what happens in a rowing boat when two children are given a bun each, but a swan snatches the bun out of the hand of the smaller child? Auntie, operating at the fourth stage of equity, gives the smaller child another bun. Tears stop on the bunless child while the other child bursts into tears: "It's not fair, he's had two buns now and I've only had one!" The key point to note is that in this or any other model of moral development both the bottom-up and the top-down principles are interacting.

We will briefly look at two models of religious development only to demonstrate the point that there are emergent models of religious development. I shall not dwell on unnecessary details. Erikson, a psychoanalyst, suggests an emergent view of personal development through eight stages. Each stage is described by polar opposites.[3] Each person's growth will occur at different points across this "horizontal" scale of each stage. In infancy the child develops trust and hope or withdraws and mistrusts. In early childhood the child either succeeds or fails to develop will and self-control in differing degrees. In the play age, the child experiences the tension between a sense of purpose and achievement through play, or jealousy through failure. During the school age the child develops competences or feels inferior. Through adolescence the more successful individual develops a greater sense of faithfulness. The polar opposites for young adults are inclusive love towards others or exclusive isolation. Adulthood is marked by more or very little care towards people and life. Old age is marked by integrity and wisdom at one pole, or despair, disgust and disdain at the other. Qualities and virtues in the first and last stages Erikson finds safeguarded in religious traditions. Wisdom is provided by a living "religious or philosophical tradition" (Erickson 1968: 140) to which an individual aspires. The numinous threads through the stages often appear as a recurring "sense of a *hallowed presence*" (Erickson 1977: 85–6).

Fowler's stages of faith build on Piaget's work. The first stage – Intuitive-Projective Faith – is marked by the imagination's use of powerful images to picture ultimate reality, avoiding terrifying images if possible. Stage two – Mythic-Literal Faith, is linked to Piaget's concrete operational stage around the age of seven, focusing on religious story and myth taken literally, with developing associated feelings of personal goodness or badness. Stage three is called Synthetic-Conventional Faith. Linked to Piaget's stage of formal operations, the teenager can view the religious stories personally and symbolically, frequently developing a conformist religious outlook. Many adults do not progress beyond this stage.

The fourth stage is called Individuative-Reflective Faith. The individual personalizes and internalizes her or his faith stance, understanding the nature and deeper meanings of religious symbols and stories, as well as recognizing different relative faith positions. Conjunctive Faith, the fifth stage, is a time when the individual seeks to harmonize or integrate different aspects of faith into a meaningful whole. At this stage the individual is open to entertaining the truths expressed in other religious traditions, combining loyalty to one's own communal faith with loyalty to a "community of communities". The sixth stage – Universalizing Faith – focuses on two qualities developed by very mature individuals. Decentring of self or selflessness, and emptying of self or total focus on the needs of others are qualities found in people like Gandhi, Mother Teresa, Martin Luther King (Fowler 1981: 203–4; Dykstra and Parks 1986: 28–31) and Nelson Mandela, but are also found in wonderful people in local communities too.

In this section we have taken note of the fact that there are emergent, or bottom-up, views of the development of religious personhood. The purpose of this exercise is to highlight the fact that initially religious experience is *not* fundamentally like this. Religious experiences can happen to anyone at almost any age. Do they occur through 'bottom-up', 'top-down' causation or both?

Let us now study a religious experience of a five-year-old child. The analysis, which I will quote below, is a summary of the analyses of many groups of teenagers.

THE ANTS: A RELIGIOUS EXPERIENCE OF A FIVE-YEAR-OLD

When I was about five I had the experience on which, in a sense, my life has been based. It has always remained real and true for me. Sitting in the garden one day I suddenly became conscious of a colony of ants in the grass, running rapidly and purposefully about their business. Pausing to watch them, I studied the form of their activity, wondering how much of their own pattern they were able to see for themselves. All at once I knew that I was so large that, to them, I was invisible – except, perhaps, as a shadow over their lives. I was gigantic, huge – able at one glance to comprehend, at least to some extent, the work of the whole colony. I had the power to destroy or scatter it, and I was completely outside the sphere of their knowledge and understanding. They were part of the body of the earth. But they knew nothing of the earth except the tiny part of it which was their home.

Turning away from them to my surroundings, I saw there was a tree not far away, and the sun was shining. There were clouds, and blue sky that went on for ever and ever. And suddenly I was tiny – so little and weak and insignificant that it didn't really matter at all whether I existed or not. And yet, insignificant as I was, my mind was capable of understanding that the limitless world I could see was beyond my comprehension. I could know myself to be a minute part of it all. I could understand my lack of understanding.

A watcher would have to be incredibly big to see me and the world around me as I could see the ants and their world, I thought. Would he think me to be as unaware of his existence as I knew the ants were of mine? He would have to be vaster than the world and space, and beyond understanding and yet I could be aware of him – I was aware of him, in spite of my limitations. At the same time he was, and he was not, beyond my understanding.

Although my flash of comprehension was thrilling and transforming, I knew even then that it was no more than a tiny glimmer. And yet, because there was this glimmer of understanding, the door of eternity was already open. My own part, however limited it might be, became in that moment a reality and must be included in the whole. In fact, the whole could not be complete without my own particular contribution. I was at the same time so insignificant as to be almost non-existent and so important that without me the whole could not reach fulfilment.

Every single person was a part of a Body, the purpose of which was as much beyond my comprehension now as I was beyond the comprehension of the ants. I was enchanted. Running indoors, delighted with my discovery, I announced happily, "We're like ants running about on a giant's tummy!" No one understood, but that was unimportant. I knew what I knew.

It was a lovely thing to have happened. All my life, in times of great pain or distress or failure, I have been able to look back and remember, quite sure that the present agony was not the whole picture and that my understanding of it was limited as were the ants in their comprehension of their part in the world I knew.[4]

Let me summarize what the teenagers who analysed this passage discovered.

1   *The five-year-old child*:
   (a) The child *knew* that she was an unseen, unknown, large and powerful force who could destroy the ants and the work of their whole colony.
   (b) The child *knew* that, seeing the tree, the sun shining and the blue sky that went on for ever and ever, that she was so little, weak and insignificant, that it did not matter if she did not exist. Yet she knew she was a part of the world, and could not understand it all.
   (c) The child *knew* that she was aware of a watcher, vaster than the world and space.
   (d) The child went indoors and announced happily, "We're like ants, running about on a giant's tummy." No one understood. The child didn't mind. The child *knew* what she *knew*.

However, the groups of teenagers all recognized that much of the language was not that of a five-year-old, and discovered at least two more sophisticated layers of language in the passage.

2   *Teenage level*: The concept of the "Body" must have been discovered later by the child – from Pauline theology (e.g. 1 Corinthians 6:15 "your bodies are part of the body of Christ") – perhaps she learnt it at church or in school.
3   *Mature years*: Some language is fairly sophisticated, showing mature reflection, e.g. "the door of eternity was already open" which is hardly the language of a small child or a teenager.

So is religious experience knowledge? The answer for the five-year-old is "yes". The *use* of language states it is knowledge: the child *knew*. Yes, the child appropriated concepts learnt from the social environment, which have *emerged* into the child's consciousness (upward causation) to describe experiences read or heard about earlier, but the fact that religious language was *chosen* means that the child *wishes* the *interpretation* to be religious. The child *purposely*, even if unconsciously, selected her religious *intuition* to make the connections between her sense experience and the religious mode of discourse she uses to articulate her new-found knowledge. The experience of the "watcher" almost certainly exercises "downward causation" joyfully embraced, i.e. this knowledge influenced this person's behaviour.

The interpretation is elaborated over time, becoming increasingly significant and important for this person. However, a personal religious experience, to be knowledge, must pass James's tests for the individual. Does this experience bring delight? Does this experience bring good fruits? The account offers positive answers to both questions.

But does the possibility of this reinterpretation process not make this and any other religious experience an imaginative mishmash, and totally untrustworthy? No. Why not? Take any experience, and you will find you are doing this all the time with many experiences you have. An example: you rush into the office when everyone is very busy, and a colleague rushes out of the door at the same time and bumps your shoulder very hard. You both rush on, no time to speak, and think no more about it. However, at the end of the day you reflect on this incident. He is not your favourite colleague. He did it on purpose! You are angry . . . The next day you change your mind!

After an analysis of the account of *The Ants* was completed by different student groups, I told them that the person was a female, fifty-five years of age at the time she recorded this experience.[5]

Here I must draw one thread already described into this discussion. In Chapter 9, we discussed the 'Spirit of the Child', reporting that Rebecca Nye, in analysing the experiences of children between six and eleven years, discovered what she calls *relational consciousness*. You can discern all the four conscious forms of perceptiveness Nye mentions in *The Ants* experience above: "I–Others", "I–Self", "I–World" and "I–God".

WHEN DOES THE INSIGHT WITHIN AN EXPERIENCE BECOME PERSONAL KNOWLEDGE AND THEN COMMUNAL KNOWLEDGE?

This last question must be faced. When does an experience, or, more accurately, the insight within an experience, become knowledge? That is the question faced by a scientist, who is advancing a new theory, who recognizes the gap of time between their own conviction backed by evidence and the slower development of conviction in the scientific community. The same question faces a composer who hopes that his opera will be accepted into the repertoire. This question, as a moral question, is faced by reformers, like Wilberforce, who looked to the ending of slavery in Britain. This question haunts the convictions of Hosea that God is a God of love, shared by Jesus of Nazareth and promoted as "Treat others as you want them to treat you. This is what the Law and Prophets are all about" (Matthew 7:12 CEV). There was an interval of time before communities accepted the inner authority of the teacher and the teaching, both morally and religiously as of divine origin.

This process of dialogue between individuals and their communities, and now between members of World Faiths living in our global village, illustrates the principle of dialogue identified by Revd Canon Professor John Bowker, who repeats that "our concepts, pictures and imaginings about God, as about anything else, are necessarily incomplete, provisional, approximate, and corrigible" (Bowker 1987: 74).

THE INTUITIVE EMOTIONAL SELECTION OF A FORM OF KNOWLEDGE

Let me try to summarize pictorially the tentative conclusions of our explorations so far, based on Wynn's insights. Imagine a clock face with two hands on the central spindle. Remove the two hands and the central spindle and replace the spindle with a much longer one on which is mounted ten hands. Instead of 12 at the top and 6 at the bottom of the clock face, cover the numbers with a circular blob of white paint. Paint over the numbers 1–5 and 7–11 with different colours each representing a form of knowledge, such as physical, chemical, biological, aesthetic, moral, personal, religious.

One colour represents one form of knowledge and one hand has a matching colour. Pretend for a moment that there are only ten forms of knowledge, for which you need ten hands painted with each form of knowledge colour. At rest, five hands are pointing to the white blob at the top of the clock face (12) and five pointing towards the white blob at the bottom (6). This is a monitoring dial inside your brain, to show it is working properly. If your brain is fully wired for all forms of knowledge (mine is not – no good for wine or art discussions), what happens is that when you start an exploration, (see Chapter 2) your senses focus on the matter of interest. This involves percepts, thoughts, then particular feelings or emotions which will alert the emotional magnet distinct for each form of knowledge and fixed at the head of each hand. This will alert that arrow which will then point to its form of knowledge. The brain will then select concepts in that form of knowledge to articulate first the reflections and then the findings of the host person. Redesign the "clock face" dial to include all forms of knowledge (cf. Wynn, pp. 134 and 310 above, and pp. 370–2 below).

In the last century, whether we were scientists or artists, moralists or religionists, we all tended to use the monocular vision of a primary mode of discourse, e.g. science or religion. But now in the 21st century, is it not time for us all to recognize, that in all such disciplines, we need to use a range of forms of discourse to give a more accurate picture of our area of study? Fundamentalist science or religion fails to communicate the rich, multi-discourse understandings of the human achievements of science, or the scientific or human insights which inform a more credible understanding of the nature of religious experience and religion.

SCIENCE AND RELIGIOUS EXPERIENCE: ARE THEY SIMILAR FORMS OF KNOWLEDGE?

We have already seen in Chapter 18 that Plantinga has already demonstrated that three kinds of basic beliefs are fundamental beliefs which can *practically* establish knowledge – percept beliefs, memory beliefs, and beliefs about other peoples thoughts. They need no supporting arguments, just supporting examples to give adequate evidence that this *pragmatic* procedure works. Swinburne makes the same point using different language (ch. 16) when outlining his Principle of Credibility – a rational *practical* principle that "when a person believes something to be present in normal sense, it usually is" and it would be inconsistent not to apply this to religious experience. Likewise his Principle of Testimony – we think that what others say they perceived, probably happened. This has statistical support, already quoted (ch. 18), that between a third and a half of United States and UK citizens report such experiences, making his principle self evidently reliable.

We have seen (ch. 16) from Alston's analysis of "perceiving God" that both science and religion begin with the formation of knowledge from the eight elements involved in processing perception. These elements are: the **experience, perception** – *intuition crossing the logical gap between "object" and "percept"*, **memory** *adding associated ideas, and* **concepts** *which help to classify the experience, followed by conscious and unconscious* **reflections, reason, interpretation, and evaluation**. This is based on the claim that sense perception "is reliable and the argument for that claim is that our success in predicting sense experiences on the basis of sense perception . . . is best explained by supposing sense perception . . . to be reliable" (Alston 1993: 141). Likewise, 'personal perception' and 'mystical perception' are pragmatically justified, but on different bases, as outlined in the summary of Alston's work in Chapter 16.

Dr Mark Wynn seeks to explore how he can support Alston's work by taking note of those who are studying the relationship of aesthetic values to a physical object, or moral values to physical behaviour, or humour to physical behaviour and who tend to conclude that one cannot relate values to a physical cause.

Wynn begins his exploration by quoting a situation where John and Joan are seated in an over full subway train where there are no empty seats. One standing passenger, a woman in her thirties, uncomfortably holds two almost full shopping baskets. John is aware of her presence but not sensitive to her discomfort. Joan is aware of the woman's discomfort and is morally uncomfortable unlike John who lacks the necessary moral sensitivity (quoted from Blum 1994: 31–3, by Wynn 2005: 1). Wynn notes that emotions are evoked in Joan through her focused attention and sensitivity to the situation. So there are situations in which our response is affectively toned.

Wynn wonders how this could affect one's reflections on whether one can identify values in a religious experience "as a kind of affectively toned sensitivity to the values that 'make up' God's reality?" (Wynn 2005: 5). For example, McDowell, who has studied the relationship of aesthetic or moral values to physical events recognizes that "the property of being morally wrong . . . has a normative dimension, and its extension is therefore only visible in the light of a normative perspective, rather than the perspective of empirical science." "McDowell urges that we cannot trace value experience to qualities in the world that are value free" (7).

Alston considers the possibility that the state of consciousness through which we perceive God is purely affective in terms of its phenomenal content and in this connection he writes:

> One nagging worry is the possibility that the phenomenal content of mystical perception wholly consists of affective qualities, various ways the subject is feeling in reaction to what the subject takes to be the presence of God. No doubt such experiences are strongly affectively toned; my sample is entirely typical in this respect. The subjects speak of ecstasy, sweetness, love, delight, joy, contentment, peace, repose, bliss, awe, and wonder. Our inability to specify any other sort of non-sensory phenomenal qualities leads naturally to the suspicion that the experience is confined to affective reactions to a believed presence, leaving room for no experiential presentation of God or any other objective reality. (Alston 1991: 49–50 cited by Wynn 2005: 8)

Wynn's work is focused on this topic, that "a great deal of religious experience is affectively toned." Alston has hinted above that a genuine perception of God is a "nagging worry" as it might just be "purely affective". Wynn notes that at the beginning of the passage Alston portrays the affective component as a "feeling in reaction to what the subject takes to be the presence of God". Wynn comments, that "this formulation assumes that the element of feeling in a mystical perception is a 'reaction' to (what is presumably) a feeling neutral thought" (8). But this way of talking portrays feelings as sensations, which are responses, like a bruise on your arm caused by a bump. This would certainly suggest that a religious experience of this kind would lack evidential value to sustain belief in an 'objective God'. So Alston's picture of God can seem to be an "affective reaction" to a "believed presence" (9).

Wynn suggests that if we adopt McDowell's conception of affective experience (rather than Alston's) "it will be easier to see how a religious experience whose phenomenal content may be purely affective may still be veridical". McDowell's conception of affective experience is that "the 'real' source of that (religious) experience is a set of 'objective value properties'" (10). While Alston downplays this view, preferring his causal view, that "perception of objective external realities" causes a religious experience, "he admits the possibility of affects playing this sort of role" (10). So Wynn is encouraged to prefer McDowell's model, which holds that a religious experience contains two components, a thought component and an affective component.

Another advantage of McDowell's account is that it is not possible to identify the stimulus conditions, which lead to a religious experience. Alston does think of this possibility, when he suggests that these "have to do with God's purposes and intentions . . . " (12–13). Perhaps God brings about such conditions miraculously. Put another way it is not possible to pair off any set of stimulus conditions with a mystical perception. Thus the supervenience of a religious experience can take place in circumstances which might involve a particular 'eureka', an apparently random set of thoughts, but in a physical situation which may or may not be significant.

Wynn finds that this McDowell-inspired approach illustrates the difficulty of sometimes finding language appropriate to record the religious experience.

In summary form, Wynn hopes that his exploration of McDowell's view and Alston's view can be fruitfully combined in some of the ways mentioned here. First, there is no way of identifying any patterned correlation between the supervenience of a mystical perception with any particular kinds of thought or physical stimulus or context. Secondly it is possible that a mystical perception may be purely affective in some situations. Thirdly, the combination of the two theoretical approaches mentioned is more acceptable to "the thought that the affective dimension of religious experience is cognitively significant" (17).

In conclusion Wynn has shown that "affective experience can disclose values". In the first example the values were moral. Wynn has fruitfully related Alston's "affectively toned theistic experiences" to McDowell's "cognitivist reading of value experience" (28). We should not expect any correlation therefore, within the supervenience of "mystical perception" with "non-evaluative features of the world", which even if affective may be evidence to justify belief as knowledge. It follows that theistic experience filled with deep feelings can lead to a mode of value perception even if purely affective, yielding religious knowledge. Wynn is clear: emotional feelings can constitute thoughts. Wynn emphasizes that in matters of religion we do not need to opt for objective content over emotional form or vice versa, for form and content are not properly separable. He is clear that perception and conception are often infused by feeling. So an *emotional* element can trigger, or transmute, into a *thought* component.

If eighteen chapters summarized the findings of the twentieth century, then this chapter, with Wynn at the helm, even though he has drawn on elements from Alston, Wildman and Brothers, and others we have not mentioned, heralds in the twenty-first century. For us, this suggests that God speaks to a person from within.

We have already explored in earlier chapters the question of whether science and religious experience have similar forms of knowledge from the evidence of the subject specialists. We now note the judgements of a distinguished epistemologist, Professor Robert Audi (Audi 2005).

If we ask, "What is the nature of knowledge?" Audi gives us a clear answer: "Knowledge arises in experience. It emerges from reflection. It develops through inference . . . knowledge is at least true belief" (Audi, 2005: 220–1). Audi also reminds us that "the concept of knowledge is not itself precise" (230). At the end of his detailed analyses he concludes that, "Knowledge is *true belief based in the right way on the right kind of ground*" (251), after he has described many different ways of establishing knowledge.

Audi views scientific, moral and religious knowledge as distinct forms of knowledge. "Scientific knowledge . . . emerges only after we use imagination, both in formulating questions and in framing hypotheses to answer them . . . (a) place where scientific *invention* occurs" (Audi, 2005: 260). After exploring science he concludes: "I am inclined to say that in spite of both scientific error and the fallibility of scientific attitudes, we do have much scientific knowledge, even if it is all only approximate knowledge . . . (but) we are quite some distance from the artificial picture one might have of scientific knowledge as a set of beliefs of precisely formulated and strictly true generalizations, arrived at by inductive transmission of knowledge from its basic sources in experience and reason" (267). He notes that, "Scientific generalizations . . . are inductively known . . . on the basis of the facts, such as observations data, which we use to confirm them. If there can be scientific knowledge on this basis, then there can be knowledge based on inductive grounds, grounds that do not entail the proposition we know on the basis of them. Why then should there not be inductively grounded moral knowledge even if no moral knowledge is deductively grounded?" (273). In other words, the same kind of argument justifying science can be applied to moral knowledge to justify it, for "We certainly appeal to facts to justify moral judgements. I might justify my judgement that I ought to meet Jane by citing the simple fact that I promised to" (273).

Concerning religious knowledge, Audi makes it clear that nothing in his analysis of the nature of knowledge "implies either that there can or that there cannot be cogent arguments for God's existence. For instance, nothing said about the basic sources of knowledge . . . implies that those sources could not in some way lead to arguments yielding knowledge of God or of some other spiritual reality" (278).

Concerning "the important arguments for God's existence – of direct (non-inferential) knowledge of God", which is the focal point of our exploration of religious experience, Audi makes it quite clear that the framework of his book on epistemology does not rule out this possibility (279). "One may formulate an argument that proceeds from premises describing the *occurrence* and character of the experience" (283). Audi recognizes that a special sensitivity is needed for seeing the beauty in a painting, so it follows that when Gerard Manley Hopkins says "The world is charged with the grandeur of God", such a response may also need a special sensitivity (282). Audi confirms that "there are times when a belief is justified not by grounding in one or more conclusive arguments, but by its support from . . . many sets of independent premises none of which alone, would suffice to justify it" (283). It is worth noting that, "whatever the grounds needed for justified theistic belief, weaker ground will suffice for theistic faith" (286).

So we may conclude that it is just as inaccurate to talk about scientific *proof*, as requesting *proof* for the existence of God, so the question need never again be threatening in a Religious Education classroom or anywhere else! Audi puts it this way: "It seems a mistake to talk of scientific proof at all if it means (deductive) proof of scientific hypotheses or theories from observational or other scientific evidence" (286). Scientific knowledge "is commonly either approximate knowledge, often known to need refinement, or knowledge of approximations, formulated with the appropriate restrictions left unspecified. There is good reason to think that we also have, and certainly have not been shown not to have, moral knowledge. And there is apparently no cogent reason to deny the possibility of religious knowledge" (286).

I think it is possible to draw some other conclusions from Audi's analyses relevant to our explorations. He has just told us that, "Scientific generalizations . . . are inductively known . . . on the basis of the facts, such as observable data, which we use to confirm them. If there can be scientific knowledge on this basis, then there can be knowledge based on inductive grounds, grounds that do not entail the proposition we know on the basis of them" (273). From this Audi has shown us that moral knowledge is supported by facts, in his example, the fact was "the promise made". That is not visually observable – it is a *value* judgement – a *moral* value – a promise. But a beautiful scene, a memorable song which others thought was just noise, a sympathetic person, and a religious experience, are all affirmed on the basis of *value* judgements – evaluations of experiences of a painting or a view, music not noise, a person, a religious situation, and in each case evidence within that mode of discourse can be cited, and we have cited a huge amount evidence supporting the importance of religious experience – at least a third of the population of the English-speaking countries each side of the Atlantic. We can borrow Wynn's insight here, that religious experience contains feelings or emotions and thoughts, and apply it to all the most important human forms of knowledge, like aesthetic, moral, personal, as well as religious knowledge – which are based on value judgements for their evidence. That does not devalue the importance of science, but complements it, for all these modes of thinking use inductive thinking as their foundations. Does that not mean that they all have similar forms of knowledge, and are all equally valuable?

So it seems that it may be that, apart from Logic and Mathematics, all forms of knowledge are pragmatically justified through their use, e.g., making a chair, formulating a hypothesis, loving our neighbour, making poverty history, worshipping God.

If so is it not reasonable to suggest that science and religious experience are similar forms of knowledge?

As well as similarities in the basic structure of these forms, there are differences of which the first difference needs commenting on again, that is the different degrees of freedom in interpretation possible in the different forms of knowledge. In recognizing objects described by a common language there are no degrees of freedom. There are very few in sciences whereas in religion, there is in theory complete "intellectual freedom". Belief and experiences, indicated by Alston above, influence the embraced orientation of the religious, but also of the agnostic and the atheist.

One might be led to ask the question, why all this freedom of interpretation in "choosing" your lifeway? Professor John Hick provides the best answer I know. "If we were from the beginning set 'face to face' with God we would never be able to make a free response to the Deity. There could be no question of freely loving and choosing to worship One whose very presence overwhelms us" (Hick 2004: 44).

The second key difference can show how science (i.e. all sciences) and religious experience are similar forms of knowledge. Science becomes knowledge through the main route of knowledge-formation. "Knowledge arises in experience. It emerges from reflection . . . " (Audi, 2005: 220). This is *conscious* knowledge formation, which occupied a time span of varying lengths *after* the initial experience. But "direct experiential awareness of God" often confronts a person in a moment of time as conceptually formed knowledge. It may have been growing in a *pre-conscious* or *unconscious* mode, a possible time span *before* the experience. (Alston 1993:35; Watts 2002: 96; Ch. 18 above; pp. 360–3 above; cf. Audi 2005: 281–3). This direct apprehension happens in other situations for some people, 'at first sight': falling in love, deciding to become a dancer, singer, builder, driver, after being instantly inspired by some meeting or event. This can inspire, for example, an instant new aesthetic judgement in music or art, or moral judgement on patterns of human behaviour.

At the end of Chapter 16 (314) Ward noted that within the global village a converging account of the qualities of God will emerge. He focuses attention on "converging spirituality" (2004: 226f.) and says, that "Global faith will then not be seen as a new world religion, superseding all others. It will rather be an attitude which can be taken by an adherent of any tradition, but which seeks to deepen its understanding of faith by attention to other traditions, and by reformulating its own principles in the light of the criticisms other traditions may make" (231). He centres on "the experiential view of religion" and "inclusivism", suggesting that "the ultimate spiritual Reality . . . is genuinely known in various partial ways in diverse religious traditions" (227). "The global view . . . has come to see the many religions of the world as ways of learning to transcend self by participating in a Supreme Objective Good" (232). Ward notes that "there is a general descriptive core, which fits many ideas of the Transcendent. In many traditions there is said to be a supreme reality, which embodies perfection, and that perfection includes such characteristics as wisdom, freedom, compassion and bliss" (233). Hopefully these will "favour forms of religious faith which are critical, experiential, personalist and global" (235).

We now have greater understanding of distinctive forms of knowledge, insights into the hierarchy of forms of knowledge, and have discovered common methods of constructing forms of knowledge. This includes knowledge derived through insights from and about religious and scientific experiences, and revelations from other forms of knowing. There is common ground between Humanists, whether atheist, agnostic, or religionist. There are some fundamental values and beliefs shared by different World Faiths. Through dialogue and congenial social contact between Faith Communities we hope for peace and harmony in our multi-faith global village. Perhaps the new model of 'religious experience' explored by Wynn and others may provide a model for all World Faiths to take seriously – one possible starting point for inter-faith dialogue (Forward 2001: 119f).

It is fitting that Professor Putnam provides a clue as to how dialogue may work. His admission regarding the importance of the religious dimension of life (Putnam 1998:1, 179; see page 155 above), and indeed his reminding us that Wittgenstein, no less, had a very respectful attitude to religious belief, provides support for the necessity of developing trust and fostering compassion.

NOTES

1   For example, *Interreligious Insight: A Journal of Dialogue and Engagement* 1 (1) (January 2003), from Interreligious Insight, c/o World Congress of Faiths, 2 Market Street, Oxford OX1 3RF. See Van de Weyer (2003).
2   Their term for transcendental, religious or other spiritual experiences.
3   Erikson (1963). For a comprehensive summary and evaluation cf. Wulff (1991: 369–98, 403–10).
4   From the archives of the Religious Experience Research Centre, Lampeter, quoted in Robinson (1977a: 12–13). Used by Miles (1983: 436).
5   Miles (1983: 224–5a, 246, 250, 436). Copy at Alister Hardy Research Centre, Department of Theology and Religious Studies, University of Wales, Lampeter, Ceredigion, Wales SA48 7ED.

REFERENCES

Alston, W. P. (1993), *Perceiving God* (1991), Ithaca, NY, and London: Cornell University Press.
Audi, R. (2005) *Epistemology*, 2nd edn, New York: Routledge.
Blum, L. (1994), *Moral Perceptions and Particularity*, Cambridge: Cambridge University Press.
Bowker, J. (1987), *Licensed Insanities*, London: Darton, Longman & Todd.
Dykstra, C. and Parks, S. (eds), (1986), *Faith Development and Fowler*, Birmingham, AL: Religious Education Press.
Erikson, E. H. (1963), *Childhood and Society*, 2nd edn, New York: W. W. Norton.
Erickson, E. H. (1968), *Identity, Youth and Crisis*, New York: W. W. Norton.
Erickson, E. H. (1977), *Toys and Reasons: Stages in the Ritualisation of Experience*, New York: W. W. Norton.
Forward, M. (2001), *Inter-religious Dialogue*, Oxford: One World.
Fowler, J. W. (1981), *Stages of Faith*, San Francisco, CA: Harper & Row.
Hick, J. (2004), *The Fifth Dimension: An Exploration of the Spiritual Realm*, Oxford: One World.
Kaufman, G. D. (1993), *The Face of Mystery*, Cambridge, MA: Harvard University Press.
McDowell, J (1981), "Non-Cognitivism and Rule Following" in S. H. Holtzmann and C. M. Leich (eds), (1983), *Wittgenstein: To Follow a Rule*, pp. 141–62; London: Routledge and Kegan Paul.

Miles, G. B. (1983), "A Critical and Experimental Study of Adolescents' Attitudes to and Understanding of Transcendental Experience", PhD dissertation, School of Education, University of Leeds.

Piaget, J. (1932), *The Moral Judgement of the Child*, London: Routledge & Kegan Paul.

Putnam, H. (1998), *Renewing Philospphy* (1992), Cambridge, MA and London: Harvard University Press.

Robinson, E. (1977a), *The Original Vision*, Oxford: Religious Experience Research Unit, Westminster College, Oxford.

Van de Weyer, R. (2003), *A World Religions Bible*, Alresford: O Books, John Hunt Publishing.

Ward, K. (1992), *Defending the Soul*, Oxford: One World.

Ward, K. (2000), *Christianity: A Short Introduction*, Oxford: One World.

Ward, K. (2004), *The Case for Religion*, Oxford: One World.

Watts, F. and Williams, M. (1988), *The Psychology of Religious Knowing*, Cambridge: Cambridge University Press.

Wildman, W. T. and Brothers, L. A. (1999), "A Neurophysical-Semiotic Model of Religious Experience', in R. J. Russell, N. Murphy, T. C. Meyering and M. A. Arbib (eds), *Neuroscience and the Person*, Notre Dame, IN: University of Notre Dame Press.

Wynn, M. R. ((2005), *Emotional Experience and Religious Understanding*, Cambridge: Cambridge University Press.

Glossary

Agnosticism means 'not known' or 'I don't know'. In a religious context an 'agnostic' is one who does not wish to deny or affirm that God exists. The word was invented by Thomas Huxley in 1869. We know that there are no proofs for basic assumptions, and it is not possible to mount any. We can only give reasons. As Huxley found no proof, not realising that all systems have intuitive unsupported assumptions, which communities of like-minded people 'find sensible' and accept, he wrote, "I neither affirm nor deny the immortality of man. I see no reason for believing it, but on the other hand, I have no means of disproving it. I have no . . . objection to the doctrine."[1]

atheism is defined as "disbelief in God or gods" (Gk. *atheos* without God, *a* not, *theos* God).[2] When expressed as a personal statement of belief, e.g. 'Bert <u>does not</u> *believe in* God', the *use* could be atheistic or agnostic, and needs further clarification. In fact, Bert <u>does not</u> *believe in* God and also <u>does not</u> *believe that* 'gods or God exist'. Jim <u>does not</u> *believe in* God, but *believes that* God <u>does</u> exist (cf. C. S. Lewis, *The Screwtape Letters*, HarperCollins). '*Belief that*' is a doctrinal, cognitive or knowledge claim; '*belief in*' is about 'trust in' or 'faith in' – an emotional attachment, a relationship. Bert holds **the most common *use* of 'atheism'**, i.e., **gods do not, or God does not exist.**

bad faith is an idea found in existentialism meaning 'self-deception' so one avoids 'facing up to one's true self' (e.g. 'smoking is O.K.' even though I have lung cancer). **Bad faith** means the same as another term existentialists use, **inauthentic existence**. It means you are not 'true to your better self', that you are drifting along with the crowd (e.g. of smokers), rudderless, and you are not exercising your freedom to become the person you should be and want to become deep down (see **existentialism** and **good faith** below, and Chapters 8 & 9).

behaviourism expresses the *outlook on life* that statements must be open to demonstrable proof. In practice, this is an alternative term to logical positivism, which is explored below. It is the belief that a human being is the sum of her or his actions, their stimuli and responses, or behaviours and their effects.

Christian Faith. A model used by some Christians in exploring their faith-stance is to draw a pentagon on a piece of paper, with one label at each corner: Reason; Scripture, encoded revelation; Tradition, a denomination's interpretation and practice; Experiences of life, empirical and personal; Religious experience, personal enrichment or direct revelation. They can then plot their position within the pentagon to indicate the relative importance of each mode of knowing to them. The pattern may change more than once as the child grows older. Members of other Faiths can use this model if they wish.

coherence and **correspondence** theories of truth: see **truth** below.

combinatorial analysis means that you combine different facts, ideas and reasons in a sensible logical way. Piaget uses the term as a characteristic of formal operational thinking. See the discussion of Piaget in Chapter 4.

critical realism takes note of relevant observations, facts, empirical evidence, reasons, theories or hypotheses, and ideological beliefs to give an account of the way the world is, so that its overall view **corresponds** with relevant data, often from different forms of knowledge, which are related together in a **coherent** way. See **truth** below.

deduction is the method of thinking which begins with *assumptions*. For example: If a + b = 5, and a + c = 7; if c = 4 what is the value of 'b'? Don't think about the answer so much as the method you use to work out the answer: that is deduction. Look up the section on 'rational knowledge' in Chapter 2 and look again at the incident involving the Wayfarer dinghy. It starts from two *assumptions*, from which the results are *deduced*. Also see **hypothetico-deductive thinking**.

deism is the view that true religion is natural religion, that is, natural religion is deduced from nature, as in the argument from design. Natural religion finds expression in nature as in beauty, consistency of natural laws and the interactive harmony of nature. True religion is an expression of mature human nature responding to the benevolence of God expressed through the fruits of creation. Deism, then, defends natural religion, but rejects revelation: the idea that God can reveal wishes or will to human beings. So deists frequently reject revealed religions like Christianity or Judaism, though members of revealed religions often embrace natural theology, that is ideas of God, and beliefs about God, derived from reflections on nature.

determinism is the view that every event is the result of a previous series of events, so that everything that happens is the result of independent natural (and human) causes which interact, and over which an individual has no control. It implies that the future is as unalterable as the past, for everything is linked together in a causal chain. Newtonian physics has been viewed in this light, while quantum theory admits degrees of indeterminacy, and chaos theory accepts 'degrees of freedom' as do the biological and human sciences (see Chapters 5 & 6). Our understanding of degrees of freedom is a dimension of downward causation, with increasing degrees of freedom the higher up the hierarchy of forms of knowledge one is working in.

direct realism is the theory that what you see is unmediated, that is a spade is a spade! **naïve realism** is the view that every object has a one to one correspondence with a concept. I have argued throughout that all perceptions are interpreted with concepts, which are evaluated, and gathered into schemes of thought, in order to distinguish information from knowledge (see Chapter 2). Some of these schemes of thought which are 'read off' from data, reasons, and hypotheses, are very complicated as we have seen, hence the preference for the philosophical position of **critical realism**.

dualism exists where there is a polarity between two objects of ideas which some find impossible to hold together, e.g. finite man and infinite God, mind and matter. It is most often used to refer to the tension between mind and matter, and how they relate (see Chapters 17 &19).

emotivism is the view that moral judgements, aesthetic statements, beliefs about persons you know, religious beliefs are not knowledge – a view advanced by **logical positivists.** Emotivists, including the logical positivists, say that all these kinds of statements are just expressions of my feelings or emotions, hence the term emotivism. All those who hold this view are **non-cognitivists.** That is they say such views express feelings but not knowledge. Those who say that moral judgements recognize the moral or immoral properties of actions are **cognitivists.** That is they say they are knowledge claims. 'You *know* that boy is honest.' The hierarchical model of forms of knowledge, advanced in this book, asserts the commonly held view that there are different kinds of knowledge, which have been distinguished. They include empirical, moral, aesthetic, personal and religious knowledge (see Chapter 4).

empiricism is the view that experience is the primary data of human knowledge. Logical Positivists take the view that experience must be largely based on the five senses – sight, hearing, touch, taste and smell, though other recognizable experiences are included as they relate to concepts, such as dizziness. Beliefs are upheld if they can be shown to relate to experiences. Logical positivists are empiricists (see Chapter 2). Do note that **critical realists** accept the knowledge recognized by logical positivists, but rejects their narrow definition of knowledge, recognizing other kinds of knowledge outlined in the hierarchy of knowledge outlined earlier (see Chapter 6).

epistemology simply means the study of knowledge. When someone says 'the epistemological foundations of logical positivism are vested in the verification principle' they are saying in a complicated way that the knowledge base of logical positivism is their test 'the meaning and truth of a sentence depends on the steps taken to show it is true'. In Chapter 4 the defenses of different kinds of knowledge were epistemological. That is they were trying to show what criteria were acceptable to make aesthetics or morality, for example, forms or kinds of knowledge.

existentialism a mid Twentieth Century philosophy focused 'on being human', well expressed in novels by Sartre (e.g. *Nausea*) and Camus (e.g. *The Plague* and *The Stranger*). It centres on four questions: What is existence? What is human existence? What is 'authentic existence' or 'being true to your (better) self"? What is 'inauthentic existence' or self deception, which holds you back from being your true self? (see **bad faith** above and **good faith** below; Chapters 8 & 9).

foundationalism is concerned with a two tier structure of knowledge, arguing that there are *foundation beliefs*, often called *basic beliefs*. These basic beliefs can be epistemologically justified (i.e. accepted as knowledge claims) if they are based on emotional or rational intuition, empirical observations and organizations, or a process of induction or deduction. Intuition is often based on induction, which means from a particular observation one can make a generalization. There are three kinds of basic beliefs: *perceptual beliefs*, e.g. you see one duck, so all other birds, which look the same, you call ducks, *memory beliefs*, and *beliefs about other people's thoughts* (see pp. 352 & 363, Plantinga on *foundational beliefs*). Science when it moves from the observation of particular objects or situations to scientific theory is a model of inductive thinking. The second tier of the structure of knowledge consists of inferred beliefs or truths, which are based on basic beliefs, e.g. the radiators are off (basic belief); "you must be cold" (inferred belief).

freewill relates to freedom. It is also the opposite of **determinism**. Briefly, it asserts that to be fully human you constantly exercise your choice of action according to your wishes within the environmental, sociological, psychological and personal constraints, which surround each of us. Put another way, we are limited by where we live, our parents, teachers, our abilities and temperament, but there is no reason why you and I cannot exercise our freedom and try and remove or weaken these limitations.

Simply freewill can be defined as 'sanctioned choice'. The sanctions set our limits, we are free inside them. But freedom is a moral concept, you are responsible for what you freely do. Licence involves acting without responsibility, perhaps immorally. So freewill and freedom are part of moral knowledge and operate in the mode of moral discourse – arguments in which you are making moral judgements. The exercise of freedom is for existentialists an act of **good faith** as you act in the mode of authentic existence, responsibly, and therefore morally.

good faith or **authentic existence** are terms used in existentialism and are now in wider use, meaning living life according to what it means to be your true or 'better' self. That involves using your freewill responsibly (see **freewill** above), and, in Camus' eyes, developing your talents fully for your personal satisfaction and self-esteem and to help others as seems needful (see **bad faith** and **existentialism** above). Authentic existence for the Christian means working creatively and in dialogue with God, in order to live joyfully within the ethos of God's love, a view shared by members of other World Faiths, alongside humanists, existentialist agnostics, and other people of goodwill.

holism involves looking at a system or situation as a whole, rather than focusing alone on the basic elements of matter, parts of a system, or isolated pieces of behaviour, insisting that in the end the whole is greater than the sum of the parts. Two examples, one from biology, and one from world views. Dawkins insists that the centre of human biology is the selfish gene, a part. Rose insists that the organism is the central unit of study in human biology, the whole biological system and so he takes a holistic view of the organism.

Monod, Dawkins, Rose and Wilson are all self-declared materialists. That is their world-view. Materialism focuses on a part, the bits of matter, which make up systems. Barbour, Davies, Polanyi and Polkinghorne are critical realists, who accept the hierarchy of forms of knowledge, and their different modes of being and becoming as a result of complex and inter-related systems within the embracing outlook provided by the religious mode of discourse. They focus on the whole, and take a holistic view of life.

hypothetico deductive thinking is a characteristic of Piaget's most mature stage of thinking, formal operational thinking (see Chapter 4). It describes the structure of thinking which begins with a theory or a hypothesis from which a person seeks to deduce what follows from it. An extended example discussed earlier is Darwin's theory of evolution (see Chapters 2 and 5).

Let's take an example of a seventeen or eighteen year old. Before I look at this example, be clear! We are not looking for 'correct answers'. We are looking at kinds of thinking. Let's take two (conflicting) hypotheses of God's action in the world:

Hypothesis 1 – God works through natural processes in nature and through such processes in sentient creatures including humans.

Hypothesis 2 – God intervenes in history to support his chosen race.

This student opts for hypothesis 2, and engages in an excellent piece of *hypothetico deductive thinking*.

The topic of conversation is 'Crossing the Red Sea' (Exodus 14).

Q. 'How would you explain the dividing of the waters of the Red Sea?
A. 'God intervened in history to work the good of his chosen race.'
Q. 'For what purpose?'
A. 'To show his race, his chosen people, his covenant people, he was   working on behalf of them, and that what he said he would carry out.'[3]

You or I may not like the theology of this answer, but that is irrelevant. The key point is that this student answered the question, not in empirical discourse as many do, but in theological discourse correctly, postulating a theological hypothesis and deducing some implications from it. It is an excellent piece of formal operational reasoning (see Chapter 4).

idealism as a philosophy is the belief that reality is found in the *ideas* in the human mind. Only ideas are real. So the external world only exists in so far as ideas describe objects and events in the outside world. It was of some importance at the end of the nineteenth century, and has reappeared in the work of some living philosophers. It does not make sense of life to many today. Critical realism is the most widely held outlook on life, which often covers many different religious or secular belief systems.

implicit religion is defined as one's key commitments which have the greatest effects on one, and may well describe their *outlook on life*. For most people their sense of *Self* is a sacred focus, held in awe. One commitment may be the daily round of tasks, and people are very pleased with any day filled with good relationships and jobs well done. Another commitment might be a regular drink at the local public house, with its ritual dimension – having the drink, social dimension – communal solidarity, and mythic and ethical dimensions – being a man.[4] See Chapter 11 for fuller discussion.

induction – inductive thinking involves the task of 'intuitively recognising' that a collection of objects belong to one class. For example, a child might suddenly 'discover' that all the small and big black and white, and brown and white animals are cows. However, more advanced examples of inductive thinking are used by engineers and scientists, like the rest of us, to make a claim which goes beyond the objects or data observed.

Darwin, for example, in looking at the variations in similar animals and birds, proposed the hypothesis of evolution to account for the small changes. He and other biologists have been testing that *inductive theory* ever since. See the example of induction in Darwin's work quoted in Chapter 2, in the section called, 'Nineteenth Century Scientific Knowledge'. In Chapter 5 there are more details on Darwin's theory. Often the thinking of a mature mind in dealing with a problem postulates a theory or guess and then tests it out. If a hypothesis, or *an intuitive informed guess,* is tested, that kind of thinking is called *hypothetico-deductive thinking.* Turn to the description of this above.

intuition is a 'connecting or creative faculty' which makes the link between the object or perception of the object and the interpretation of the perception or concept which comes into your mind without reasoning. We know that the universe exists through an intuition. There is no proof.[5]

Intuition is what connects our minds to 'what is outside of them'. Intuition is not based on a perception, a fact, a sentence, an object, or a memory. It operates in various ways. For example it tells you $1 + 1 = 2$, which is an eternal mathematical truth;[6] it helps you recognise a colour – 'that's red'! You intuitively know how to ride a bicycle by watching others (see Polanyi's tacit knowledge in Chapter 6). You intuitively know that God is present (over a third of the population of the USA and UK. have had this experience). You know that that is the right thing to do in this situation – you can't give a reason (a moral intuition). Handel intuitively knew what to write down when writing the Hallelujah chorus in his oratorio, *Messiah.*

With their clear understanding of the earlier underlying models of the structures of their subjects, Einstein intuitively grasped relativity, and Dirac intuitively grasped quantum mechanics.

You intuitively knew that that person was the right girl or man for you, even if your specifications were not met, because when another person previously met your specifications nothing 'clicked'! Intuition is linked to the creative imagination, which can lead to an idea coming to you out of the blue. Does intuition, via evolution, come out of the swamp, from human complexity, from human creativity, from God? Are they all linked?

knowledge by acquaintance is the knowledge we obtain by being aware of objects which we can see or examine, or the knowledge we obtain by being in the presence of a person with whom you converse, but you may 'click' with him, and feel you know him, as Williams knew Fred (see Chapter 6). You can decide after you have met a person whether you feel you *know* something *about* him, or whether you have begun to get to *know him.* The former you may describe as knowledge by description, and the latter as knowledge by acquaintance. You can make the same distinction of objects, if you only know, for example, two things about that fountain, or you installed it and know everything about it.

logical positivism see Chapter 2.

materialism is the *philosophical world view* that all that exists are material bits like quarks, atoms, molecules, which occur in random combinations, whether in rock formations or human beings. The world exists accidentally, pointlessly and without any overall purpose.

meaning is disclosed by a speaker or writer by the way she *uses* language. See Chapter 2 for a fuller discussion.

metaphor is "an application of name or descriptive term or phrase to an object or action to which it is imaginatively but not literally applicable *(e.g. a glaring error, food for thought, leave no stone unturned.*"[7] Psychotherapists have outlined a view of how metaphors can operate. First, a person strains inwardly to grasp an aspect of an experience of which s/he had hardly been aware. Next comes a stage of 'unfolding' when 'new' feelings are put into words. Third, there is then a phase when relevant associations emerge into conscious awareness. Lastly, the experience is expressed in ways in which the individual can better understand both the experience and her or himself. Perhaps this is one way in which religious ideas are put into words.[8]

metaphysics is the philosophical investigation of reality in all its aspects – its nature, constitution and structure. It is broader that all subjects in the hierarchy of knowledge, because it examines the *assumptions*, called by different names like axioms, principles, or foundation beliefs, upon which every form of knowledge is based. The task of metaphysics is to identify them, examine them and state them.

Metaphysics examines concerns like the nature of existence, experience, concepts, knowledge, truth, goodness, mind, matter, the existence of God, and justification – that is ways of constructing a valid argument. What do those terms mean and how do we use them? Metaphysics has been defined as, "a sustained and rational study of what there is and the ultimate nature of what there is."[9]

No activity of exploration can begin, in the sciences or the arts, without making *assumptions*. Often they are not stated, particularly in science, for many practicing scientists are unaware of them. For example, "physics presupposes the following three things:

1. that there exists a physical reality independent of our mental states;
2. that the interactions of the stuff constituting this reality conform to certain general laws;
3 that we are capable of grasping physical laws and obtaining evidence that favors or disfavors specific proposed laws."[10]

The first two assumptions are concerned with 'what ultimately exists', that is, the nature of reality: these are metaphysical assumptions. The third assumption is concerned with the nature of knowledge: philosophers call this an epistemological assumption (ology = 'study of', episteme = the Greek word for 'knowledge').

One key question is whether there are only material things – **materialism**, or only entities in the mind called *ideas*, hence **idealism**. Metaphysics examines the basic assumptions of **dualism** the belief that both mind and matter exist and are functionally interrelated, **naïve realism, critical realism**, and all other metaphysical, or ultimate, outlooks on life, religious or secular.

Metaphysics, in the hands of some philosophers, however can have its limits. For example, in exploring the question of the 'existence of God', metaphysical enquiry frequently focuses on the notion of God as an 'entity', on the arguments for the existence of God, and thus producing what has been called 'the God of the philosophers'. For most people the existence of God, the existence of music, morality and relationships, for example, emerge from experience, for which rational analysis and argument from the philosophers provide some additional data for consideration in the light of their experience.

mind is a subject of universal philosophical and religious philosophical study. A central problem is the relationship of the mind and body. Plato, a Greek philosopher, said the soul and body were distinct. Greek philosophical thought followed this dualistic picture. Descartes said the mind and body were distinct: this forms the doctrine of mind–body dualism (see Chapters 2, 6, 17).

Hebrew thought knows of no such dualism. The New Testament is centrally Hebrew thought with Greek influences important in some books like John's Gospel. It can distinguish between 'spirit' and 'flesh', but the word 'soma', often unhelpfully and misleadingly translated 'body', is an integrated, unitary concept. It indicates the total personality of a human being in the form of 'spirit filled flesh'. In English we talk of a person as a psychosomatic (= mind + soma) unity. Actually the word 'psychosomatic' was formed, thinking it meant 'mind + flesh' unity. Now you know the meaning of soma, psychosomatic = 'mind + spirit + flesh'.

When the New Testament is talking about the resurrection of the body, I think it is not talking about the resurrection of flesh, but of the whole personality of a human, and St Paul says that resurrection takes place in a spiritual body (Acts 9:1–18; 1 Corinthians 15:8, 44). Other Christians sincerely believe in a physical resurrection. Either way, belief in the resurrection involves a psychological transformation, just like falling in love, becoming a Newcastle United supporter, or becoming wild about sailing! That's the psychological mechanism: this is not the time or place to talk of content and its meaning.

naturalism is the belief that everything is composed of natural substances, studied by the methods of the sciences. Ethical naturalism presupposes the autonomy of moral discourse and the appropriateness of introducing moral concepts into empirical discourse, e.g. Fred picked up that apple without paying for it.

pantheism is the belief that God is identical with everything. This is different from **panentheism**, which is the view that God is in everything.

personal knowledge has been used in two senses in this book. The first sense is Polanyi's use. In objecting to scientists claiming that scientific knowledge was objective, he chose the term 'personal knowledge' to apply to all knowledge, on the grounds that all knowledge has been formulated and known by persons. Polanyi rightly argued that all knowledge is personal even though controlled by agreed criteria or rules. I use 'personal knowledge' to apply to that form of knowledge and mode of discourse related to *knowing persons* (see 'knowing Fred' in Chapter 6). From this starting point, or assumption, it is possible to use the term 'personal knowledge' to apply analogically to 'personal knowledge of God', as a metaphor 'pointing towards God'.

phenomenology literally means the study of things. So the phenomenology of religion is concerned with viewing religion as 'a thing' which can be analysed 'from outside'. So, for example, an objective or 'outside' study of Christianity can be made by a Christian or an atheist. The phenomenological study of religion involves the 'bracketing out' of my beliefs, as I study the beliefs of another faith, trying to stand in the shoes of believers within the faith, and represent their views as 'seen by them' (see Chapter 8 for a fuller discussion).

philosophical analysis see Chapter 2.

rationalism is the belief that reason is the most important way of acquiring knowledge. In its strongest form it rejects other approaches such as 'revelation', 'tradition' or 'empiricism' based on sense perception (see **Christian Faith** above).

relativism mainly relates to knowledge or morality.

cognitive relativism is the belief that there are no universal truths about the world as there are different ways of interpreting it. Someone holding this view feels he cannot say "something is true for all", only that "something is true for me". For example relativism can relate to many personal, social, religious or aesthetic issues.

ethical relativism is the belief that there are no universal moral values or principles to establish agreed moral judgments. Such a person holds that "something is good for me". Those holding this view contrast the moral codes of different ethnical, religious or social communities as illustrations of their position. Those who oppose this position note how much common ground there is between different moral codes (see p. 150).

subjectivism is usually used to mean that the judgements expressed represent the views, opinions, attitudes, beliefs and feelings of the speaker. This contrasts with the shared judgements in society – e.g. in empirical, scientific, moral, personal or religious forms of knowledge. It is occasionally loosely used to defend a belief in relativism, which is that one person's view is as good as the view of anyone else.

theism. In antiquity this referred to belief in many gods, each with different attributes in different cultures, e.g. Roman or Greek.

Today, in monotheistic Faiths, it usually relates to the existence of God, who created the Universe, and is regarded as all powerful, present everywhere, loving to all, and who therefore is the source of moral goodness. One can be in communion with God through prayer and worship.

truth can be approached in different ways.

1. The **correspondence theory of truth** is the view that a belief or statement is true if it corresponds to the object or reality to which it relates. A strict or literal form states that each item or object must correspond to a term: this is **direct or naïve realism**. For example, if the statement that 'Jesus was born of a virgin' is treated literally, it claims that he had no human father. 'A human being can walk' is likewise an observational truism: it corresponds with what one sees. Both statements are in the mode of empirical discourse.

2. The **coherence theory of truth** is the view that beliefs should relate to each other in an integrated or complementary way, rather than in a contradictory way, both within a form of knowledge, and even if those truths are from different forms of knowledge. For example, if you look at a Gospel account of Jesus, you may test its truth against criteria like:

• human behaviour is limited by natural laws;

• balanced human behaviour will tend to be consistent in similar circumstances.

So part of the 'truth' of the virgin birth would emerge from recognising that Jesus was born from two parents like any other child. This is expressed within the mode of scientific discourse. The main part of the truth of the virgin birth story comes from the disciples' assessment of Jesus, which is expressed *poetically by telling this story* to affirm *his significance*. That is done, after their experience of his life and teaching. They 'saw' God's spirit working in him more fully than in anyone they had ever known either in the flesh, or through Scripture. This insight into the *significance* of Jesus is expressed in the mode of religious discourse, using the poetic model of a virgin birth understood as expressing *significance*. So the 'truths' of biology are not violated by the 'truths' of theology, but found to be complementary: one searches for coherence between them. This example employs a literary convention in the Bible which often attributes 'a special birth', to a person who is regarded of 'prophetic stature'.

This is an example of **critical realism** which sees that 'overall different statements from within one form of knowledge, or even from different forms of knowledge can 'hang together'. See the quantum theory of light mentioned in Chapter 5 for an example from within one form of knowledge. In both religion and in science, models, metaphors and analogies are used which are partial, adequate, revisable, but necessary for picturing as well as possible what is real.

3. The **pragmatic** theory of truth of James has been described in Chapter 3 (pp. 42–3). Briefly, true assumptions are said to be those which work, and which develop behaviour which produces desirable results.

value is the significance or worth of something, some idea or someone. I have argued that all empirical knowledge is based on perception, which is interpreted and valued as significant. Thus value is a constituent element in the structure of empirical knowledge. Equally it is a significant element in all other forms of knowledge for an interpretation in any mode of knowing is only significant if it is of value, hence value is an ingredient of all forms of knowledge. The distinction between 'information' and 'knowledge' is that the former has no value unless it is an element in a 'knowledge structure'. A simple example: 'The train arrives at 11.13 p.m.' So what?! Useless *information*! But if your friend is coming home, and the train arrives at the station at that time, you need that piece of information to complete your *knowledge* of when to meet the train.

The **verification principle** exists in two forms. The verification principle of the logical positivists states that the meaning of a proposition depends on the steps by which it can be verified, or tested to see if it is true.

When that was shown to be a metaphysical statement by Wittgenstein, he replaced the former test for truth by his own verification principle: the meaning of words in a proposition or sentence depends on the use to which it is put. Put another way, the meaning of words depends on the way they are used. This is the 'use' rule.

REFERENCES

In the compilation of this brief list of simple definitions I have made use of, and sometimes quoted, words or phrases from entries in:

Audi, R. (ed.), (1995), *The Cambridge Dictionary of Philosophy*, Cambridge: Cambridge University Press.

Craig, E. (2000), *Concise Routledge Encyclopedia of Philosophy*, New York: Routledge.

Flew, A. (1979), *A Dictionary of Philosophy*, London: Pan Books.

Honderich, T. (ed.), (1995), *The Oxford Companion to Philosophy*, Oxford: Oxford University Press.

Reese, W. L. (1996), *Dictionary of Philosophy and Religion: Easter and Western Thought*, New Jersey: Humanities Press.

Sykes, J. B. (ed.), (1983), *The Concise Oxford Dictionary*. Oxford: Oxford University Press.

Any errors are not from these dictionaries: they are mine alone.

NOTES

1   As Chairman of the London Schools Board of Education he insisted that Religious Instruction should be part of the school curriculum.
2   J. B. Sykes (ed.), (1983), *The Concise Oxford Dictionary*, Oxford: Oxford University Press. A helpful comment to me concerning 'believing that' and 'believing in' led me to formulate the following ilustration.
3   G. B. Miles (1971), 'A Study of Logical Thinking and Moral Judgements in G.C.E. Bible Knowledge Candidates', M.Ed. dissertation, University of Leeds, p.179.
4   E. Bailey (1997), *Implicit Religion in Contemporary Society*, The Netherlands: Kok Pharos; E. Bailey (1998), *Implicit Religion: An Introduction*, London: Middlesex University Press.
5   P. F. Strawson (1959), *Individuals*, London: Methuen; P. R. Baelz (1968), *Christian Theology and Metaphysics*, London: Epworth.
6   R. Penrose (1999), *The Emperor's New Mind*, Oxford: Oxford University Press, p. 151.
7   J. B. Sykes (ed.), *The Concise Oxford Dictionary*, Oxford: Oxford University Press.
8   This is a suggestion from F. Watts and M. Williams (1988), op. cit., p. 148.
9   G. Heimar and M. Losonsky (1998), *Beginning Metaphysics*, Oxford: Blackwell, p. 1.
10   M. Jubien (1997), *Contemporary Metaphysics*, Oxford: Blackwell, p. 1.

Bibliography

Acton, H. B. (1970), *Kant's Moral Philosophy*, London: Macmillan.

Allen, R. T. (1992), *Transcendence and Immanence in the Philosophy of Michael Polanyi and Christian Theism*, Edinburgh: Rutherford House.

Allen, R. T. (2005), "Polanyi and the Rehabilitation of Emotion", in S. Jacobs and R. T. Allen (eds), (2005), *Emotion, Reason and Tradition*, Aldershot: Ashgate.

Allport, G. W. (1969), *The Individual and His Religion* (1950), London: Macmillan,

Allport, G. W. and Ross, J. M. (1967), "Personal Religious Orientation and Prejudice", *J. Pers. Soc. Psychol.* 5, 432–3.

Alston, W. P. (1983), "Christian Experience and Christian Belief", in A. Plantinga and N. Wolterstorff (1983), *Faith and Rationality*, Notre Dame, IN: University of Notre Dame Press.

Alston, W. P. (1991), *Perceiving God: The Epistemology of Religious Experience*, Ithaca, NY: Cornell University Press.

Alston, W. P. (1993), *Perceiving God* (1991), Ithaca, NY, and London: Cornell University Press.

Alston, W. P. (2004), "Does Religious Experience Justify Religious Beliefs?" in M. L. Peterson and R. J. Vanarragon (eds), (2004), *Contemporary Debates in Philosophy of Religion*, Oxford: Blackwell Publishing.

Ameriks, K. (1995), "Kant, Immanuel", in R. Audi (ed.), (1995).

Argyle, M. (1967), *The Psychology of Interpersonal Behaviour*, London: Penguin.

Argyle, M. (2000), *Psychology and Religion*, London and New York: Routledge.

Armstrong, K. (1987), *Tongues of Fire: An Anthology of Religious and Poetic Experience* (1985), Harmondsworth and New York: Penguin.

Ashby, R. W. (1985), "Logical Positivism", in D. J. O'Connor (ed.), (1985), *A Critical History of Western Philosophy* (1964), Glencoe, NY: Free Press.

Audi, R. (ed.), (1995), *The Cambridge Dictionary of Philosophy*, Cambridge: Cambridge University Press.

Audi, R. (2005), *Epistemology*, 2nd edn, New York and London: Routledge.

Ayer, A. J. (1982), *Language, Truth and Logic* (1946), Harmondsworth: Pelican.

Baelz, P. R. (1968), *Christian Theology and Metaphysics*, London: Epworth Press.

Bailey, C. (1984), *Beyond the Present and the Particular*, London and Boston: Routledge & Kegan Paul.

Bailey, E. (1997), *Implicit Religion in Contemporary Society*, Kampen: Kok Pharos.

Bailey, E. (1998), *Implicit Religion: An Introduction*, London: Middlesex University Press.

Barbour, I. G. (1966), *Issues in Science and Religion*, Englewood Cliffs, NJ: Prentice-Hall.

Barbour, I. G. (1974), *Myths, Models and Paradigms*, London: SCM Press.

Barbour, I. G. (1990), *Religion in an Age of Science*, London: SCM Press.

Barbour, I. G. (1998), *Religion and Science*, London: SCM Press.

Barbour, I. G. (1999), "Neuroscience, Artificial Intelligence and Human Life: Theological and Philosophical Reflections", in R. J. Russell *et al.* (eds.), (1999).

Barnes, K. C. (1960), *The Creative Imagination*, London: Allen & Unwin.

Bartel, T. W. (ed.), (2003), *Contemporary Theology: Essays for Keith Ward*, London: SPCK.

Barton, J. (1997), "Biblical Commentaries", *Epworth Review* 24 (3), 35–6.

Barton, J. (ed.), (1998), *The Cambridge Companion to Biblical Interpretation*, Cambridge: Cambridge University Press.

Batson, C. D. (1971), "Creativity and Religious Development", PhD thesis, University of Princeton.

Batson, C. D. and Ventis, W. L. (1982), *The Religious Experience*, Oxford: Oxford University Press.

Batson, C. D., Schoenrade, P. and Ventis, W. L. (1993), *Religion and the Individual*, New York and Oxford: Oxford University Press.

Beardsworth, T. (1977), *Living the Questions*, Oxford: Religious Experience Research Unit, Westminster College.

Beit-Hallahmi, B. (1995), "Object Relations Theory and Religious Experience", in R. W. Hood Jr. (ed.), (1995), ch. 12.

Beit-Hallahmi, B. and Argyle, M. (1997), *The Psychology of Religious Behaviour, Belief and Experience*, London and New York: Routledge.

Berger, P. L. (1963), *Invitation to Sociology: A Humanistic Perspective*, New York: Doubleday.

Berger, P. L. (1971), *A Rumour of Angels* (1969), London: Pelican.

Berger, P. L. (1973), *The Social Reality of Religion* (1969), Harmondsworth: Penguin.

Berger, P. L. (1980), *The Heretical Imperative* (1979), London: Collins.

Berry, R. J. (ed.), (1991), *Real Science, Real Faith*, Eastbourne: Monarch.

Best, D. (1992), *The Rationality of Feeling*, London and Bristol, PA: Falmer Press.

The Bible (1997), The Contemporary English Version (C.E.V.), Swindon: British and Foreign Bible Society.

Blackham, H. (1968), *Humanism*, Harmondsworth: Penguin.

Blum, L. (1994), *Moral Perceptions and Particularity*, Cambridge: Cambridge Univesrsity Press.

Bolton, N. (1972), *The Psychology of Thinking*, London: Constable.

Bonnett, M. (1994), *Children's Thinking*, London: Cassell.

Bouquet, A. C. (1968), *Religious Experience*, Cambridge: Heffer.

Bowker, J. (1978), *The Religious Imagination and the Sense of God*, Oxford: Oxford University Press.

Bowker, J. (1987), *Licensed Insanities*, London: Darton, Longman & Todd.

Bowker, J. (1995), *Is God a Virus?*, London: SPCK.

Bowker, J. (1998), *The Complete Bible Handbook*, London: Dorling Kindersley.

Braithwaite, R. B. (1955), "An Empiricist's View of the Nature of Religious Belief" in I. T. Ramsey (ed.), (1966), *Christian Ethics and Contemporary Philosophy*, London: SCM Press.

Braithwaite, R. B. (1955), *An Empiricist's View of the Nature of Religious Belief*, Cambridge: Cambridge University Press.

Brown, R. E., Fitzmyer, J. A. and Murphy, R. E. (eds), (1995), *The New Jerome Biblical Commentary*, London and New York: Geoffrey Chapman.

Buber, M. (1958), *I and Thou*, Edinburgh: T. & T. Clark.

Bucke, R. M. (1969), *Cosmic Consciousness: A Study in the Evolution of the Human Mind* (1901), New York: Dutton.

Burhenn, H. (1995), "Philosophy and Religious Experience", in R. W. Hood Jr. (ed.), (1995), ch. 7.

Burke, C. D. (1998), "Evolution and Creation", in F. Watts (ed.), (1998), *Science Meets Faith*, London: SPCK.

Burke, P. A. (1999), "The Healing Power of the Imagination", *International Journal of Children's Spirituality* 4 (1), 9–17.

Butler, D. C. (1967), *Western Mysticism*, London: Constable.

Byatt, A. S. (1997), *Unruly Times: Wordsworth and Coleridge in Their Time*, London: Vintage.

Campbell, A. (ed.), (1987), *A Dictionary of Pastoral Care*, London: SPCK.

Camus, A. (1984), *The Plague* (1960), Harmondsworth: Penguin.

Carroll, L. (1998), *Alice's Adventures in Wonderland*, London: Macmillan.

Chadwick, H. (1986), *Augustine*, Oxford: Oxford University Press.

Chapman, M. (1988), *Constructive Evolution*, Cambridge: Cambridge University Press.

Chisholm, R. M. (1966), *Theory of Knowledge*, Englewood Cliffs, NJ: Prentice-Hall.

Clark, W. H. (1967), *The Psychology of Religion* (1958), New York: Macmillan.

Clarke, P. B. and Byrne, P. (1993), *Religion Defined and Explained*, London: St. Martin's Press.

Clayton, P. (1999), "Neuroscience, the Person and God", in R. J. Russell *et al.* (eds.), (1999).

Conford, P. (ed.), (1996), *The Personal World: John Macmurray on Self and Society*, Edinburgh: Floris Books.

Coplestone, F. C. (1974), *Religion and Philosophy*, New York: Gill & Macmillan.

Costello, J. E. (2002), *John Macmurray: A Biography*, Edinburgh: Floris Books.

Craig, E. (2000), *Concise Routledge Encyclopedia of Philosophy*, New York: Routledge.

Cupitt, D. (1979), *Jesus and the Gospel of God*, London: Lutterworth.

Cupitt, D. (1980), *Taking Leave of God*, London: SCM Press.

Cupitt, D. (1985), *Only Human*, London: SCM Press.

Cupitt, D. (1998), *After God: The Future of Religion*, London: Phoenix.

D'Aquili, E. G. and Newberg, A. B. (1998), "The Neuropsychology of Religion" (1996), in R. Watts (ed.), (1998), *Science Meets Faith*, London: SPCK.

Darwin, C. (1968), *Origin of Species* (1859), Harmondsworth: Penguin.

Davies, P. (1984), *God and the New Physics*, Harmondsworth: Pelican.

Davies, P. (1992), *The Mind of God*, Harmondsworth: Penguin.

Davies, P. (1998), *The Fifth Miracle*, Harmondsworth: Penguin.

Dawkins, R. (1981), *The Selfish Gene* (1976), London: Granada.

Dawkins, R. (1991), *The Blind Watchmaker* (1986), Harmondsworth: Penguin.

Dawkins, R. (1998), *Unweaving the Rainbow: Science, Delusion and the Appetite for Wonder*, London and New York: Allen Lane, Penguin.

Dawkins, R. (1999), *River out of Eden* (1995), London: Orion Books, Phoenix.

De Duve, C. (1995), *Vital Dust*, New York: Basic Books.

De Selincourt, E. (ed.), (1965), *Wordsworth's Collected Works*, Vols 2 and 5, Oxford: Oxford University Press.

Dearden, R. F. (1968), *The Philosophy of Primary Education*, London: Routledge & Kegan Paul.

Delaney, C. F. (1995), "Personalism", in R. Audi (ed.), (1995).

Dickson, A. (ed.), (1990), *The Penguin Freud Library*, vol. 13, *The Origins of Religion* (1985), London: Penguin.

Dickson, A. (ed.), (1991), *The Penguin Freud Library*, vol. 12, *Civilisation, Society and Religion*, London: Penguin.

Dixon, T. (2003), *From Passions to Emotions*, Cambridge: Cambridge University Press.

Donovan, P. (1979), *Interpreting Religious Experience*, London: Sheldon Press.

Dostoevsky, F. (1982), *The Brothers Karamazov* (1880), Harmondsworth: Penguin.

Durkheim, E. (1976), *The Elementary Forms of the Religious Life* (1915), London: Allen & Unwin.

Dyer, W. W. (1990), *You'll See It When You Believe It*, New York: Avon Books.

Dykstra, C. and Parks, S. (eds), (1986), *Faith Development and Fowler*, Birmingham, Al: Religious Education Press.

Ellis, G. F. R. (1999), "Intimations of Transcendence: Relations of the Mind and God", in R. J. Russell *et al.* (eds.), (1999).

Erickson, E. H. (1963), *Childhood and Society*, 2nd edn, New York: W. W. Norton.

Erickson, E. H. (1968), *Identity, Youth and Crisis*, New York: W. W. Norton.

Erickson, E. H. (1977), *Toys and Reasons: Stages in the Ritualisation of Experience*, New York: W. W. Norton.

Eyre, A. (1997), *Football and Religious Experience: Sociological Reflections*, Oxford: Religious Experience Research Unit, Westminster College.

Ferre, F. (1968), *Basic Modern Philosophy of Religion*, London: George Allen & Unwin.

Fishbein, M. and Ajzen, I. (1975), *Beliefs, Attitude, Intentions and Behavior*, Reading, MA: Addison-Wesley.

Fisher, J. W. (1999), "Helps to Fostering Students' Spiritual Health", *International Journal of Children's Spirituality*, 4 (1), pp. 29–49.

Flew, A. (1979), *A Dictionary of Philosophy*, London: Pan Books.

Ford, D. F. (ed.), (1997), The Modern Theologians, 2nd edn, Oxford: Blackwell Publishers.

Forward, M. (2001), *Inter-religious Dialogue*, Oxford: One World.

Fowler, J. W. (1981), *Stages of Faith*, San Francisco, CA: Harper & Row.

Francis, L. J. (1991), "The Personality Characteristics of Anglican Ordinands: Feminine Men and Masculine Women?", *Personality and Individual Differences* 12, 1133–40.

Frankl, V. E. (1963), *Man's Search for Meaning*, New York: Washington Square Press.

Frankl, V. E. (1977), *The Unconscious God*, London: Hodder & Stoughton.

Franks Davis C. F. (1989), *The Evidential Force of Religious Experience*, Oxford: Clarendon Press.

Freud, S. (1907), *Obsessive Actions and Religious Practices*, in A. Dickson (ed.), (1990).

Freud, S. (1913), "Totem and Taboo", in A. Dickson (ed.), (1990).

Freud, S. (1927), *The Future of an Illusion*, in A. Dickson (ed.), (1991).

Freud, S. (1935a), "Postscript" to *Autobiographical Study*, in A. Dickson (ed.), (1991).

Freud, S. (1939), "Moses and Monotheism", in A. Dickson (ed.), (1991).

Fromm, E. (1971), *Psychoanalysis and Religion* (1950), New Haven and London: Yale University Press.

Fromm, E. (1978), *To Have or to Be*, London: Jonathan Cape.

Fuller, A. R. (1994), *Psychology of Religion*, Lanham, MA, and London: Littlefield Adams.

Gaarder, J. (1995), *Sophie's World*, London: Phoenix House.

Gadamer, H. G. (1991), *Truth and Method*, New York: Crossroads.

Gale, R. (1994), "The Overall Argument of Alston's 'Perceiving God'", *Religious Studies* 30: 2, 135–49, 150–81, Cambridge: Cambridge University Press.

Garber, D. (1995), "Rationalism", in R. Audi (ed.), (1995).

Gelwick, R. (1977), *The Way of Discovery: An Introduction to the Thought of Michael Polanyi*, New York: Oxford University Press.

Ghiselin, B. (1952), *The Creative Process*, Berkeley and Los Angeles, CA: California University Press.

Gleick, J. (1988), *Chaos*, London: Sphere Books, Cardinal.

Glock, C. Y. and Stark, R. (1965), *Religion and Society in Tension*, Chicago: Rand McNally.

Glock, C. Y. and Wuthnow, R. (1979), "Departures from Conventional Religion: the Nominally Religious, the Non-religious, and the Alternatively Religious", in R. Wuthnow (ed.), (1979), *The Religious Dimension*, New York: Academic Press.

Goldman, R. (1964), *Religious Thinking from Childhood to Adolescence*, London: Routledge, Kegan Paul.

Goleman, D. (1996), *Emotional Intelligence*, London: Bloomsbury.

Goleman, D. and the Dalai Lama (2003), *Healing Emotions*, Boston, MA: Shambhala Publications.

Goleman, D. and the Dalai Lama (2004), *Destructive Emotions*, London: Bloomsbury.

Goleman, D., Boyatzis, R. and McKee, A. (2003), *The New Leader*, London: Time Warner.

Graves, D. (1996), *Scientists of Faith*, Grand Rapids, MI: Kregel Resources.

Gray, J. (1993), *Men Are from Mars and Women Are from Venus*, London: Thorsons.

Greeley, A. M. (1974), *Ecstasy: A Way of Knowing*, Englewood Cliffs, NJ: Prentice-Hall.

Greeley, A. M. (1975), *The Sociology of the Paranormal*, London: Sage.

Greeley, A. M. (1992), "Religion in Britain, Ireland and the U.S.A.", in R. Jowell, L. Brook, G. Prior and B. Taylor (1992), *British Social Attitudes: The 9th Report*, Aldershot: Dartmouth.

Greenwood, S. F. (1995), "Transpersonal Theory and Religious Experience", in R. W. Hood Jr. (ed.), (1995), ch. 21.

Gross, R. and McIlveen, R. (1998), *Psychology: A New Introduction*, London: Hodder & Stoughton.

Halligan, F. R. (1995), "Jungian Theory and Religious Experience", in R. W. Hood Jr. (ed.), (1995: ch 11).

Halloran, J. D. (1967), *Attitude Formation and Change*, Leicester: Leicester University Press.

Hamlyn, D. W. (1970), *The Theory of Knowledge*, London: Macmillan.

Happold, F. C. (1975), *Mysticism: A Study and An Anthology* (1963), Harmondsworth: Penguin.

Harding, R. (1940), *The Anatomy of Inspiration*, London: Cassell.

Hardy, A. (1978), *The Divine Flame* (Gifford Lectures of 1965 [London and Glasgow: Collins, 1966]), Oxford: Religious Experience Research Unit, Westminster College. Available from Alister Hardy Research Centre, Department of Theology and Religious Studies, University of Wales, Lampeter, Ceredigion, Wales SA48 7ED.

Hardy, A. (1979), *The Spiritual Nature of Man*, Oxford: Clarendon Press.

Haughton, R. (1972), *The Knife Edge of Experience*, London: Darton, Longman & Todd.

Hay, D. (1990), *Religious Experience Today*, London: Mowbrays.

Hay, D. with Nye, R. (1998) *The Spirit and the Child*, London: Fount, HarperCollins.

Hebblethwaite, B. L. (1972), "The Appeal to Experience in Christology", in S. Sykes and J. P. Clayton (eds), (1972), *Christ, Faith and History*, Cambridge: Cambridge University Press.

Hebblethwaite, B. L. (1983), "The Experiential Verification of Religious Belief in the Theology of Austin Farrer", in J. C. Eaton and A. Loades (eds), (1983), *For God and Clarity*, Allison Park, PA: Pickwick Publications.

Hebblethwaite, B. L. (1987), "Religious Language and Religious Pluralism", *Anvil* 4 (2), 101–11.

Hebblethwaite, B. L. (1987), *The Problems of Theology* (1980), Cambridge: Cambridge University Press.

Hebblethwaite, B. L. (1994a), *Modern Theology*, vol. 10, Oxford: Blackwell.

Hebblethwaite, B. L. (1994b), "The Communication of Divine Revelation", in A. G. Padgett (ed.), (1994), *Reason and the Christian Religion*, Oxford: Clarendon Press.

Heimar, G. and Losonsky, M. (1998), *Beginning Metaphysics*, Oxford: Blackwell.

Heisig, J. W. (1979), *Imago Dei: A Study of C. G. Jung's Psychology of Religion*, Cranbury, NJ: Associated Universities Press.

Hepburn, R. W. (1957), "Poetry and Religious Belief ", in S. E. Toulmin, R. W. Hepburn and A. MacIntyre (1957), *Metaphysical Beliefs*, London: SCM Press.

Hessayon, D. G. (1981), *The Rose Expert*, Waltham Cross: PBI Publications.

Hick, J. (1968), *Evil and the God of Love*, London: Fontana.

Hick, J. (1983), Philosophy of Religion, Englewood Cliffs, NJ.

Hick, J. (1989), *An Interpretation of Religion: Human Responses to the Transcendent*, Basingstoke: Macmillan.

Hick, J. (2004), *The Fifth Dimension: An Exploration of the Spiritual Realm* (1999),Oxford: One World.

Hill, C. C. (1979), *Problem Solving*, London: Pinter.

Hill, P. C. (1995), "Affective Theory and Religious Experience", in R. W. Hood Jr. (ed.), (1995), ch. 15.

Hill, P. C. and Hood Jr., R. W. (eds), (1999), *Measures of Religiosity*, Birmingham, AL: Religious Education Press.

Hirst, P. H. (1965), "Liberal Education and the Nature of Knowledge", in P. H. Hirst (ed.), (1974), *Knowledge and the Curriculum*, London: Routledge & Kegan Paul.

Hirst, P. H. (1999), "The Nature of Educational Aims", in R. Marples (ed.), (1999), *The Aims of Education*, London: Routledge.

Hirst, P. H. and Peters, R. S. (1970), *The Logic of Education*, London and Boston: Routledge & Kegan Paul.

Holm, J. (1975), *Teaching Religion in School*, Oxford: Oxford University Press.

Holm, J. (1977), *The Study of Religion*, London: Sheldon Press.

Holm, J. (ed.) with Bowker, J. (1994), Themes in Religious Studies: *Human Nature and Destiny*; *Worship*; *Making Moral Decisions*; *Myth and History*; *Attitudes to Nature*; *Picturing God*; *Sacred Writings*; *Women in Religion*; *Rites of Passage*; and *Sacred Place*, London and New York: Pinter Publishers (ten volumes).

Holm, N. G. (1995), "Role Theory and Religious Experience" in R. W. Hood Jr. (ed.), (1995), ch. 17.

Hood Jr., R. W. (ed.), (1995), *Handbook of Religious Experience*, Birmingham, AL: Religious Education Press.

Hood Jr., R. W. (1999), "American Psychology of Religion and the Journal for Scientific Study of Religion", Paper presented at the annual convention of the Society for the Scientific Study of Religion, Boston.

Hooker, M. (1982), "Myth, Imagination and History", *Epworth Review* 9 (1), 50–6.

Honderich, T. (ed.), (1995), *The Oxford Companion to Philosophy*, Oxford: Oxford University Press.

Horne, J. R. (1978), *Beyond Mysticism*, Ontario: Laurien University Press.

Hostie, R. (1957), *Religion and the Psychology of Jung*, London: Sheed & Ward.

Howatch, S. (1997), *A Question of Integrity*, London: Little, Brown and Co.

Howcroft, K. G. (1998), "Reason , Interpretation and Postmodernism – Is There a Methodist Way?", *Epworth Review* 25 (3), 28–9.

Hudson, W. D. (1974), *A Philosophical Approach to Religion*, London: Macmillan.

Hudson, W. H. (1936), *From Far Away and Long Ago*, London: Dent.

Hughes, E. J. (1986), *Wilfred Cantwell Smith: A Theology for the World*, London: S.C.M.

Hull, J. (1991), *Touching the Rock*, London: Arrow Books.

Huxley, A. (1954), *The Doors of Perception*, London: Chatto & Windus.

Huxley, A. (2004), *The Perennial Philosophy* (1945), New York: Perennial.

Hyde, K. E. (1990), *Religion in Childhood and Adolescence*, Birmingham, AL: Religious Education Press.

Jacobs, S. and Allen, R. T. (eds), (2005), *Emotion, Reason and Tradition*, Aldershot: Ashgate.

Jakobsen, M. J. (1999), *Negative Spiritual Experiences: Encounters with Evil*, Oxford: Religious Experience Research Unit, Westminster College.

James, W. (1902), *The Varieties of Religious Experience*, London and New York: Longmans, Green and Co.

James, W. (1928), *The Varieties of Religious Experience* (1902), New York and London: Longmans Green and Co.

James, W. (1968), *The Varieties of Religious Experience* (1902), London and Glasgow: Fontana.

Jones, C., Wainwright, G., and Yarnold, E. (eds), (1987), *Study of Spirituality*, 4 vols, Oxford: Oxford University Press.

Jowell, R., Brook, J., Prior, G. and Taylor, B. (1992), *British Social Attitudes: The 9th Report*, Aldershot: Dartmouth.

Jubien, M. (1997), *Contemporary Metaphysics*, Oxford: Blackwell.

Jung, C. G. (1957), *The Undiscovered Self* (1933), trans. R. F. Hull, New York: New American Library.

Jung, C. G. (1973a), *Psychology and Religion* (1938), New Haven: Yale University Press.

Jung, C. G. (1973b), *Memories, Dreams, Reflections* (1963), London: Fontana.

Jung, C. G. (1967), "The Theory of Psychoanalysis" (1913), in H. Read, M. Fordham and G. Adler (eds), *The Collected Works of C. G. Jung*, vol. 4, Princeton, NJ: Princeton University Press.

Jung, C. G. (1977), *Psychology of the Unconscious* (1911), in H. Read, M. Fordham and G. Adler (eds), *The Collected Works of C. G. Jung*, vol. 7, London: Routledge & Kegan Paul.

Kant, E. (1960), *Religion Within the Limits of Reason Alone* (1793), New York: Harper & Brothers.

Katz, S. T. (1978), "Language, Epistemology and Mysticism", in S. T. Katz (ed.), (1978), *Mysticism and Philosophical Analysis*, London: Sheldon Press.

Kaufman, G. D. (1993), *The Face of Mystery*, Cambridge, MA: Harvard University Press.

Kee, A. (1971), *The Way of Transcendence*, London: Pelican.

Kerr, F. (1997), *Immortal Longings*, London: SPCK.

Kirkpatrick, L. A. (1995), "Attachment Theory and Religious Experience", in R. W. Hood Jr. (ed.), (1995), ch. 19.

Kung, H. (2003), "A Global Ethic: Development and Goals" in *Interreligious Insight: A Journal of Dialogue and Engagement* 1(1): 8–19. C/o World Congress of Faiths 2, Market Street, Oxford, OX1 3EF.

Lapsley, R. and Westlake, M. (1988), *Film Theory: An Introduction*, Manchester: Manchester University Press.

Lash, N. (1988), *Easter in Ordinary*, London: SCM Press.

Laski, M. (1961), *Ecstasy*, London: Cresset Press.

Laski, M. (1980), *Everyday Ecstasy*, London: Thames & Hudson.

Lee, J. M. (ed.), (1985), *The Spirituality of the Religious Educator*, Birmingham, AL: Religious Education Press.

Leuba, J. H. (1978), *The Psychology of Religious Mysticism* (1925), New York: Harcourt Brace Publishers.

Leuba, J. H. (1994), *A Psychological Study of Religion: Its Origins, Functions and Future* (1912), New York: Macmillan.

Loder, J. E. (1981), *The Transforming Moment*, San Francisco, CA: Harper & Row.

Lovell, K. (1965), *Educational Psychology and Children*, London: University of London Press.

Luckman, T. (1967), *The Invisible Religion*, New York: Collier-Macmillan.

Luscombe, P. (2000), *Groundwork of Science of Religion*, Peterborough: Epworth Press.

Lyttkens, H. (1979), "Religious Experience and Transcendence", *Religious Studies* 15, 211–12.

Mackinnon, D. (1978), "Some Epistemological Reflections on Mystical Experience", in S. T. Katz (ed.), (1978), *Mysticism and Philosophical Analysis*, London: Sheldon Press.

Macmurray, J. (1961), *Religion, Art and Science*, Liverpool: Liverpool University Press.

Macmurray, J. (1972), *Reason and Emotion* (1935), London: Faber & Faber.

Macquarrie, J. (1975), *Thinking about God*, London: SCM Press.

Macquarrie, J. (ed.), (1971), *A Dictionary of Christian Ethics*, London: SCM Press.

Maslow, A. H. (1970), *Religions, Values and Peak Experiences* (1964), New York: Viking Press.

Maslow, A. H. (1987), *Motivation and Personality*, 3rd edn, New York: Harper & Row.

Maslow, A. H. (ed.), (1970), *New Knowledge in Human Values*, 3rd edn, New York: Harper.

Maxwell, M. and Tschudin, V. (1990), *Seeing the Invisible*, London: Arkana.

McCallister, B. J. (1995), "Cognitive Theory and Religious Experience", in R. W. Hood Jr. (ed.), (1995), ch. 14.

McDowell, J. (1981), "Non-Cognitivism and Rule Following" in S. H. Holtzmann and C. M. Leich (eds), (1983), *Wittgenstein: To Follow a Rule*, London: Routledge and Kegan Paul.

McFadyen, A. I. (1990), *The Call to Personhood*, Cambridge: Cambridge University Press.

McGrath, A. (2004), *The Twilight of Atheism*, London: Rider.

McGrath, A. (2005), *Dawkin's God*, Oxford: Blackwell.

Meadow, M. J. and Kahoe, R. D. (1984), *Psychology of Religion*, New York: Harper & Row.

Meissner, W. W. (1984), *Psychoanalysis and Religious Experience*, New Haven: Yale University Press.

Methodist Worship Book (1999), Peterborough: Methodist Publishing House.

Miles, G. B. (1971), "A Study of Logical Thinking and Moral Judgements in G.C.E. Bible Knowledge Candidates", MEd dissertation, University of Leeds.

Miles, G. B. (1981), "Transcendental and Religious Experiences of Sixth Form Pupils: An Analytic Model", Paper presented at Implicit Religion Conference, Denton Hall, Yorkshire, May 1981. Available from Canon Professor Edward Bailey, The Old School, Church Lane, Yarnton, Oxford OX5 1PY <www.implicitreligion.org>

Miles, G. B. (1983), "A Critical and Experimental Study of Adolescents' Attitudes to and Understanding of Transcendental Experience", PhD dissertation, School of Education, University of Leeds.

Miles, G. B. (1993), "Transcendental and Religious Experiences of Sixth Form Pupils: An Analytic Model", Paper presented at RIMSCUE Conference, September 1993; printed in SPES: Newsletter No. 1 of RIMSCUE Centre, October 1994, Faculty of Education, University of Plymouth, Douglas Avenue, Exmouth, Devon EX8 2AT.

Miles, G, B. (1996) "'I am David': A Study in the Development of Primary School Pupils' Understanding of the Concept of God." Journal of Beliefs and Values, Vol. 17, No. 1: 13–17; Vol. 17, No. 2: 15.

Miles, T. R. (1972), *Religious Experience*, London: Macmillan.

Monod, J. (1977), *Chance and Necessity*, Glasgow: Fount.

Mooney, P. (1996), "Macmurray's Notion of Love for Personal Knowing", *Appraisal* 1 (2): 57–67.

Moore, J. R. (1979), *The Post Darwinian Controversies*, Cambridge: Cambridge University Press.

Moore, P. (1978), "Mystical Experience, Mystical Doctrinal, Mystical Technique", in S. T. Katz (ed.), (1978), *Mysticism and Philosophical Analysis*, London: Sheldon Press.

Moorman, M. (1963), "Wordsworth, William", in *Collier's Encyclopaedia*, New York: Crowell-Collier.

Morea, P. (1990), *Personality: An Introduction to the Theories of Psychology*, Harmondsworth: Penguin Books.

Morgan, R. (1998), "The Bible and Christian Theology", in J. Barton (ed.), (1998), *The Cambridge Companion to Biblical Interpretation*, Cambridge: Cambridge University Press.

Mott, N. (1991), *Can Scientists Believe?*, London: James & James.

Murphy, N. (1999), "Supervenience and the Downward Efficacy of the Mental: A Nonreductive Physicalist Account of Human Nature", in R. J. Russell *et al.* (eds.), (1999).

Neitz, M. J. (1995), "Feminist Theory and Religious Experience", in R. W. Hood Jr. (ed.), (1995), ch. 22.

Newton-Smith, W. H. (1986), *The Rationality of Science*, London: Routledge & Kegan Paul.

Nielsen, M. E. (2000), "Psychology of Religion in the USA" (pp. 1–12). Internet: 'Back to the Psychology of Religion' home page.

Oakeshott, M. (1971), "Education: the Engagement and Its Frustrations", in R. F. Dearden,

P. H. Hirst and R. F. Peters (eds), (1972), *Education and the Development of Reason*, London and Boston: Routledge & Kegan Paul.

Oakeshott, M. (1975), *On Human Conduct*, Oxford: Clarendon Press.

Onions, C. T. (ed.), (1975), *Shorter Oxford Dictionary*, Oxford: Oxford University Press.

Otto, R. (1959), *The Idea of the Holy* (1917), Harmondsworth: Pelican.

Otto, R. (1932), *Mysticism East and West: A Comparative Analysis of the Nature of Mysticism*, New York: Macmillan.

Oxford English Dictionary (1979), The Compact Edition, Compact Editor, A. Boni. London: Book Club Associates.

Paffard, M. (1973), *Inglorious Wordsworths*, London: Hodder & Stoughton.

Park, J. (1999), "Emotional Literacy: Education for Meaning", *International Journal of Children's Spirituality* 4 (1), 19–28.

Parkes, P. L. (ed.), (1906), *John Wesley's Journal (abridged)*, London: Pitman.

Peacocke, A. (1993), *Theology for a Scientific Age*, London: SCM Press.

Peacocke, A. (1996), *God and Science: A Quest for Christian Credibility*, London: SCM Press.

Peacocke, A. (1999), "The Sound of Sheer Silence: How Does God Communicate with Humanity?", in R. J. Russell *et al.* (eds.), (1999).

Penrose, R. (1999), *The Emperor's New Mind*, Oxford: Oxford University Press.

Peterson, M., Hasker, W., Reichenbach, B. and Basinger, D. (1991), *Reason and Religious Belief*, Oxford: Oxford University Press.

Phenix, P. (1964), *Realms of Meaning*, New York: McGraw-Hill.

Piaget, J. (1932), *The Moral Judgement of the Child*, London: Routledge & Kegan Paul.

Piaget, J. (1951), *Play, Dreams and Imitation in Childhood*, London and Boston: Routledge & Kegan Paul.

Piaget, J. and Inhelder, B. (1958), *The Growth of Logical Thinking from Childhood to Adolescence*, London and Boston: Routledge and Kegan Paul.

Pirouet, L. (1989), *Christianity World Wide: AD 1800 Onwards*, London: SPCK.

Pittenger, N (1970), *Christology Reconsidered*, London: SCM Press.

Plantinga, A. (1983), "Reason and Belief in God", in A. Plantinga and N. Wolterstorff (eds), (1983).

Plantinga, A. and Wolterstorff, N. (eds), (1983), *Faith and Rationality*, Notre Dame, IN: University of Notre Dame Press.

Polanyi, M. (1958), *Personal Knowledge: Towards a Post-Critical Philosophy*, Chicago, MI: University of Chicago Press.

Polanyi, M. (1960), *Knowing and Being*, London: Routledge and Kegan Paul.

Polkinghorne, J. (1986), *One World*, London: SPCK.

Polkinghorne, J. (1991), *Reason and Reality*, London: SPCK.

Polkinghorne, J. (1994), *Quarks, Chaos and Christianity*, London: SPCK.

Polkinghorne, J. (1998), *Science and Theology*, London: SPCK.

Poloma, M. M. (1995), "The Sociological Context of Religious Experience" in R. W. Hood Jr. (ed.), (1995), ch. 8.

Popper, K. (1974), *Unending Quest*, London: Fontana.

Popper, L. (1959), *The Logic of Scientific Discovery*, London: Hutchinson.

Power, M. and Dalgleish, T. (1997), *Cognition and Emotion: From Order to Disorder*, Hove: Erlbaum.

Price, H. R. (1969), *Belief*, London: Allen & Unwin.

Prigogine, I. and Stengers, I. (1988), *Order out of Chaos*, London: Fontana, Flamingo.

Proudfoot, W. (1985), *Religious Experience*, Berkeley, CA: University of California Press.

Pruyser, P. W. (1977), "The Seamy Side of Current Religious Beliefs", *Bulletin of the Menninger Clinic* 35, 77–97.

Pruyser, P. W. (1983), *The Play of the Imagination: Towards a Psychoanalysis of Culture*, New York: International Universities Press.

Putnam, H. (1998), *Renewing Philosophy* (1992), Cambridge, MA: Harvard University Press.

Quinn, P. L. and Taliaferro, C. (eds), (1999), *A Companion to the Philosophy of Religion* (1997), Oxford: Blackwell.

Quinton, A. M. (1985), "Contemporary British Philosophy", in D. J. O'Connor (ed.), (1985) *A Critical History of Western Philosophy* (1964), Glencoe, NY: Free Press.

Race, A. (ed.), (2003–), *Interreligious Insight: A Journal of Dialogue and Engagement*, c/o World Congress of Faiths, 2 Market Street, Oxford, OX1 3EF.

Ramsey, I. T. (1967), *Religious Language*, London: SCM Press.

Ramsey, I. T. (1973), *Religious Language*, London: SCM Press.

Rayburn, C. A. (1995), "The Body and Religious Experience", in R. W. Hood Jr. (ed.), (1995), ch. 20.

Reardon, B. M. G. (1966), *Religious Thought in the Nineteenth Century*, Cambridge: Cambridge University Press.

Reese, W. L. (1996), *Dictionary of Philosophy and Religion: Easter and Western Thought*, New Jersey: Humanities Press.

Reiss, M. J. and Straughan, R. (1996), *Improving Nature? The Science and Ethics of Genetic Engineering*, Cambridge: Cambridge University Press.

Rhys Jones, G. (1996), *The Nation's Favourite Poems*, London: BBC Worldwide.

Richardson, A. and Bowden, J. (eds), (1991), *A New Dictionary of Christian Theology*, London: SCM Press.

Richardson, N. (1997), "Biblical Interpretation and Christian Experience", *Epworth Review* 24 (1), 75–6.

Rizzuto, A.-M. (1979), *The Birth of the Living God*, Chicago: University of Chicago Press.

Roberts, J. D. (1991), *A Philosophical Introduction to Theology*, London: SCM Press; Philadelphia: Trinity Press International.

Roberts, M., King, T. and Reiss, M. (1994), *Practical Biology for Advanced Level*, London: Nelson

Robinson, D. (1983), "Spirit", "Spiritual", *Concordance to the Good News Bible*, Swindon: The British and Foreign Bible Society, *c.* 1113–16.

Robinson, E. (1977a), *The Original Vision*, Oxford: Religious Experience Research Unit, Westminster College, Oxford.

Robinson, E. (1977b), *This Time Bound Ladder*, Oxford: Religious Experience Research Unit, Westminster College.

Robinson, E. (1978), *Living the Questions*, Oxford: Religious Experience Research Unit, Westminster College.

Robinson, J. A. T. (1963), *Honest to God*, London: SCM Press.

Rogers, C. R. (1980), *A Way of Being*, Boston: Houghton Mifflin.

Rose, S. (1997), *Lifelines: Biology, Freedom, Determinism*, Harmondsworth: Penguin.

Russell, B. (1937), *The Scientific Outlook*, London: Allen and Unwin.

Russell, B. (1995), "Intuition", in R. Audi (ed.), (1995).

Russell, R. J., Murphy, N., Meyering, T. C. and Arbib, M. A. (eds), (1999), *Neuroscience and the Person*, Vatican City State: Vatican Observatory Publications and Berkeley, CA: Center for Theology and Natural Sciences; Notre Dame, IN: University of Notre Dame Press.

Sarap, M. (1993), *Post-Structuralism and Postmodernism*, New York and London: Harvester Wheatsheaf.

Scheffler, I. (1978), *Conditions of Knowledge* (1965), Chicago: University of Chicago Press.

Schillebeeckx, E. (1980), *Interim Report on the Books Jesus and Christ*, London: SCM Press.

Schleiermacher, F. E. D. (1928), *The Christian Faith*, Edinburgh: T. & T. Clark.

Schleiermacher, F. E. D. (1988), *On Religion: Speeches to Its Cultured Despisers* (1821), trans. R. Crouter, Cambridge: Cambridge University Press.

Scott, D. (1996), *Michael Polanyi* (1985), London: SPCK.

Scott, P. and Spencer, C. (1998), *Psychology: A Contemporary Introduction*, Malden, MA, and Oxford: Blackwell.

Scott Peck, M. (1990a), *The Road Less Travelled: A New Psychology of Love, Traditional Values and Spiritual Growth* (1978), London: Arrow Books.

Scott Peck, M. (1990b), *People of the Lie* (1983), London: Arrow Books.

Shafranske, E. P. (1995), "Freudian Theory and Religious Experience", in R. W. Hood Jr (ed.), (1995), ch. 10.

Smart, N (1973), *The Science of Religion and the Sociology of Knowledge*, New Haven: Princeton University Press

Smart, N. (1966), "Myth and Transcendence", *The Monist* 50 (4): 475–87.

Smart, N. (1968), *Secular Education and the Logic of Religion*, London: Faber & Faber.

Smart, N. (1973), *The Science of Religion and the Sociology of Knowledge*, Princeton, NJ: Princeton University Press.

Smart, N. (1978), "Understanding Religious Experience", in S. T. Katz (ed.), (1978), *Mysticism and Philosophical Analysis*, London: Sheldon Press.

Smart, N. (1997), *Dimensions of the Sacred* (1996), London: Fontana.

Soskice, J. M. (1997), "Religious Language", in P. L. Quinn and C. Taliaferro (eds), (1999), *A Companion to the Philosophy of Religion*, Oxford: Blackwell.

Soskice, J. M. (1985), *Metaphor and Religious Language*, Oxford: Clarendon Press.

Speake, J. (ed.), (1981), *A Dictionary of Philosophy* (1979), London: Pan Books.

Spilka, B. and McIntosh, D. N. (1995), "Attribution Theory and Religious Experience", in R. W. Hood Jr. (ed.), (1995), ch. 18.

Spinks, G. S. (1963), *Psychology and Religion*, London: Methuen.

Stace, W. T. (1973), *Mysticism and Philosophy* (1960), London: Macmillan.

Stannard, R. (1996), *Science and Wonders: Conversations about Science and Belief*, London: Faber & Faber.

Starbuck, E. D. (1899), *The Psychology of Religion*, New York: Scribner's.

Steiner, G. (1989), *Real Presences*, London: Faber and Faber.

Stiver, D. R. (1998), *The Philosophy of Religious Language*, Oxford: Blackwell.

Strawson, P. F. (1959), *Individuals*, London: Methuen.

Streng, F. J. (1978), "Language and Mystical Awareness", in S. T. Katz (ed.), (1978).

Surin, K. (1986), *Theology and the Problem of Evil*, Oxford: Blackwell.

Swinburne, R. (1979), *The Existence of God*, Oxford: Clarendon Press.

Sykes, J. B. (ed.), (1983), *The Concise Oxford Dictionary*. Oxford: Oxford University Press.

Tamminen, K. and Nurmi, K. E. (1995), "Developmental Theory and Religious Experience", in R. W. Hood Jr. (ed.), (1995), ch. 13.

Tart, C. (ed.), (1992), *Transpersonal Psychologies*, 3rd edn, New York: HarperCollins.

Thiselton, A. C. (1992), *New Horizons in Hermeneutics*, London: Harper Row.

Thiselton, A. C. (1997). "Theology and Hermeneutics", in D. F. Ford (ed.), (1997), *The Modern Theologians*, 2nd edn, Oxford: Blackwell.

Thiselton, A. C. (1998), "Biblical Studies in Theoretical Hermeneutics", in J. Barton (ed.), (1998).

Thouless, R. H. (1971), *An Introduction to the Psychology of Religion* (1923), Cambridge: Cambridge University Press.

Tillich, P. (1951), *Systematic Theology, Vol. 1*, Chicago: University of Chicago Press.

Tillich, P. (1962), *Courage To Be* (1952), London: Collins, Fontana.

Tillich, P. (1968), *A History of Christian Thought*, New York: Simon & Schuster.

Tillich, P. (1969), *Morality and Beyond* (1964), London: Fontana.

Trusted, J. (1997), *An Introduction to the Philosophy of Knowledge*, 2nd edn, London: Macmillan.

Turner, D. (1995), *The Darkness of God*, Cambridge: Cambridge University Press.

Turner, D. (2004), *Faith, Reason and the Existence of God*, Cambridge: Cambridge University Press.

Van de Weyer, R. (2003), *A World Religions Bible*, Alresford: O Books, John Hunt Publishing.

Vardy, P. (1992), *The Puzzle of Evil*, London: HarperCollins, Fount.

Vardy, P. and Grosch, P. (1994), *The Puzzle of Ethics*, London: Fount.

Vernon, P. E. (ed.), (1970), *Creativity: Selected Readings*, Harmondsworth: Penguin.

Vinacke, V. E. (1952), *The Psychology of Thinking*, New York: McGraw-Hill.

Wallas, G. (1926), *The Art of Thought*, New York: Harcourt.

Walsh, R. and Vaughan, F. (1993), *Paths Beyond Ego: The Transpersonal Vision*, Los Angeles: Tarcher.

Walsh, W. (1960), *The Use of Imagination*, London: Chatto & Windus.

Ward, G. (1997), "Postmodern Theology", in D. F. Ford (ed.), *The Modern Theologians*, 2nd edn, Oxford: Blackwell.

Ward, K. (1974), *The Concept of God*, Oxford: Blackwell.

Ward, K. (1987), *Images of Eternity*, London: Darton, Longman & Todd.

Ward, K. (1992), *Defending the Soul*, Oxford: One World.

Ward, K. (1994), *Religion and Revelation*, Oxford: Clarendon Press.

Ward, K. (1996), *God, Chance and Necessity*, Oxford and Rockport, MA: One World.

Ward, K. (2000), *Christianity: A Short Introduction*, Oxford: One World.

Ward, K. (2002), *God, Faith and the New Millenium* (1998), Oxford: One World.

Ward, K. (2002), *God: A Guide for the Perplexed*, Oxford: One World.

Ward, K. (2004), *The Case for Religion*, Oxford: One World.

Watson, R. A. (1995), "Dualism", in R. Audi (ed.), (1995).

Watts, F. (ed.), (1998), *Science Meets Faith*, London: SPCK.

Watts, F. (1999), "Cognitive Neuroscience and Religious Consciousness", in R. J. Russell *et al.* (eds.), (1999).

Watts, F. (2002), *Theology and Psychology*, Aldershot: Ashgate.

Watts, F. and Williams, M. (1988), *The Psychology of Religious Knowing*, Cambridge: Cambridge University Press.

Weil, S. (1973), *Waiting for God*, New York: Harper & Row.

Westphal, M. (1997), "The Emergence of Modern Philosophy of Religion", in P. L. Quinn and C. Taliaferro (eds.), (1999), *A Companion to Philosophy of Religion*, Oxford: Blackwell.

White, J. (2003), "Five Critical Stances towards Liberal Philosophy of Education in Britain", *Journal of Philosophy of Education* 37 (1): 147–84.

Wildman, W. T. and Brothers, L. A. (1999), "A Neurophysical-Semiotic Model of Religious Experience", in R. J. Russell *et al.* (eds.), (1999).

Wiles, M. (1994), *The Remaking of Christian Doctrine* (1974), London: SCM Press.

Wilkinson, D. (1993), *God, the Big Bang and Stephen Hawking*, Tunbridge Wells: Monarch.

Williams, R. (1979), *The Wound of Knowledge*, London: Darton, Longman & Todd.

Wilson, E. O. (1978), *On Human Nature*, Cambridge, MA: Harvard University Press.

Wilson, J. (1961), *Philosophy and Religion*, Oxford: Oxford University Press.

Wilson, J. (1971), *Education in Religion and the Emotions*, London: Heinemann.

Winnicott, D. W. (1988), *Human Nature*, London: Free Association Books.

Wittgenstein, L. (1967), *Philosophical Investigation* (1953), Oxford: Blackwell.

Wordsworth, W. (1979), *The Prelude* (1850), J. Wordsworth, M. N. Abrams, S. Gill (eds), New York and London: Norton and Co.

Wright, D. (1967), "A Review of Empirical Studies in the Psychology of Religion", *Catholic Psychology Group Bulletin* 10, 11–37.

Wulff, D. M. (1991), *Psychology of Religion: Classic and Contemporary Views*, New York: Wiley.

Wulff, D. M. (1995), "Phenomenological Psychology and Religious Experience", in R. W. Hood Jr. (ed.), (1995), ch. 9.

Wynn, M. R. (2003), "Religion and the Revelation of Value: the Emotions as Sources for Religious Understanding", in T. W. Bartel (ed.), (2003), *Comparative Theology: Essays for Keith Ward*, London: SPCK.

Wynn, M. R. (2005), *Emotional Experience and Religious Understanding*, Cambridge: Cambridge University Press.

Young, F. (1990), *The Art of Performance: Towards a Theology of Holy Scripture*, London: Darton, Longman & Todd.

Zachry, W. (1990), "Correlation of abstract religious thought and formal operations in high school and college students", *Review of Religious Research* 31: 405–412, quoted in R. W. Hood Jr. (ed.), (1995).

Zaehner, R. C. (1973), *Mysticism, Sacred and Profane* (1957), Oxford: Oxford University Press

Index